Global Perspectives on Health Geography

Series editor

Valorie Crooks, Department of Geography, Simon Fraser University, Burnaby, BC, Canada

Global Perspectives on Health Geography showcases cutting-edge health geography research that addresses pressing, contemporary aspects of the health-place interface. The bi-directional influence between health and place has been acknowledged for centuries, and understanding traditional and contemporary aspects of this connection is at the core of the discipline of health geography. Health geographers, for example, have: shown the complex ways in which places influence and directly impact our health; documented how and why we seek specific spaces to improve our wellbeing; and revealed how policies and practices across multiple scales affect health care delivery and receipt.

The series publishes a comprehensive portfolio of monographs and edited volumes that document the latest research in this important discipline. Proposals are accepted across a broad and ever-developing swath of topics as diverse as the discipline of health geography itself, including transnational health mobilities, experiential accounts of health and wellbeing, global-local health policies and practices, mHealth, environmental health (in)equity, theoretical approaches, and emerging spatial technologies as they relate to health and health services. Volumes in this series draw forth new methods, ways of thinking, and approaches to examining spatial and place-based aspects of health and health care across scales. They also weave together connections between health geography and other health and social science disciplines, and in doing so highlight the importance of spatial thinking.

Dr. Valorie Crooks (Simon Fraser University, crooks@sfu.ca) is the Series Editor of Global Perspectives on Health Geography. An author/editor questionnaire and book proposal form can be obtained from Publishing Editor Zachary Romano (zachary.romano@springer.com).

More information about this series at http://www.springer.com/series/15801

Svetlana Malkhazova
Varvara Mironova • Natalia Shartova
Dmitry Orlov

Mapping Russia's Natural Focal Diseases

History and Contemporary Approaches

With Contributions by Vadim Rumyantsev and Mikhail Soldatov

Springer

Svetlana Malkhazova
Department of Biogeography,
Faculty of Geography
Lomonosov Moscow State University
Moscow
Russia

Natalia Shartova
Department of Biogeography,
Faculty of Geography
Lomonosov Moscow State University
Moscow
Russia

Varvara Mironova
Department of Biogeography,
Faculty of Geography
Lomonosov Moscow State University
Moscow
Russia

Dmitry Orlov
Department of Biogeography,
Faculty of Geography
Lomonosov Moscow State University
Moscow
Russia

Translated by Alexandra White, Biological Sciences and Comparative Literature & Society, Columbia College, Columbia University, New York City, NY, USA

ISSN 2522-8005 ISSN 2522-8013 (electronic)
Global Perspectives on Health Geography
ISBN 978-3-030-07814-0 ISBN 978-3-319-89605-2 (eBook)
https://doi.org/10.1007/978-3-319-89605-2

Orginal publication: Медико-географический атлас России «Природноочаговые болезни» / под ред. С.М. Малхазовой. 2-е изд., перераб. и доп. – М.: Географический факультет МГУ, 2017 – 216 с. Medico-geographical Atlas of Russia "Natural Focal Diseases" / edited by S.M. Malkhazova. 2nd revised edition. – Faculty of Geography, Lomonosov Moscow State University, Moscow, 2017. – 216 p.

Softcover re-print of the Hardcover 1st edition 2019

This Springer imprint is published by the registered company Springer Nature Switzerland AG
The registered company address is: Gewerbestrasse 11, 6330 Cham, Switzerland

Foreword

Numerous natural focal (or natural nidal) diseases, the pathogens of which inhabit natural landscapes, pose a serious danger to human health. In the last few decades, increasing anthropogenic activity (agricultural development in new regions, suburban real estate development and increasing recreational pressure) has led to a substantial increase in contact between the population and natural foci, as well as the creation of epidemiological preconditions for the spread of such diseases. In addition, the migration of people from epidemiologically unsafe areas, an enormous increase in the scale of tourism, including international travel, and increasing international shipments have raised the risk of infection with diseases that find Russia's environmental conditions favourable. The geography of natural focal diseases, the socio-economic preconditions of their spread and the spatial structure of focal territories on different spatial and temporal levels must be studied in order to support epidemiological monitoring and the development of sanitary and preventative measures that protect the population from disease.

Maps, literature, archival materials, government statistics (collected by the Russian Federal State Statistics Service for socio-economic measures and the Russian Federal Service for Surveillance on Consumer Rights Protection and Human Wellbeing for natural focal disease morbidity rates in the population) and data collected as part of field research served as sources for this book. The study was conducted based on the administrative and territorial divisions in the Russian Federation as of 2015.

The monograph includes the following thematic chapters:

1. Introduction
2. Natural and Socio-economic Conditions
3. Natural Focal Disease Hosts and Vectors
4. Major Natural Focal Disease Distribution
5. Spatial Organization of the Sanitary and Epidemiological Service

The book is unique in the combination of cartographic and textual information, infographics, photos and analytical material that it contains. It consists of about 80 original maps on a scale between 1:25,000,000 and 1:60,000,000 for Russia and 1:120,000,000 for the world. A number of model regions are shown on a scale between 1:3,000,000 and 1:5,000,000. The bibliography, which lists more than 400 sources on natural focal diseases and research methodologies, most by Russian authors, is valuable in its own right.

The prepared series of maps allows the reader to gage the amount of information available on natural focal diseases in Russia, to highlight regions that necessitate additional targeted research and to identify natural and socio-economic disease preconditions, the geographic ranges of the main pathogen hosts and vectors, and the spectrum of most commonly diagnosed natural focal diseases.

Nosogeographic maps depict population morbidity rates and perennial disease dynamics for 85 administrative subjects of the Russian Federation. For the majority of maps, the temporal depth of analysis is approximately 20 years (1997–2015). Twenty-three disease units have been mapped. Taken together, the maps allow for the evaluation of the epidemiologic danger posed by a specific territory and the stability of a disease's behaviour in a particular region, followed by conclusions on the necessity of developing and carrying out specialized preventative and therapeutic measures.

The book is aimed at public health, medical, science and education professionals, as well as a wide range of readers with an interest in issues of human health and the environment.

Moscow, Russia

Prof. Svetlana M. Malkhazova
Chair of the Department of
Biogeography of the Faculty of Geography
Lomonosov Moscow State University

Acknowledgements

The monograph was compiled based on materials collected and produced by the Federal Service for Surveillance on Consumer Rights Protection and Human Wellbeing (Rospotrebnadzor) with financial support from Russian Geographical Society (RGS) and Russian Foundation for Basic Research (RFBR). Section 3 was prepared with support from the following grant: Russian Science Foundation (RSF) Grant # 14-50-00029 'A Scientific Basis for the Creation of a National Depository Bank of Living Systems'. Geographical analysis of the distribution of the West Nile Fever (Section 4) has been supported by the grant programme from Russian Science Foundation (RSF) (Project № 17-77-20070 'An initial assessment and projection of the bioclimatic comfort in Russian cities in XXI century against the context of climate change').

The book's authors thank their colleagues and graduate and undergraduate students of the Faculty of Geography at Moscow State University for their professional consultations and help in data compilation and processing and preparing the Book for print. The authors are grateful to members of the Department of Biogeography of the Faculty of Geography at Lomonosov Moscow State University and the members of the Medical Geography and Human Ecology commission of the Moscow Centre of the RGS for their valued advice and suggestions throughout their discussion.

While preparing this monograph, we received very helpful comments and recommendations from our colleagues. We would like to express our most sincere gratitude to P. V. Pestina, N. V. Ryabova, T. V. Vatlina, O. S. Adischeva, L. V. Ezerova, A. E. Beljaev and O. P. Chernyavskaya.

Contents

About the Authors

Svetlana Malkhazova, Doctor in Geography, is Professor and Chair in the Department of Biogeography of the Faculty of Geography at Lomonosov Moscow State University. She specializes in medical geography and human ecology. She is the author or co-author of over 300 scientific publications, including 15 monographs and several textbooks and medico-geographical atlases.

Varvara Mironova, Ph.D. in Geography, is Senior Research Associate in the Department of Biogeography of the Faculty of Geography at Lomonosov Moscow State University. Her interests include nosogeography, natural focal and naturally endemic disease mapping, and ecological parasitology. She is the author of 50 scientific publications.

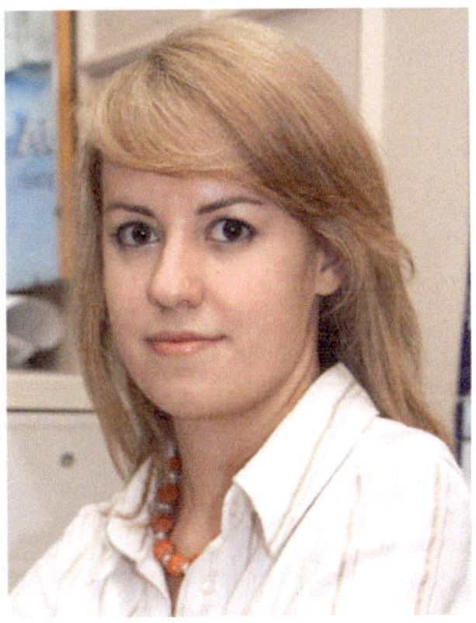

Natalia Shartova, Ph.D. in Geography, is Senior Research Associate in the Department of Landscape Geochemistry and Soil Geography of the Faculty of Geography at Lomonosov Moscow State University. Her interests include medical geography, human ecology and urban population health.

Dmitry Orlov is Research Associate in the Department of Biogeography of the Faculty of Geography at Lomonosov Moscow State University. His interests include medical geography, human ecology and ecological epidemiology.

List of Figures

1 Introduction

Abstract

This chapter contains a short summary of natural focal disease theory, which was developed by Evgeny Pavlovsky, a Soviet scientist, in the 1930s. This theory generalized material and data that had been gathered on natural focal infectious and parasitic diseases and the complex networks of mutually linked and mutually dependent pathogen, animal host and arthropod vector populations, which make up a biotic component that is organically linked to natural landscapes. The history of research on natural focal diseases in Russia, the contributions of Russian and Soviet scientists to the development of the theory of the emergence and spread of natural focal diseases and modern achievements in this discipline are also described. The understanding of what makes a natural focus is discussed, along with the theory's main tenets. The level of cartographical research on the spread of natural focal diseases in Russia is briefly characterized, and the methodology of map compilation is also explained. The chapter includes a geographical map of Russia and a map of its federal administrative divisions, which simplifies the later connections between the thematic content, geographic units and units of territorial divisions.

1.1 Natural Focal Disease Theory

The notion of diseases having natural foci was first proposed by Danylo Zabolotny (Fig. 1.1) at the turn of the twentieth century.

In 1939, Evgeny Pavlovsky, a member of the Academy of Sciences (Fig. 1.2), presented the main tenets of natural focal disease theory, which were based on his own research and that of his predecessors, at the General Conference of the Academy of Sciences of the USSR (Pavlovsky 1946). A final version of his theory is presented in his 1964 monograph (Pavlovsky 1964, 1966), for which he received the country's highest honour—the Lenin Prize.

Over the following decades, natural focal disease theory was continuously developed both by Pavlovsky himself and several generations of his students and followers (Korenberg 2010). Natural focal disease theory, which explained the causes for the emergence and continued existence of a large group of diseases, made a hefty contribution to the development of many disciplines: epidemiology, parasitology, medical geography, biogeography and ecology. Despite the fact that the phenomenon has not been sufficiently studied, today it is utterly clear that this is one of the greatest scientific achievements of the twentieth century.

In the English-language literature, along with the term focus, the term nidus is more widely used, and the science studying the phenomenon of natural foci is commonly called the landscape epidemiology (Meade and Emch 2010; Kanaroglou et al. 2016). In the works of the Russian authors, both terms are found with the former being used more often. Therefore, in this book, which is devoted to the study of natural focal diseases in Russia, we will adhere to the term focus/focal/focality.

At the turn of the twentieth century, it was widely believed that humans could only contract infections directly or indirectly from other humans. One of the main theoretical accomplishments of natural focal disease theory is that it discovered the possibility of disease transmission between a single individual and wild animals. The theory's essence is in the acknowledgment of the fact that pathogens that cause a number of diseases emerged and exist in nature independently of humans and are members of natural ecosystems. This new perspective on the possibility of contracting an infectious disease pathogen from a natural source had a significant influence on the further development of epidemiology and preventative medicine.

As shown by numerous studies, natural focal diseases can be caused by viruses, various bacteria, Protozoa and helminths (Natural Focality of Diseases… 2003). The exact number of such diseases is currently unknown. Still recently

S. Malkhazova et al., *Mapping Russia's Natural Focal Diseases*, Global Perspectives on Health Geography, https://doi.org/10.1007/978-3-319-89605-2_1

Fig. 1.1 Danylo Zabolotny (1866–1929). *Credit* https://fotki.yandex.ru/next/users/humus777/album/436485/view/1081424

Fig. 1.2 Evgeny Pavlovsky (1884–1965). *Credit* http://www.ras.ru/nappelbaum/ddb81522-eb5b-4490-87cd-969c267859af.aspx#

experts estimated that there were tens of natural focal infections. Today, it is clear that it was a vast underestimate. There are several hundred entries in International Catalog of Arboviruses alone, many of which have medical significance (Atlas of Pathogen Distribution… 2001b). The overall list of natural focal diseases will undoubtedly grow.

Pathogens typically inhabit various animals and periodically live in the water or soil, meaning that the diseases that they cause are zoonotic or sapronotic. Usually, humans are accidental hosts for these pathogens; moreover, they are commonly biological dead ends for these diseases. The pathogens' main characteristic is their ability to survive in natural ecosystems. In the vast majority of cases, natural focal disease pathogens cannot be passed from one person to another, and humans are infected only when they inadvertently enter a natural focus.

Blood-sucking pathogen vectors were first identified in the late nineteenth century. Today, the list includes blood-sucking dipterous insects: mosquitoes, blackflies, biting midges (Ceratopogonidae), sandflies and horseflies. It has been established that various tick and flea species, among others, also play an important role in spreading the pathogens that cause natural focal diseases.

The vector passes the pathogen to an animal, i.e. the recipient of the infection. However, the animal that acts as the vector's nutrient source does not always serve as the infection reservoir as well. In some cases, the bitten organism is a dead end for the infection. In other cases, the recipient is asymptomatic and the disease is latent. Finally, in some cases, the recipient can either become a host or become sick, and only then does it become a source of infection. By the same token, blood-sucking arthropods cannot always spread the infection. Pathogens survive for a short time inside many of these animals, especially in lice, bed bugs, fleas and mosquitoes.

Natural focal disease theory initially only applied to vector-borne zoonotic diseases, as vector-borne disease theory gave rise to it in many ways. This theory formed in the late nineteenth century after a series of discoveries of pathogens and their vectors, described in the works of Charles Louis Alphonse Laveran, Patrick Manson, Joseph Bancroft, Giovanni Batista Grassi and Ronald Ross, Edmond and Etienne Sergent, David Bruce, Carlos Finlay, Walter Reed and others. According to this theory, disease pathogens are transmitted by arthropod vectors in nature.

However, by the mid-1950s, it had become clear that the sphere of natural focal diseases is not limited to vector-borne disease. It was proven that natural focus theory also applied to non-vector-borne zoonotic diseases, such as hemorrhagic fever with renal syndrome, leptospirosis, rabies, etc.

The understanding that sapronotic diseases, whose reservoirs are soil and water, not animals, also have natural foci began to form in the mid-1980s (Litvin 1986; Litvin and Korenberg 1999). Today, it has been established that soil serves as the reservoir for pseudotuberculosis, tetanus and listeria, while bodies of water serve as the reservoir for diseases like Legionnaire's disease, cholera and melioidosis. Sapronotic disease natural foci are conditional upon numerous and varied contacts with organisms that inhabit the soil, many of which serve as natural hosts for these microorganisms, while land-dwelling warm-blooded animals are usually accidental hosts. For instance, cholera pathogens have a symbiotic relationship with hydrobionts, especially zooplankton, in which the pathogen attaches itself to the organism in order to survive in the water (Lipp et al. 2002). *Vibrio cholerae* has been found in the larvae of water insects, molluscs, crabs, fish and frogs (Islam et al. 1994). In addition, data point to a biocenotic link between cholera pathogens and certain aquatic plants (Spira et al. 1981). Birds and mammals, including humans, do not play a determining role in the existence of sapronotic disease pathogens in nature, and the warm-blooded carrier is not an essential structural member of the natural focus (Litvin et al. 1998). Therefore, as natural focal disease theory evolved, sapronotic diseases were added to the list, removing first vectors and then warm-blooded hosts from the list of a focus' necessary components.

The natural focus is the central and most crucial concept in natural focal disease theory. There are a number of interpretations of this term. Pavlovsky (1964, 1966) defined the natural focus as a portion of the geographic landscape in which the pathogen circulates among species in a typical biocenosis. Later, as the theory was developed and explored in greater depth, a natural focus was taken to mean the minimal portion of land in which an independent zoonotic process can exist for a maximally prolonged period of time without pathogen introduction from an external source (Kucheruk 1972). Anatoly Voronov, the founder of the medical geography school at M. V. Lomonosov Moscow State University, emphasized the connection between landscapes and natural foci and saw the natural focus as a natural and territorial complex along with the pathogens, vectors and hosts that inhabit it (Voronov 1981).

Vladimir Beklemishev and his followers had a slightly different vision. A natural focus is the combination of the pathogen population with the vertebrates that support its existence (Beklemishev 1970), or according to Nikiforov's similar explanation (1968), a biocenosis inhabited by the pathogen's population. Beklemishev believed that, because every population is always associated with a specific territory, so every focus of infection is also associated with a specific territory. Olsufiev and Dunaeva (1970) had similar points of view, underscoring along with Beklemishev that an infectious focus is a host–parasite system that is part of the biocenosis and is characterized by a specific epizootic process, i.e. the circulation of the infection among animal populations via characteristic methods of transmission.

The development of natural focal disease theory by the end of the twentieth century led to the conclusion that the only mandatory and specific component of any focus is a pathogen population (Litvin and Korenberg 1999). The presence of a pathogen in an ecosystem is what makes it a natural focus. According to this perspective, an infectious disease focus is any natural ecosystem that contains a pathogen population (Korenberg 2010).

While acknowledging the theoretical significance of current scientific achievements, we nonetheless consider the natural focus to be a territorial unit with corresponding spatial and temporal characteristics, in accordance with the opinions of E. N. Pavlovsky, V. V. Kucheruk and A. G. Voronov.

Natural foci vary in their contours, sizes and epidemiological characteristics. The size of a focus depends on the type of pathogen, the environmental situation, the social and living conditions of the population, the nature of the infection, the pathogen–host interactions and the availability of natural barriers. It can range from 100 km^2 [such as one of the leptospirosis foci on Lake Nero, which has been extensively studied by Karasyova (2000)] to tens or hundreds of thousands of square kilometres (plague foci in the Caspian region).

According to Voronov (1981), the smallest territorial unit in nature that can contain a focus is a landscape, as smaller and simpler morphological components of a landscape (like a landmark or facies) do not contain all the features that are necessary for the prolonged survival of a pathogen population with its supporting animal host and arthropod vector populations. However, landscape and focus borders do not always coincide, and each natural focus is considered to be a unique and singular phenomenon.

Foci have a specific structure that can be more or less defined, and consist of the following elements: areas where the infection is relatively stable (the focal nuclei), areas to which the infection is spreading (epizootic fields) and areas that are always free of pathogens (Kucheruk 1972; Voronov 1981).

Two focal structure types are identified depending on how sharp the differences are between the morphologic components of the focus: homogenous and heterogeneous. In homogenous foci, the pathogen is spread across the entire territory, and people are at risk of infection at nearly any point in the focus, because nuclei and epizootic fields are weakly differentiated. Such foci are typically found in forest landscapes (tick-borne encephalitis). Heterogeneous foci are characterized by sharp differences between each morphological element. In these cases, risk of infection is linked to nuclei at all times and to epizootic fields only during mass reproduction of the pathogen, host and vector populations. Such foci are found in arid territories (plague, leishmaniasis).

These natural focal diseases have nuclei in river floodplains and epizootic fields in desert landscapes.

The natural focus' sphere of influence can spread beyond its borders, at times quite far. This depends on the distance that the vectors or pathogen reservoirs can travel. The amount of time a person needs to spend in the focus to become infected also varies greatly depending on the characteristics of the reservoir animals and transmission mechanisms.

Many researchers have shown how people become infected when they interact with a member of the host–parasite system (pathogen–vector–host) (Voronov et al. 2014).

- As a result of contact with vectors. This also means that a person in unpopulated territory can gather vectors around herself/himself even previously they were homogeneously spreaded over it.
- As the result of contact with animals who are infection reservoirs. In this case, the reservoir seldom plays an active role (for example, in the case of rabies); sometimes humans become infected from holding a killed or dead animal.
- As the result of human behaviour in the focus. The concentration of lumberjacks in forests, the creation of settlements and homesteads with a majority non-immune population in regions endemic for a specific natural focal disease, troop relocation, border patrol work in endemic regions, etc. play a major role in this process.

The emergence of disease in a new focus can be the result of the following:

- Migration of infected vectors to biotopes that were created due to human activity, especially in cases when the vector finds nutrient sources in the new focus. These nutrient sources can become infection reservoirs and, subsequently, donors.
- Migration of infected host animals, which bring parasites with them when they move from a natural environment to a human settlement or infect local species in a new setting. The invasion of human living spaces by 'wild' infected animals is especially dangerous when housing is built in a natural focus.
- Migration of humans from the focus where they were infected to a home environment. In a number of cases, human-to-human transmission is impossible (tularaemia), however, in some diseases, when new vectors and host can be found in buildings, human infection is possible within a settlement.

Evgeny Pavlovsky listed the following characteristics of natural foci for classification purposes:

- *In terms of origin,* they can be primary natural, secondary (daughter) natural or anthropurgic (whose emergence and maintenance is linked to various forms of human activity).
- *In terms of location,* they can be autochthonous (i.e. their emergence and existence are directly linked to their location) and heterochthonous (i.e. they shifted from the location of the mother focus due to the activity of their vectors, hosts or sometimes people) in terms of their location.
- *In terms of age,* they can be ancient or modern.
- *In terms of pathogen specificity,* they can be zoonotic (where animals are the main carriers of diseases pathogenic to humans) or anthroponomics (also known as physioanthropic, according to Pavlovsky, or foci of diseases that are usually found in humans but whose vectors have a connection to the natural environment).
- *In terms of reservoir diversity,* they can have one host species or many.
- *In terms of vector genera diversity,* they can have vectors that all belong to the same genus or vectors that belong to different genera.
- *In terms of territorial limits,* they can be diffuse (without clearly defined borders) or limited (with more or less clear borders).
- *In terms of their association with location,* they can be stationary or errant (wandering).
- *In terms of their epidemiological risk status and intensity,* they can be potentially dangerous or latent (no humans means no morbidity, 'obscuring' the focus); epidemiologically valent, weakened or fading (depending on natural or anthropogenic conditions); and faded or eradicated (as the result of targeted efforts).
- *In terms of evolution,* they can be progressing or regressing.
- *In terms of stability,* they can be stable or ephemeral.

Natural focal diseases are unique in their territorial association with specific geographic landscapes. The most important effect of this association is the possibility of predicting the latent epidemiological danger posed by a territory based on the types of landscapes and biocenoses (which include disease pathogens, vectors and hosts) found within it. Research on the geography of natural foci is thus a relevant and productive pathway within the disciplines of medical geography and biogeography.

A wide range of natural focal diseases is found in the Russian Federation. Some of them (tick-borne encephalitis, Lyme borreliosis, tularaemia, hemorrhagic fever with renal syndrome) occupy vast territories. Others (North Asian tick typhus, Crimean-Congo hemorrhagic fever, Omsk hemorrhagic fever) are found in limited regions.

Like plant communities, natural foci are classified by zone based on the representative typical landscape conditions of a specific zone: tick-borne encephalitis, Lyme borreliosis and hemorrhagic fever with renal syndrome (forest zone), North Asian tick typhus and tick-borne spirochetosis (forest-steppe and desert-steppe landscapes), plague (arid zones: steppe, semi-desert, and desert), yellow fever (equatorial and rainforest zones), etc. Intrazonal foci, which do not occupy typical areas in any of the zones, usually contain tularaemia, leptospirosis and Omsk hemorrhagic fever. Various helminthiases associated with river systems and transmitted via fish—opisthorchiasis, diphyllobothriasis, etc. —are also quite dangerous. Many diseases have extrazonal foci, i.e. those that spread beyond the borders of 'their' zone. For instance, tick-borne encephalitis foci are found in extrazonal conditions, namely in small 'islands' of birch trees commonly found in steppes, etc.

If the host–parasite system of a natural focal disease (pathogen—vector—host) is seen as a biotic component that has organic ties to its ecosystem, then this combination can serve as a unit for the evaluation of the degree of epidemiological risk in a natural environment (Fig. 1.3). On a global scale, there is a fairly clear connection between the spread of a disease and the geographic zoning of an environment. The number of natural focal diseases typical in a specific population and closely linked to natural factors increases significantly from cooler latitudes to warmer ones and reaches its peak in the subequatorial climates zone. This conclusion corresponds with the increased biodiversity and ecosystem complexity in the tropics compared to the arctic (Malkhazova 2001).

Vershinsky's 1964 map *Diseases with Natural Focality. A Nosogeographical Map of the USSR* on a scale of 1:25,000,000 (Fig. 1.4) was the first and only such work to reflect the combination of all infections found on Soviet territory and to base this depiction on geographical zones. The map shows territories along with a comprehensive set of factors that includes preconditions that lead to human disease. Three territorial categories are identified: those with natural preconditions and disease foci within a landscape specific to the zone type, those with natural preconditions and zonal disease foci that are outside of the disease's main range and those with natural preconditions and disease foci that do not have a clear connection with specific landscape types.

The landscape-based epidemiological approach was further developed during the compilation of the *Nosoecosystems of Asian Russia* map on a scale of 1:7,500,000, which was part of the *Maps of the Nature, Population, and Agriculture of Asian Russia* series, provided in both digital and print form (Koneva et al. 2012). It reflects the patterns of animal and human pathogen spread on various spatial levels (planetary, regional, local) and pathogen connections to various hosts and natural ecosystems.

Human influence on the environment facilitates the growth of focal territories beyond their characteristic environmental conditions. For instance, plague, which is found in arid landscapes in nature, attacked cities in extremely varied parts of the world thanks to the activity of synanthropic rats. Because of this, as previously mentioned, anthropurgic foci are defined as foci that emerge as the result of human activity or that exist in environments that have been modified by human activity. In these foci, domesticated

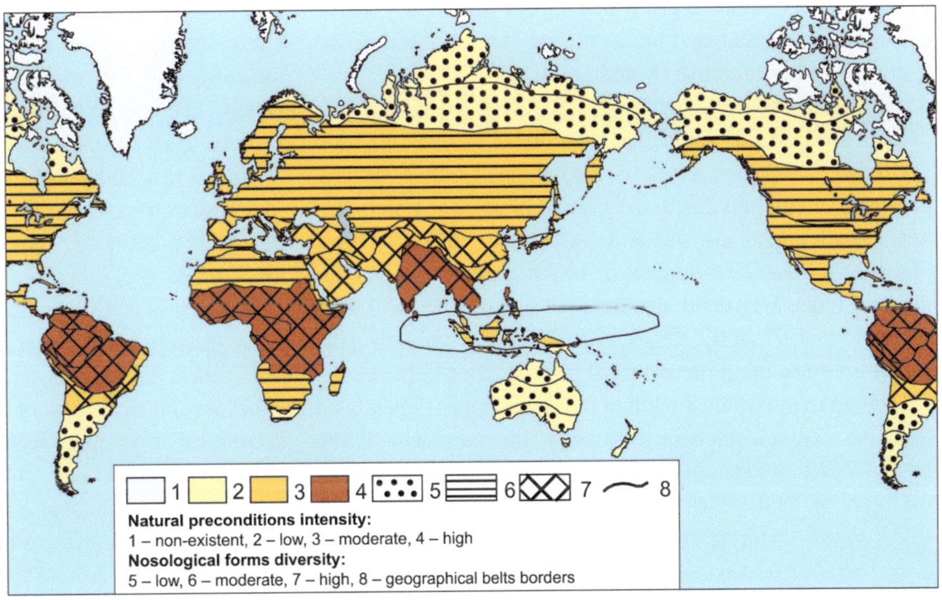

Fig. 1.3 A medical geography assessment of the Earth's geographic zones (based on a series on natural endemic diseases). *Credit* Malkhazova 2001

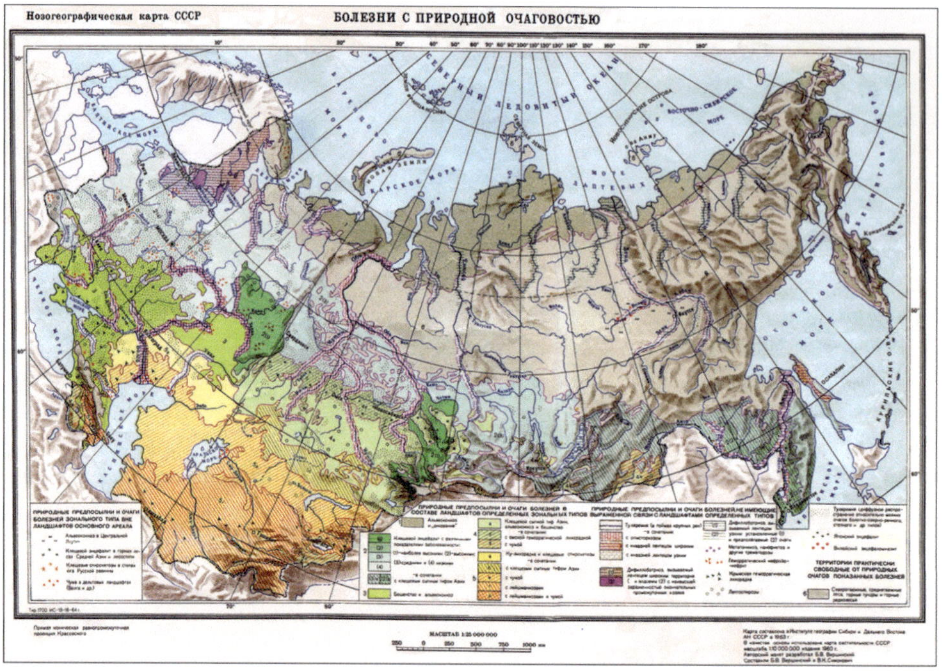

Fig. 1.4 Diseases with natural focality. A nosogeographical map of the USSR. *Credit* Vershinsky 1964

and synanthropic animals are the main pathogen carriers. Natural–anthropurgic foci are an intermediate type of focus, within which wild, domesticated or synanthropic animals can be pathogen hosts.

One of the main issues that occurs as a result of anthropogenic changes to ecosystems is the transformation of disease foci that exist in natural landscapes and changing forms of population contact with pathogen vectors and hosts. However, these issues have yet to be sufficiently explored.

Two types of medico-geographical consequences of anthropogenic changes to natural ecosystems can be distinguished: direct (short-term), which appear in the early stages of land development, and indirect (long-term), which are the result of further land use (Malkhazova 2001).

Regardless of landscape type, early stages of land development always lead to increased patchiness in disease host and vector habitats, which is usually accompanied with increasing pathogen circulation due to the growth of ecotone communities. Development also leads to increased contact between humans and disease natural foci, which also raises epidemiological risks. Migration also facilitates growth in morbidity rates. It is at the early stages of land development that the morbidity rates of all natural focal diseases tend to increase.

The direction of focal evolution in later stages depends on both the ecological characteristics of the members of the host–parasite system and the nature of human activity and the type of land development in the specific territory. In some cases, anthropogenic interference may lead to the disappearance of a natural focus. In others, members of the host–parasite system adapt to the new conditions. Moreover, the pathogen can be spread beyond the borders of the primary natural foci, leading to the formation of anthropurgic foci. Secondary (natural–anthropurgic and anthropurgic) tularaemia, plague and leptospirosis foci, among others, emerged thanks to this process.

Later on, under specific conditions, an anthropurgic focus can return to being a natural focus. For instance, Evgeny Pavlovsky believed foci in areas of ancient civilization to be an example of this phenomenon. However, not all diseases have reversible anthropurgic foci. Many vector species can easily cohabitate with humans. The foci of *Ornithodors papillipes, O. lochoreus, Rhipicephalus sanguineus,* some *Hyalomma* species and *Aedes aegypti* are non-reversible. Other species cannot survive for long in human dwellings, meaning that the foci of the diseases that they transmit cannot become fully anthropurgic (tick-borne encephalitis vectors, like *Ixodes persulcatus,* etc., are an example of this).

In the last 100 years, the principles of natural focal disease research have acquired new theoretical approaches and practical techniques, most of which come from adjacent disciplines: geography, cartography, biology, ecology, systems analysis and mathematics. In turn, natural focal disease epidemiology had an effect on the development of the above disciplines. The use of an ecological approach had a major effect on the evolution of natural focal disease theory. The accumulation of facts based on this approach allowed not only to elucidate natural focus structure and connections between foci and local conditions but also to identify the necessity of natural focal disease research.

In Russia and around the world, discovery of new natural foci and research on their epidemiology and epizootology are underway. Hypotheses about the existence of natural foci whose pathogens have existed for a long time but only became active recently, thanks to increased contact between the focus and the human population, have been proven. This lays the conditions to increased risk and periodic morbidity spikes of well-known natural focal diseases (for instance, tick-borne encephalitis, leptospirosis, tularaemia) and the appearance of emerging and re-emerging infectious diseases and diseases that recently appeared or that had been around for a long time but recently underwent a morbidity increase or spread to a new geographical region (Morse 1996). In the late twentieth century, highly contagious hemorrhagic fever pathogens were first described: Marburg virus (1967), Lassa virus (1969) and Ebola virus (1976). The search for a natural reservoir for the Marburg and Ebola viruses among vertebrates and arthropods has yet to produce a result.

Russia's natural focal disease conditions are considered to be unfavourable (On the Status… 2014). In 2013, 18,107 cases of various natural focal diseases were reported (for comparison, the number was 23,365 in 2012 and 24,496 in 2011). Sanitation services conduct various preventative campaigns directed at decreasing host and vector populations and also specifically prevent some infections. Climate and weather conditions in specific years influence morbidity rate dynamics and cyclical annual variations in vector and carrier population sizes. Despite an insignificant decrease in overall morbidity, natural focal diseases remain a relevant public health and social concern. This is related to growth in the scale and intensity of the development of territory of natural foci, the emergence of substantial territories of unploughed and abandoned agricultural lands, the growth and spread of suburban gardening, and, as a result, increasing contacts between the non-immune population and natural foci.

The geographic spread or increasing epidemiological risk of certain dangerous diseases that had not been previously widespread, such as Crimean-Congo hemorrhagic fever or West Nile fever, is a phenomenon that warrants special mention. From time to time, outbreaks of disease thought of as old or exhaustively researched occur on Russian territory. For instance, in 2013, a tularaemia outbreak affecting over 1000 people occurred in Yugra in Khanty-Mansi Autonomous Okrug.

Therefore, natural focal diseases are a serious issue in Russia, making it necessary to take data on the geography of natural focal diseases into account when making decisions about land development and the construction of recreational buildings, tourism development and conservation efforts.

Cartography of natural focal diseases is the first step in a medico-geographical evaluation and appraisal of Russia, with its enormous environmental and geographic diversity and intense economic growth in previously untouched areas. This publication consolidates and presents temporal and spatial information on the most common natural focal diseases found in Russia.

1.2 Natural Focal Diseases in Russian Atlas Cartography

Research on the geographic characteristics of natural focal diseases on Russian territory is inextricably linked with cartography. Today, a fairly substantial number of maps have been compiled. They have been published either independently or as part of an atlas, the latter of which provides the most complete depiction of this natural phenomenon. Atlas cartography, which is characterized by its systematized, comprehensive, and interdisciplinary approach, provides a multifactorial perspective on the subject in question, while the atlases themselves vary in purpose, conceptual approach, territorial breadth and scale.

Atlases edited by Prof. E. Rodenwald are a classic example. The *Atlas of Infectious Diseases*, which includes maps of the European part of the USSR, was published before World War II. This atlas was provided to German doctors that fought on the Eastern Front. After the war ended, E. Rodenwald and H. Jusatz created the *Welt Seuchen Atlas* (1952–1957). Among modern publications, the *World Atlas of Epidemic Diseases* (Cliff et al. 2004) and, from a methodological standpoint, *Cartographies of Disease. Maps, Mapping and Medicine* (Koch 2005) are of particular interest.

Three atlas categories are examined in this chapter. They are presented chronologically, as the topic developed cartographically. First, maps of natural focal diseases appeared in comprehensive general atlases, then in various comprehensive thematic atlases—medical geography, medical demography, medical ecology and ecology atlases, etc.—and finally in atlases of natural focal diseases.

Today, practically, the entire territory of Russia is represented in comprehensive atlas cartography. At the same time, the depth, detail and accuracy of this research vary significantly from one region to another and depend on a combination of objective (geographic features, the history of development, modern agricultural and industrial potential) and subjective (public interest, availability of scientific infrastructure) factors. This note extends to medical geography maps, including maps of natural focal diseases.

1.2.1 Natural Focal Disease Maps in Comprehensive General Atlases

Relatively few maps dedicated specifically to natural focal diseases and related topics can be found in atlases. Most of these maps can be found primarily in regional comprehensive atlases meant for scientific and reference purposes (Fig. 1.5).

The *Transbaikal Atlas* (1967) was one of the first comprehensive atlases to include a series of maps of natural focal

Fig. 1.5 Atlases containing maps of natural focal diseases

diseases. The series appeared thanks to the medical geography school of the Institute of Geography of Siberia and the Far East, which was formed under the leadership of Evgeny Ignatiev (Medical Geography… 1964; Principles and Methods… 1968). A map entitled *Diseases with Natural Foci* on a scale of 1:5,000,000 can be found in the section called 'Medico-Geographical Evaluation of a Territory. Public Health Services'. It is accompanied by an inset map entitled *A Plague Natural Focus in the South-Eastern Transbaikal* on a scale of 1:3,500,000, which shows the population density of the Tarbagan marmot and the plague epizootic before 1947 and after 1959. Along with two other maps—*Ixodidae Ticks as Pathogen Vectors* and *Mammals as Pathogen and Vector Hosts and Nutrient Sources* on a scale of 1:5,000,000—this map reflects the actual and potential danger of specific geosystems with regard to a number of natural focal diseases (tick-borne encephalitis, North Asian tick typhus, alveolar echinococcosis, rabies, tularaemia, etc.). Another map, *Medico-Geographical Regional Zoning,* on a scale of 1:4,000,000, is the result of comprehensive medico-geographical research on the country's territory and classifies regions by their combinations of natural preconditions for human disease and the conditions of their potential emergence based on how local resources are used by humans. This medico-geographical analysis takes into account natural focal diseases that are found in specific types of territories. The maps were created based on scientific literature, data collected by foundations and the results of comprehensive field work.

In the *Atlas of Sakhalin Oblast* (1967), published at the same time as the *Transbaikal Atlas*, maps entitled *Human Helminthiases and Natural Focal Diseases* on a scale of 1:4,000,000 and *Likelihood of Infection with a Natural Focal Disease* on a scale of 1:2,000,000 were included in the 'Medico-Geographical Conditions and Public Health Services' section. Perennial data collected by regional sanitary and epidemiological stations and medical institutions were widely used in map compilation.

The first one shows the human infection rate (as a percentage) with ascariasis, diphyllobothriasis and trichuriasis. The second map shows no reported cases of infection, but the varying degree of likelihood (high, moderate, low) of the formation of human and livestock disease foci containing tularaemia, Q fever, leptospirosis or tick-borne encephalitis. The population size of taiga ticks is also mentioned. The map was to serve as a foundation for future research and the planning of preventative measures.

The second edition of the *Atlas of Tyumen Oblast* (1976) contains a markedly larger number of maps. 12 maps make up the 'Medical Geography Conditions' section, which concludes the first part of the atlas, 'Nature and Resources'. The majority of maps in this section was prepared by scientists belonging to the Irkutsk school of medical geography, so they are conceptually close to those in the *Transbaikal Atlas*. The main map of the 'Diseases Shared by Humans and Animals (Zooanthroponoses)' section, on a scale of 1:6,000,000/1:4,000,000 (the southern part of the Oblast), reflects types of territories based on the combination of pathogens in the region and the risk of human infection, based on the specifics of the environmental conditions. The legend of infections that circulate within geosystems shows the level of risk posed to the population (high, moderate, low), the agricultural and household factors that facilitate disease, and a nature analysis of the territory's natural conditions. Regions with unfavourable situations with regards to anthrax and brucellosis are highlighted. Localized diagrams reflect places where the pathogens were discovered and characterize 12 disease forms and research methods (serological analysis, virology and helminthology). Two maps, *Opisthorchiasis* and *Diphyllobothriasis,* despite their small scale of 1:8,000,000, are extraordinarily rich in data. They contain information on the intermediate, additional and definitive hosts of the disease and the degree of its prevalence (in % of examined animals). They also reflect the degree of incidence in the population in large inhabited areas and the rural population by administrative district (in % of examined individuals).

Three maps are devoted to Ixodidae ticks, which transmit pathogens that cause natural focal diseases. They show tick locations, ranges (on a scale of 1:12,000,000), taxonomy and habitats, including population sizes in different locations (on a scale of 1:6,000,000). Another five maps on a scale of 1:16,000,000 convey the patterns in the spread and attack intensity of blood-sucking insects that participate in the transmission of a number of diseases.

A synthetic map, *Regional Medico-Geographical Zoning,* on a scale of 1:6,000,000, conveys a systemic characterization of trends and patterns in the spread of natural focal diseases and takes specific features of the environment (climate, landscapes, the soil's capacity for self-purification, surface runoff, etc.) and kinds of human land use into account.

In the *Altai Krai Atlas* (1978), maps of natural focal diseases make up their own subsection, entitled 'Natural focal Diseases and Likelihood of Population Infection', in the 'Evaluation of Environmental Conditions. Nature Conservation' section, which concludes the first volume. The subsection includes four analytical maps, compiled using the same methodology on a scale of 1:4,000,000. Rich with information, they reflect natural preconditions for the existence of foci of the most significant diseases in the Krai (tick-borne encephalitis, North Asian tick typhus, tularaemia, leptospirosis). At the same time, the risk of infection among people who live very close to natural foci or periodically come into contact with them is shown using a variety of methods. A map of tick-borne encephalitis on a scale of 1:3,000,000 shows the potential risk of ticks biting humans (based on tick population size) and the actual fraction of immune individuals among the adult rural population in specific populated areas. The three other maps present three to seven types of natural or anthropurgic foci, classified based on their landscape and biocenotic preconditions, the number of years with reported cases (North Asian tick typhus, tularemia) and the actual fraction of immune individuals among the adult population (North Asian tick typhus, leptospirosis). The *Ixodidae Ticks* map (on a scale of 1:4,000,00), containing a fairly exhaustive appraisal of 19 types territorial groupings of ticks that are found in the Krai, is part of a section entitled 'The Animal World and its Resources'.

The maps are accompanied by a text that describes compilation methods, the characteristics and types of natural foci, the risk of infection and preventative measures. Materials collected by the corresponding institutions, as well as the results of a number of expeditions and special experiments, served as the basis for the mapping. The borders of the preconditions for the existence of natural foci take into account landscape segments, vegetation, the roughness of the landscape, ploughing intensity, etc.

The *Atlas of Voronezh Oblast* (1994) contains a map entitled *Natural Focal Diseases* on a scale of 1:3,000,000 in the 'Condition of the Environment' section. The table legend for nine epidemic territories mentions major natural focal diseases: rabies, leptospirosis, Q fever, tularaemia, and hemorrhagic fever with renal syndrome. The degree of epidemiological risk—high, moderate, or low—is noted. The information is split into 15 × 15 mm (45 × 45 km) segments.

The relatively short *Primorsky Krai Atlas* (1998) provides a *Medico-Geographical Map* on a scale of 1:3,000,000. The legend, entitled 'Risks Posed by Natural focal Diseases in the Krai', depicts the level of natural focal disease morbidity in the population (from low to very high) and its structure by administrative district (for six diseases: tick-borne encephalitis, pseudotuberculosis, rabies, etc.).

Extremely relevant information on risk of infection with natural focal diseases in the intensively developing Yugra (Khanty-Mansi Autonomous Okrug) is presented in the comprehensive scientific and reference *Atlas of Khanty-Mansi Autonomous Okrug/Yugra* (2006, vol. 1). Five maps on this topic are included in the 'Medico-Geographical Conditions and Population Health' section. The main map—*Nosogeographical Zoning (Zooanthroponoses)* on a scale of 1:6,000,000—reflects highly ranked disease ecosystems—faunistic associations that contain the members of host–parasite systems. The map shows a total of 15 nosoecological districts, each of which is accompanied by a list of 13 relevant diseases in order of descending risk of infection.

A map entitled *Ixodidae Ticks* on a scale of 1:6,000,000 was compiled based on long-term data. The likelihood of encountering Ixodidae ticks (from high to low) is accompanied with markers for human settlements where they are found, and tick bites and tick-borne encephalitis morbidity rates are reported. Despite its small scale (1:8,000,000), the *Blood-Sucking Diptera* map provides an understanding of the territorial distribution of insect landscapes, bite intensity and duration of their mass activity. Two maps on a scale of 1:8,000,000 characterize opisthorchiasis and diphyllobothriasis morbidity rates in the Okrug (number of cases per 100,000 individuals) by city and administrative district.

A fairly well-studied topic is presented in the *Atlas of the Kuril Islands* (2009). The 'Population' subsection of the 'Population and Economics' section includes a map with commentary, entitled *Risk of Infection with Natural Focal Diseases in the Population* on a scale of 1:2,500,000. The map characterizes the degree of risk from high (over 2.0 points) to low (less than 1.2 points). The danger was evaluated on a 5-point scale based on a combination of natural preconditions for the emergence of a specific disease when the territory is settles or visited by humans. Both qualitative and quantitative measures of the environment were taken into account (the availability of animal nutrient sources and insect vectors, climate and landscape conditions that facilitate the formation and function of natural foci, etc.). The point-based scores are explored in the '*Evaluation of Unfavourable Natural Focal Disease Preconditions on the Kuril Islands (Point-Based)*' and are accompanied with circular vector diagrams that show the average total of factor scores that characterize the risk posed by a specific territory. Calculations were conducted for 23 natural focal diseases.

The majority of popular science and school atlases, which were prepared in the late twentieth and early twenty-first centuries according to a standard and have a limited number of maps, contain essentially no information on the topic in question. One of the first maps on this topic, *Natural Focal Diseases* on a scale of 1:4,000,000, appeared in the *Atlas of Pskov Oblast* (1969) as an inset in the *Public Health Service* map. The map reflects the territorial distribution of tick-borne encephalitis and leptospirosis foci.

The National Atlas of Russia, published in 4 volumes, was a landmark publication in the field. However, medical geography information is quite limited in this collection. Volume 2, 'Nature. Ecology' (2007), contains an entire series of maps on scales ranging from 1:15,000,000 to 1:75,000,000 (*Land Vertebrates, Mammals, Birds, Rodents, Fish in Landlocked Bodies of Water*), which were the first to give such an exhaustive account of animal geography and that of specific species, some of which are the carriers of natural focal disease pathogens. Volume 3, 'Population. Economics' (2008) includes 2 maps on this topic. One is *Population Infectious and Parasitic Disease Morbidity* on a scale of 1:45,000,000, while the other is *A Sanitary and Ecological Evaluation of Russia's Territory* on a scale of 1:30,000,000, which is the same as the map in the earlier *Ecological Atlas of Russia* (2002).

1.2.2 Natural Focal Disease Maps in Comprehensive Thematic Atlases

Natural focal diseases are depicted in thematic ecological atlases. The serious danger posed by these diseases calls for the recognition of their foci as factors in the environmental danger or risk posed by a specific territory. Depending on the epidemiological characteristics of the infections and on the availability of information in each specific case, they are taken into account during the environmental evaluation of a territory on a case-by-case basis.

Among the comprehensive regional medico-geographical atlases, the *Medico-Geographical Atlas of Primorsky Krai* (1969) must be mentioned. It clearly aims to reflect the nosological potential of the territory. Approximately half of the maps (18 out of 38) touch on the topic of natural focal diseases. Despite the small scale and schematic nature of the images, the black-and-white maps provide a fairly complete understanding of the distribution of the main vectors and hosts (Ixodidae and Gamasid ticks, mosquitoes, fleas, mice, etc.) of the geography and types of disease natural foci (Japanese encephalitis, tularaemia, scrub typhus). Maps of the distribution of zoonoses can be found in the same publication: salmonella in pigs, brucellosis in livestock and anthrax in animals. The atlas is primarily based on data collected by a number of institutions and organizations.

The *Medico-Geographical Reference of the Karelian ASSR* (1990) contains seven black-and-white maps. One of them depicts parasitic and infectious disease morbidity rates (per 10,000 individuals) on average over a 4-year period. Five others (arboviral infections in the Californian encephalitis and Ukuniemi group, Pogosta disease, hemorrhagic fever with renal syndrome, diphyllobothriasis, and tick-borne encephalitis) show collective population immunity to these viruses, relative and absolute population morbidity rates, natural foci, etc. The seventh map depicts anthrax natural foci that were recorded between 1930 and 1938.

Another atlas worth mentioning is the *Ecological Atlas of Rostov Oblast* (2000). It contains five maps of natural focal diseases: anthrax, Crimean-Congo hemorrhagic fever (CCHF), plague, tularaemia and leptospirosis. The distribution of anthrax is characterized by the number of stably unfavourable locations per 100,000 ha. The epizooticity of regions (based on the duration of infection activity or association with particular types of foci—in steppes, in river deltas or the Don River, etc.) is defined for CCHF, plague and tularaemia. Territories with leptospirosis are differentiated into relatively safe zones, at-risk zones and dangerous or endemic zones. The final map reflects endemicity (epizooticity) for three natural focal diseases. All the information is associated with the Oblast's administrative districts (cartograms). The textual commentary in the atlas must be noted, as it rivals the maps in its significance. The texts, written according to a unified plan (name of disease, sources of infection, epidemiological characteristics, risk of potential infection, prevention), do not just enhance the maps but also are valuable in their own right.

The *Atlas of Irkutsk Oblast: Ecological Conditions of Development* (2004) is unique in its exhaustiveness, variety and originality. It provides an appraisal of the region from a medical geography perspective, focusing on natural focus formation patterns, their function and infection risk. 15 of the 21 atlases in the 'Medical Geography Conditions' section pertain to the topic at hand. The section is opened by *Regional Division by Landscape and Epidemiology* on a scale of 1:7,500,000. The map highlights 16 regions—territorial combination of the majority of infections and parasitoses that have been reported or found in the Oblast over a number of years. Taking into account their epidemiological significance, the diseases are grouped in two categories based on the nature and volume of primary preventative measures.

A map of ticks on a scale of 1:25,000,000, reflecting types of tick populations, their population size, the species makeup and their habitats, is crucial for this territory. The map is augmented by a table that shows changes in tick population size as a result of anthropogenic factors. Four maps, located in the order of the diseases' epidemiological significance, reflect the risk of infection (percentage of the human population that has specific antibodies) with tick-borne encephalitis, North Asian tick typhus, anthrax, Q fever, tularaemia and psittacosis. The map of agricultural zooanthroponoses reflects the spread of gastrointestinal infections common within agricultural lands.

Two synthetic maps, *The Territory's Nosoecological Potential* on a scale of 1:10,000,000 and *Ecological Preconditions of the Spread of Zooanthroponoses* on a scale of 1:5,000,000, which demonstrate the spatial units of various nosoecological entities, which are nosoecological zones, belts and regional types of zooanthroponoses, are of special interest. A special classification of zooanthroponoses, based on their links to nature networks and animal groups, serves as their foundation: ubiquitous, semiaquatic, meadow, etc. The matrix of the legend reflects the nature of pathogen circulation (spilling, patchy, local) and the nature of the host–parasite systems (predominantly natural, natural and synanthropic, predominantly synanthropic, directly involving humans).

The *Medico-Ecological Atlas of Moscow oblast* (2004) contains three cartograms in the 'Adult Population Health' section: *Salmonella Morbidity Rate* (per 100,000 individuals) in five steps, *Frequency of Diseases Caused by Animal Bites* (five stages), and *Tick Bites and Tick-Borne Encephalitis Cases*, which provides both relative and absolute data.

Population Health and Public Health Services in Sakha (Yakutia) at the turn of the century, a medico-geographical atlas (Krivoshapkin et al. 2005) contains just one map devoted to the issue of natural focality—*Number of Viliusk Encephalomyelitis Cases*—on a scale of 1:15,000,000.

The *Medico-Demographical Atlas of Kaliningrad Oblast* (2007) and the *Medico-Demographical Atlas of Moscow Oblast* (2007) are practically limited to maps of adult and child population morbidity rates for some common infectious and parasitic diseases. Only in the former there is a map of the main natural focal infections in the region. Cases reported in municipalities (tick-borne encephalitis, leptospirosis, hemorrhagic fever with renal syndrome and animal rabies) are marked on it. Areas that are unfavourable in terms of anthrax infections are marked as well.

The new edition of the *Medico-Geographical Monograph Atlas of Voronezh Oblast* (2010) contains four more maps compared to the previously published comprehensive atlas. Maps entitled *Hemorrhagic Fever with Renal Syndrome* and *Tularaemia and Leptospirosis* depict the number of cases and locations where the pathogen was found. The *Rabies* map shows rabies foci activity (the total number of cases) and the number of cases for each animal species.

The *Medico-Demographical Atlas of Smolensk Oblast* (2012) contains three maps (*Some Infectious and Parasitic Diseases* on a scale of 1:1,500,000) which reflect adult, child and teenage morbidity rates.

The *Ecological and Geographical Atlas Book of Voronezh Oblast,* published in 2013, contains the following map in a section called 'The Environment and Human Health:' *Some Infectious and Parasitic Diseases* and *Especially Dangerous Natural Focal Diseases* [rabies, hemorrhagic fever with renal syndrome, tularaemia, and leptospirosis] *Common in the Region*. The database included a 15-year time frame (1998–2012).

Comprehensive federal atlases present information on natural focal diseases in varying degrees of detail. *The Environment and Population Health in Russia* (1995) only presents a few analytical maps of population morbidity rates for brucellosis, diphyllobothriasis, opisthorchiasis, tularaemia and tick-borne encephalitis per 100,000 individuals (the average from 1991 to 1993).

The *Atlas of the Distribution of Natural Focal Viral Infection Pathogens in the Russian Federation* was published in 2001a. It depicted the geography of viruses that was known at the time of publication in great detail. Maps of the country's administrative subjects, grouped into 11 regions on a scale between 1:1,250,000 and 1:39,800,00, show the distribution of arboviruses in eleven families and eleven genera. 69 viruses were first isolated on former Soviet territory, 42 of which were listed as new.

The *Atlas of Natural and Technogenic Hazards and Risks of Emergency Situations in the Russian Federation* (2005), as well as a series of eponymous atlases, which were prepared as part of a unified program for Russia's federal districts, devote a great deal of attention to the topic in question. The 7th section of the atlas of Russia, 'Biosocial Dangers', consists of 28 maps, more than half of which are dedicated to natural focal diseases that affect humans and livestock. Biosocial dangers are types of risks such as dangerous infectious and parasitic human diseases, dangerous diseases of livestock and especially dangerous infestations and infections affecting cultivated plants. Diseases that can become epidemic in nature are seen as the most dangerous human diseases. They spread fast and wide, affect large numbers of people, have severe health effects and can even be fatal. The danger of human diseases is evaluated based on the morbidity rate, which is the number of cases per a specific portion of the population in a specific time interval.

The section opens with a map entitled *The Danger of Biosocial Emergencies* on a scale of 1:30,000,000. The danger (or risk) of such emergencies is typically reflected as the number of emergencies on Russian territory over a specific time period (usually a year), which allows the researchers to take seasonal dynamics of various influencing factors into account. Seven levels of emergency recurrence risk are shown on the map, ranging from very low (less than 1) to high (11–20). Seven maps reflect the risk of infection with tularaemia, leptospirosis, brucellosis and other natural focal diseases. The degree of risk is evaluated based on absolute or relative markers of reported disease or on the relationship between the morbidity rate in the region and the average morbidity rate in Russia overall. A somewhat lower number of maps is dedicated to the risk of disease emergence among livestock. Maps of the risk of emergency occurrence based on the appearance of brucellosis, foot-and-mouth disease, etc. are also presented. The end of the section provides a timeline of the largest emergencies in Russia. Of the 456 emergencies that occurred in Russia from 1991 to 2004, no more than 16 were related to natural focal diseases. While the section structure is preserved, the number of maps within them varies greatly.

For instance, in the *Atlas of Natural and Technological Hazards and Risks of Emergencies in the Russian Federation. Far Eastern Federal district* (2007), 22 of 28 maps are connected to natural focal diseases affecting people and livestock. It must be noted that the entire spectrum of biosocial dangers found in Russia is also found in the Far East Federal District. For a number of dangerous diseases, the morbidity rates in the District exceed the average rate in Russia. The section opens with a map called *The Risk of Biosocial Emergencies* on a scale of 1:30,000,000. Risk levels (based on annual emergency recurrence) from very low (less than 1) to high (11–20) are shown on the map. The next map—*Territories in Primorsky Krai Posing a Danger in Terms of Natural Focal Diseases* at a scale of 1:10,000,000—shows population morbidity rates per administrative district ranging from low to very high. Diagrams explaining the structure of natural focal diseases per district are also presented. Individual bars exist for tick-borne encephalitis, hemorrhagic fever with renal syndrome, pseudotuberculosis, rabies and Japanese encephalitis, while other diseases are grouped into a single bar.

The *Unfavourable Territories in Terms of Primary Population Morbidity* map highlights regions where the population morbidity rate is lower than, within the margins of, or higher than the average rate in Russia. The map is accompanied by a list of significant emergencies related to population morbidity from 1992 to 2005, which includes information on the date, location, scale and consequences of the emergency. It is followed by nine maps on a scale of 1:30,000,000 which reflect the risk of infection with pseudotuberculosis, tularaemia, brucellosis, hemorrhagic fever with renal syndrome, leptospirosis, malaria, and tick-borne encephalitis, the risk of infection with salmonella, and the risk of acute gastrointestinal infections or parasitic disease. The degree of risk is evaluated based on either the relationship between the morbidity rate in the subject and the average across Russia or on the long-term average annual number of cases from 1997 to 2005. Four maps are dedicated to the risk of disease emergence among livestock, in particular, maps of the risk of emergencies caused by epizootics of classical swine fever, foot-and-mouth disease, large cattle brucellosis, animal rabies or by the emergence of anthrax foci. *The Risk of Flooding in Anthrax Burial Grounds* is an individual map. The section concludes with a timeline of the largest emergencies in the district. No more than 2 of nearly 200 emergencies were caused by natural focal diseases.

1.2.3 Atlases of Natural Focal Diseases

There are a small number of atlases dedicated expressly to natural focal diseases. As of now, they are singular and have been prepared for regions undergoing intensive development that pose a high risk of infection with natural focal diseases (Primorsky Krai, Sakhalin Oblast, etc.). The *Medico-Geographical Atlas of Krasnoyarsk Krai* (Pospelov 1970) was one of the first and is now a rare edition. The importance of the information on natural focal diseases in an enormous and sparsely populated region of country in the atlas cannot be overstated. This research also played a large role in the development of the theoretical and methodical principles of medico-geographical cartography. These principles were furthered in the *Landscape and Epidemiology Atlas of the European Territory of the RSFSR, the Urals, and the Crimean Oblast of the Ukrainian SSR* (Pospelov 1987). It consists of two parts: 'Sect. 1. Medical Evaluation of Environmental Conditions' and 'Sect. 2. Nosogeography'. The first one depicts the natural preconditions for the emergence of natural focal diseases. The epidemiological significance of mammals and birds is discussed, along with the following topics: 'Rodents as Carriers of Natural focal Disease Pathogens' and 'Natural Carriage of Human Diseases by Ixodidae Ticks'. The second section is completely devoted to morbidity and consists of 24 maps of infectious (typhoid fever, tularaemia, rabies, etc.) and parasitic (opisthorchiasis, diphyllobothriasis) diseases. The section concludes with a comprehensive landscape- and epidemiology-based division of the area into regions, which are shown against the backdrop of forest vegetation zones. The table legend characterizes the regions based on their natural and economic conditions and on the risk of infection. The table also notes the urban and rural population size and density, the main forms of human activity and living conditions in each region.

The *Atlas of Natural Focal Diseases of Sakhalin Oblast* (1992) was based on the results of long-term, comprehensive (parasitology, zoology, virology, bacteriology, epidemiology) studies and scientific literature aimed at the study of natural foci, small mammals and arthropods—the hosts and vectors of pathogens that cause human infectious diseases.

The atlas consists of five chapters: 'Introduction' (*Vegetation* and *Field Work Territories* maps), 'Small Mammals as the Carriers of Disease Pathogens', 'Blood-Sucking Arthropods as the Vectors of Disease Pathogens', 'Natural Focal Diseases', 'Natural Focal Helminthiases'. The maps have a scale of 1:4,000,000 or 1:8,000,000. The atlas includes information on 14 infectious diseases (tick-borne encephalitis, psittacosis, Lyme disease, leptospirosis, etc.), as well as helminthiases (trichinellosis, echinococcosis, diphyllobothriasis), the natural foci of which are widespread in the Oblast. A short description of rodents, shrews, members of the Lagomorpha family, as well as specific groups and species of blood-sucking arthropods (Ixodidae ticks, fleas, horseflies, mosquitoes) that are found in Sakhalin Oblast and are the most significant pathogen vectors and carriers, is also provided. Questions of the spread of mammal and arthropod populations across the territory and across biotopes, as well as their role in epizootic processes, are discussed as well. The nosogenic territory,[1] the spread across the territory and its specific regions, data on the epizootic situation in the foci and the epidemiology of each disease are also listed. Maps of natural focal diseases show natural foci, areas with active epizootics, locations where viruses and bacterial cultures have been isolated from ticks, small rodents and other carriers, and locations where people were infected.

The *Atlas of Natural Focal Infectious and Parasitic Diseases of the Komi Republic* (2004) has practical significance for development in northern territories.

The *Medico-Ecological Atlas of Khabarovsk Krai and the Jewish Autonomous Oblast* (2005) is similar in structure to the *Atlas of Natural Focal Diseases in Sakhalin Oblast*, although it is three times greater in volume. The maps are on a scale of 1:5,400,000 and 1:9,700,000. In essence, this is an atlas of natural focal diseases. Aside from an Introduction, it includes four sections: 'Land Mammals' (23 maps), 'Blood-sucking Insects and Ticks' (9), 'Natural Focal Diseases' (32) and 'Natural Focal Helminthiases' (7). The maps and textual commentary demonstrate the degree of epizootic and epidemiological activity of the foci by district and landscape, areas of pathogen discovery, and the distribution and population size of insectivores, bats, members of the Lagomorpha family, rodents, predators, hoofed mammals, as well as blood-sucking insects and ticks. Each infection and parasitic disease is described, and its seasonality, annual oscillations in its epizootic and epidemiological activity, the category of infected individuals and the conditions of infection are also noted. A landscape-based and geographical characterization of the disease is provided; seasonality, population sizes and the role of each mammal species and arthropod in the natural foci are described. In addition, major preventative measures against natural focal diseases are listed.

The advent of the Internet allowed for the publication of atlases online. Regional institutions took advantage of this opportunity in order to publish results of long-term studies on natural focal diseases, which were compiled and generalized in atlases varying in size and depth of analysis. The availability of online publications makes it possible for this information to reach a wide audience. The *Medico-Ecological Atlas of Krasnodar Krai* (2003), the *Medico-Biological Atlas of Natural Focal Infections of Orenburg Oblast* (2007), and the *Atlas of Natural Focal Infections in Ulyanovsk Oblast…* (2009) all have online versions. One of the most recent online publications is the *Atlas of the Distribution of Natural Focal Infections in the Resort City of Sochi* (2014), which was prepared in anticipation of the 22nd Olympic Games. It is the product of the summarization, analysis and visualization of data from epizootology research and epidemiological monitoring of the territory. The atlas shows epidemic and epizootic activity caused by natural focal infections (Lyme borreliosis, hemorrhagic fever

[1]Also known as "nosoareal".

with renal syndrome, leptospirosis, infection with *Yersinia entercolitica*) based on data collected between 2009 and 2013, the risk of population infection with natural focal infections (non-existent, low, moderate, high), and maps of the *Density of Epidemiologically Significant Entities and Mosquito Species Caught in 2013*.

Therefore, medico-geographical mapping in Russia, which includes the mapping of natural focal diseases, is fairly significant in the sphere of regional and local cartography. The level of cartographic knowledge on natural focal diseases is extremely uneven in terms of both the spatial span of the territories studied and the accuracy and precision of the mapping images. Varying quality and incomplete data, as well as the use of different methods of mapping, prevent researchers from achieving a comprehensive picture of the distribution of natural foci and population morbidity on Russian territory.

Therefore, a cartographic summary that would reflect the geography of natural focal diseases throughout all of Russia on a federal level did not exist until the *Medico-Geographical Atlas of Russia 'Natural Focal Diseases'* was published in 2015 (a 2nd revised edition was published in 2017). It provided the basis for this monograph (Medico-Geographical Atlas… 2015, 2017).

1.3 Medical Geography Research Principles and Methods of Map Compilation

This edition presents the results of medical geography research on the distribution of natural focal diseases across Russia's territory. It was completed in accordance with the general methodology of thematic cartography. The main principles are relevance, evidence-based and geographically accurate images and temporal unification of maps. These principles are achieved via a systematic and comprehensive approach.

The study set the following objectives:

- The systematization and analysis of data on the role of environmental and socio-economic factors in the spread of natural focal diseases.
- Depiction of the nosogenic territories of the major natural focal diseases in Russia.
- Identification of the most active foci and evaluation of the risk posed to humans.
- Analysis of natural focal disease morbidity rates on Russia's territory overall and in greater detail, using information from model regions that have the corresponding data.
- The preparation of a foundation for the development of recommendations for the organization of differentiated preventative measures aimed at lowering natural focal disease morbidity within the framework of state-organized public health and ecological conservation methods.

These objectives are based on crucial methodological principles of medical geography research:

- Natural focal diseases are closely linked with the existing natural, socio-economic and demographic characteristics of a specific territory.
- Each natural ecosystem has a specific 'set' of natural preconditions for human disease, i.e. each landscape has typical host–parasite systems that formed under specific ecological conditions.
- The set of natural focal diseases that is characteristic for a particular territory, along with the unique features of their distribution, can be used as indices for nosogenicity, or the epidemic potential of natural ecosystems.
- Whether or not natural preconditions give rise to human disease depends on the ecological condition of the environment and on a combination of socio-economic conditions.
- A systemic approach is necessary for the identification and adequate representation of causal links between the nosogenicity (potential danger) of the environment and population health.
- Natural focal diseases and their geographic preconditions are dynamic and act in accordance with patterns of geosystem evolution and the specifics of human land use in a particular territory; accurate mapping is only possible when these dynamic changes are taken into account.
- The survey-like nature of the maps requires the use of extrapolation and analogies during analysis and synthesis of spatial information on a national (sub-planetary) level.
- The synthesis of medical geography data and the receipt of new information on the spatial patterns of the phenomenon of natural foci is made possible by mathematical modelling of maps.

Maps, texts, archival material, Rosstat socio-economic data and Rospotrebnadzor data on population natural focal disease morbidity rates, and satellite photography served as information sources throughout our research. A number of topics also drew from field work and original analysis on the local and regional levels.

Map compilation was preceded by a general preparation of a geographic basis on a scale ranging from 1:25,000,000 to 1:60,000,000 for maps of Russia and 1:120,000,000 for world maps. The maps were compiled using normal equidistant conic projection (Central Meridian 100°E and latitudes of the origin at 36°N and 56°N).

In accordance with the original conception, the monograph consists of five thematic chapters:

1. Introduction
2. Natural and Socio-economic Conditions
3. Natural Focal Disease Hosts and Vectors
4. Major Natural Focal Disease Distribution
5. Spatial Organization of the Sanitary and Epidemiological Service.

The 'Introduction' describes the contemporary understanding of natural focal diseases as complex networks of interconnected and interdependent populations of pathogens, animal carriers and arthropod vectors, which together form a biotic component that is organically linked to natural territorial networks. A brief explanation of the history of natural focal disease research, the contributions of Russian and non-Russian scientists to the development of the theory of the emergence and spread of natural focal diseases and issues faced by the field today is also provided. Existing research on natural focal disease mapping in Russia as part of both general and thematic maps is also listed. The chapter includes a physical map of Russia (Map 1.1) and a map of Russia's administrative divisions (Map 1.2). They serve as a basis for following chapters, simplifying the connection between the subject matter, geographic objects and federal and administrative land divisions.

The 'Natural and Socio-economic Conditions' chapter provides an understanding of the characteristics and conditions of the environment in which host–parasite (pathogen–vector–warm-blooded host) systems form in Russia. Natural preconditions of disease determine the possibility of the emergence, function and distribution pattern of natural foci. The structure of disease natural preconditions depends, in many ways, on the geographic location of the territory. A series of analytical, comprehensive and synthetic maps represents individual components of the landscapes and their ecological condition.

Maps of the demographic and socio-economic preconditions represent factors that have the greatest influence on the appearance and spread of natural focal diseases. Maps that reflect population distribution patterns across Russia and demographic markers—size, urban/rural population ratio—are also provided. These maps identify the territorial characteristics of population structure as one of the links in the network that forms the epidemiological risk posed by a territory. Combined with maps of land use, they allow for the identification of the sources that lead to various combinations of anthropogenic loads and the transformation of natural landscapes. Map compilation in this section was conducted in accordance with the methods and principles of geographic cartographic that have formed in the Russian scientific community (Modern Geographic Cartography 2012).

The 'Natural Focal Disease Hosts and Vectors' chapter presents maps of the ranges of animals classified by taxonomic unit (class, order, etc.) and maps of the animal population. Mapping of the main carriers' ranges was conducted in accordance with traditional methods of small-scale zoogeographical mapping. The materials that served as the basis for this map were collected by the authors of this work and are the result of long-term research in a variety of regions in Russia, as well as analysis of data in the scientific literature and in archival material. Ranges and qualitative backgrounds are used to demonstrate the distribution of epidemiologically significant birds, fish, and mammals, along with blood-sucking insects and ticks. The textual commentary illuminates the role of each species in the formation and function of natural foci and in their association with specific landscapes.

The 'Major Natural Focal Disease Distribution' map is a combination of textual commentary and a set of maps aimed at characterizing population health with regards to natural focal pathology. 23 nosogenic units were mapped. Each disease is described, and information on sources, transmission methods and disease spread, epidemiological activity, conditions of population infection and seasonal and annual oscillations in epizootic and epidemiological activity is provided. Where possible, risk of infection and the danger posed by the infection are evaluated.

The chapter adheres to the following scientific and methodological requirements:

- The data must be relevant to a medical geography analysis of the territory.
- The current state in terms of medical geography in the country (both for individual diseases and for their combination) must be based on maximally broad spatial and temporal information.
- Morbidity dynamics for major infectious and parasitic diseases must be represented.
- The conditions (favourable or unfavourable) in Russia's administrative subjects with regards to specific diseases and combinations of diseases must be evaluated.

The study of spatial and temporal patterns in disease spread among the population is central to research in medical geography. Given the specificity of the information gathered in each administrative subject of the country, these subjects serve as primary cartographic units. This method is more of a characterization of the epidemiological situation in the period in question than an examination of the spread of natural foci, whose ranges are primarily characterized by environmental conditions. During analysis of the

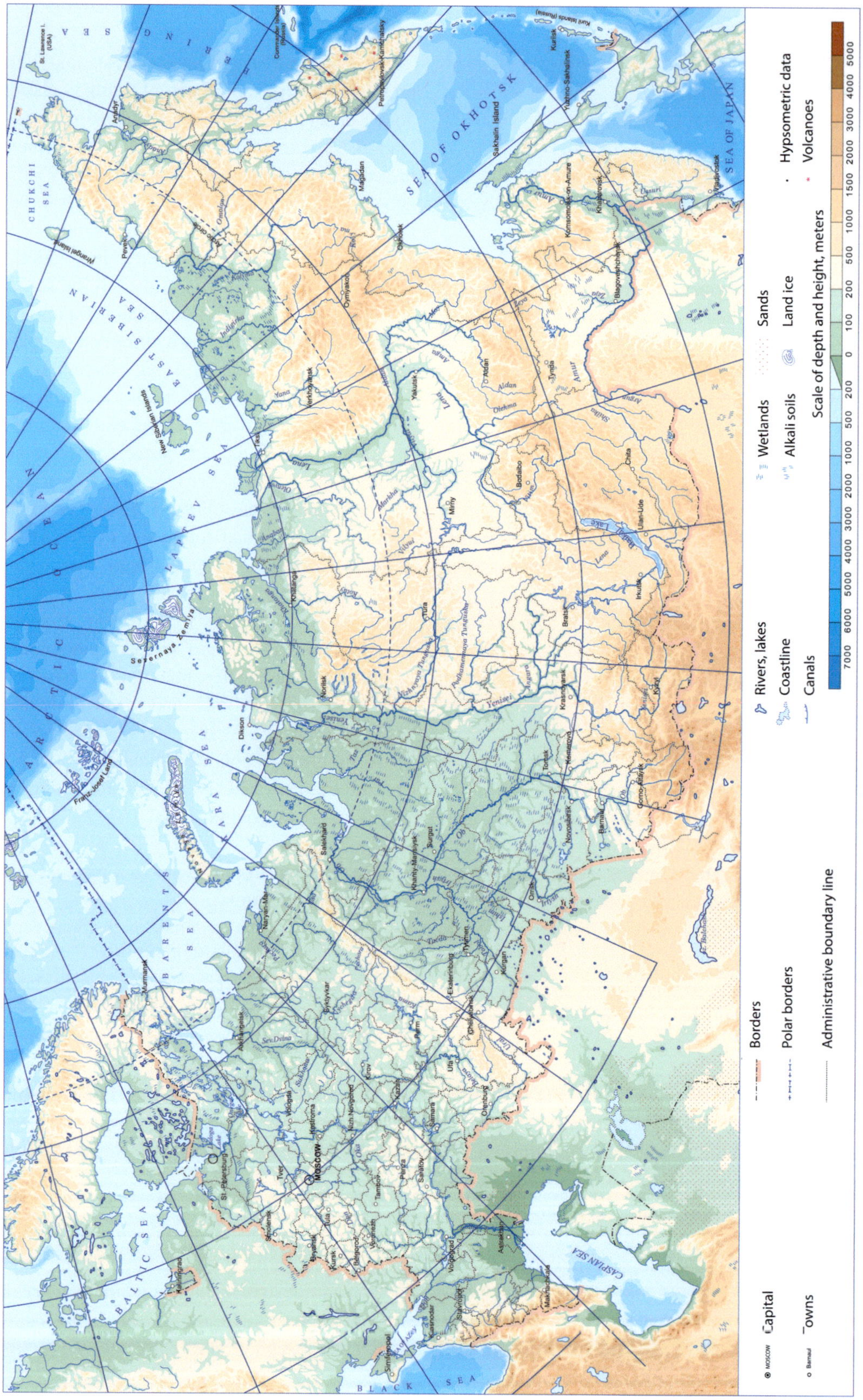

Map 1.1 Physical map of Russia

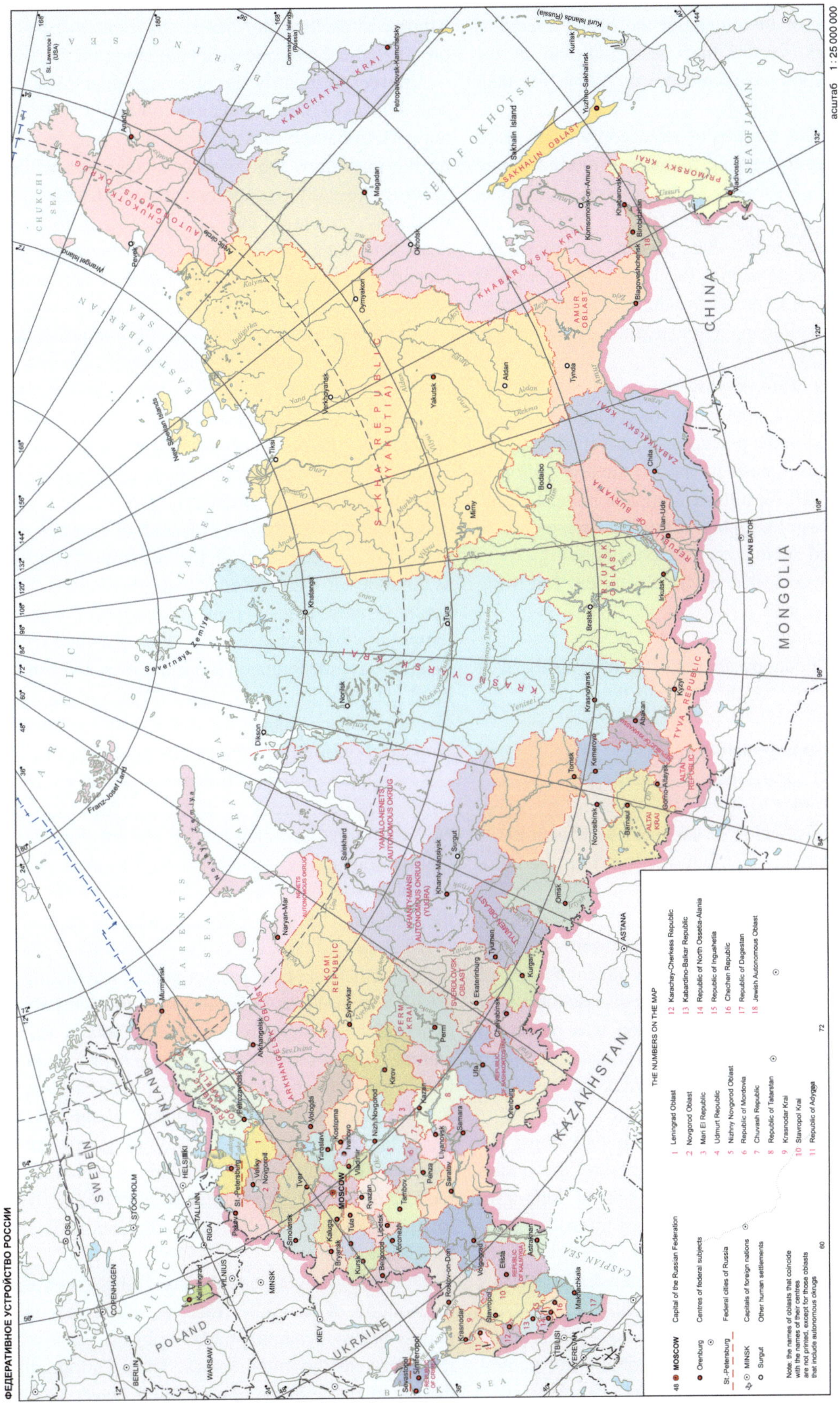

Map 1.2 Administrative divisions of Russia

geographical spread of diseases, maps of morbidity were compared with maps of the ranges of the major infectious and parasitic carriers, vectors and pathogens and of the environmental and socio-economic preconditions for their spread.

The chapter includes a series of nosogeographical maps that reflect the average perennial population morbidity and disease dynamics in 85 subjects of the Russian Federation. Analytical maps in this chapter served as a basis for the creation of integrated maps that present the morbidity structure for a network of natural focal diseases, as well as the nosological profiles of Russia's administrative subjects, which provide information on the set and intensity of various diseases in each one.

The methods that were employed in order to map the nosogenic territories and population morbidity consist of both widely used, traditional practices and novel approaches. GIS technologies were extensively used in the preparation of this work: they allowed for a more complete understanding of natural focal disease morbidity within the framework of its environmental and socio-economic preconditions.

The chapter was compiled according to the following steps:

- The preparation, updating and analysis of a database of 23 diseases, 16 of which are infectious and 7 are parasitic.
- The choice of mapping methods.
- Mathematical and cartographical modelling.
- Map analysis in order to ensure adherence to the objectives and expert evaluation of adequacy and accuracy.
- Analysis of geographic distributions based on the compiled maps.

The MS Excel and MS Access databases were compiled using statistical information on population infectious disease morbidity from 1997 to 2015 and parasitic disease morbidity from 2006 to 2015. After the information was processed (calculation of annual averages, choice of gradations, choice of scale), maps were compiled using ArcGIS, Bentley's MicroStation, and Adobe Illustrator software.

The entire process can be divided into two steps: the compilation of analytical (inventory) maps and the preparation of integrated maps. Analytical maps of long-term average annual morbidity rates were drawn up using cartodiagramming and cartogramming methods, and reflect both relative and absolute markers for each administrative subject in Russia throughout the period in question.

Maps of nosological profiles were based on the analytical maps and provide a summary of the distribution of natural focal diseases that were reported in Russia's administrative subjects. The map represents annual changes in the spectrum of natural focal diseases that are found in Russia's administrative subjects using matrix diagram (diseases are along on the vertical axis and years are along the horizontal axis).

In order to identify the most significant diseases in Russia, a map of the population morbidity structure for the network of natural focal diseases was compiled. Cartograms on the map show the diversity of diseases, while cartodiagrams represent the proportion of each disease in the overall morbidity structure. Diseases are listed in the order of decreasing long-term average annual morbidity rates in the legend. The proportions of each disease are depicted in bar charts.

An integrated map depicting the morbidity rates for the major parasitic diseases in Russia shows the annual average number of cases per 100,000 in cartograms and the number of diseases with high epidemiological activity in pie charts. Each colour marks a specific disease, while the size of the segment conveys the proportion of each parasitic disease in the overall morbidity structure.

Dynamic analysis is conducted in a series of maps of morbidity in order to ensure that the content is as informative as possible and can be used to make predictions. The dynamic nature of these maps is primarily conveyed with the help of temporally differentiated maps, the combined representation of different timepoints on the same map and a newly developed classification system for morbidity dynamics. Various diagrams—bar charts, matrices and cartograms—were employed in order to achieve maximum clarity.

Overlapping different timepoints allows for a view of the emergence and spread of dangerous infectious diseases in new territories, especially those that are rapidly expanding their nosogenic territory. West Nile fever is an example of this phenomenon. The map clearly shows the dynamics of the nosogenic territory's expansion in Russia over the course of 15 years.

Mathematical and cartographical modelling (Tikunov 1997) was used to compile maps that classify morbidity dynamics. The legends consist of average values for the annual relative morbidity rate in each of Russia's administrative subjects arranged in graphs. The advantage of such maps is that they offer the possibility of studying patterns in morbidity rate variation by way of group administrative subjects by oscillation rhythm. This classification allows for the identification of spatial trends in morbidity dynamics and to minimize the potential effect of random fluctuations on morbidity data.

Dynamic circular cartograms ('ring maps') reflect the distribution of morbidity rates for each year in a 17-year period for each administrative subject. Relative morbidity rates serve as the initial values. These cartograms allow for the visualization of the relationship between spatial data and its chronology (Visualization… 2012).

The 'Spatial Organization of the Sanitary and Epidemiological Service' chapter consists of a text describing the structure of the sanitary and epidemiological service in Russia and a series of traditional maps that depict the system of social institutions that conduct monitoring, control and prevention, thus ensuring epidemiological safety.

References

Altai Krai Atlas. vol 1. Moscow-Barnaul: GUGK; 1978, p 222 (in Russian)

Atlas of Irkutsk Oblast: ecological conditions of development. Moscow-Irkutsk; 2004, p 92 (in Russian)

Atlas of Khanty-Mansi Autonomous Okrug/Yugra. History. Population. Economy. Khanty-Mansiysk–Moscow; 2006, vol 1, p 152 (in Russian)

Atlas of natural focal infectious and parasitic diseases of the Komi Republic. Syktyvkar: SGU; CGSNL in the Komi Republic; 2004, p 60 (in Russian)

Atlas of Natural and Technological Hazards and Risks of Emergencies in the Russian Federation. Far Eastern Federal district. Moscow; 2007, p 324 (in Russian)

Atlas of Pskov Oblast. Moscow: GUGK; 1969, p 44 (in Russian)

Atlas of Sakhalin Oblast. Moscow: GUGK; 1967, p 136 (in Russian)

Atlas of Tyumen Oblast. GUGK, Moscow-Tyumen; 1976, vol 2, p 227 (in Russian)

Atlas of Voronezh Oblast. Voronezh: Voronezh University press; 1994, p 32 (in Russian)

Beklemishev VN (1970) Biocenological fundamentals of comparative Parasitology. Nauka, Moscow, p 410 (in Russian)

Chalov RS (ed) (2002) Ecological atlas of Russia. ZAO Karta, Moscow, p 163 (in Russian)

Cliff AD, Haggett P, Smallman-Raynor M (2004) The world atlas of epidemic diseases. Arnold, London, p 212p

Fedotov VI (ed) (2013) Ecological and geographical atlas book of Voronezh Oblast. Voronezh University press, Voronezh, p 512 (in Russian)

Islam MS, Drasar BS, Sack RB (1994) The aquatic flora and fauna as reservoir of *Vibrio cholera*: a review. J Diarhorreal Dis Res 12 (2):87–96

Kanaroglou P, Delmelle E, Paez A (eds) (2016) Spatial analysis in health geography. Routledge—Taylor and Francis Group, London, NY, p 344

Karaseva EV (2000) 50 years (1950–1999) of studying the natural focus of leptospirosis at the Nero Lake of the Yaroslavl region. RAT-info 4:4–8 (in Russian)

Keller AA (ed) (1990) Medico-geographical reference of the Karelian ASSR. Karel scientific center of the Academy of Sciences of the USSR, Petrozavodsk, p 195 (in Russian)

Koch T (2005) Cartographies of diseases. Maps, Mapping, and Medicine. ESRI Press, Redlands, p 388

Koneva IV, Batuev AB, Lopatkin DA (2012) Nozoecosystems of the Asian part of Russia (Map 1:7500000) Ryashchenko SV, Galyos DA (eds) Irkutsk: V.B. Sochava Institute of Geography SB RAS (1 sheet) (in Russian)

Korenberg EI (ed) (2003) Natural focality of diseases: the Gamaleya RAMN Institute Research. Coll. Moscow, Rusaki, p 254 (in Russian)

Korenberg EI (2010) Natural focality of infections: current problems and prospects of research. Zool J 89(1):5–17 (in Russian)

Kotlyakov VM, Baklanov PY, Komedchikov NN (eds) (2009) Atlas of the Kuril Islands. Vladivostok: CPI "Design. Information. Cartography", p 516 (in Russian)

Krivoshapkin VG, Timofeev LF, Lazebnik OA (2005) Population health and public health services in Sakha (Yakutia) at the turn of the century. A Medico-Geographical Atlas. FGUP "YakutAGP", Yakutsk, p 119 (in Russian)

Kucheruk VV (1972) Structure, typology and regionalization of natural foci of human diseases. In: The results of the development of the natural focality of human diseases theory and further tasks. Coll. Moscow: Medicine, p 180–212 (in Russian)

Lipp EK, Huq A, Colwell RR (2002) Effects of global climate on infectious disease: the cholera model. Clin Microbiol Rev Oct 15 (4):757–770

Litvin VYu (1986) Ecological specificity of natural focality of sapronoses. Probl Nat Focality Dis 14:114–124 (in Russian)

Litvin VYu, Korenberg EI (1999) Natural foci of diseases: the development of the concept at the close of the century. Parasitology 33(3):179–191 (in Russian)

Litvin VYu, Gintsburg AL, Pushkareva VI, Romanova YM, Boev BV (1998) Epidemiological aspects of bacterial ecology. Farmarus-Print, Moscow, p 256 (in Russian)

Lvov DK (ed) (2001a) Atlas of pathogen distribution of natural focal viral infections in Russia. Moscow: Izd-vo NLU TMG MZ RF, p 192 (in Russian)

Lvov DK (ed) (2001b) Atlas of the distribution of natural focal viral infection pathogens in the Russian Federation. Moscow: Izd-vo NLU TMG MZ RF, p 192 (in Russian)

Lur'e IK, Kravtsova VI (eds) (2012) Modern geographic cartography. Moscow: Data+, p 290 (in Russian)

Malkhazova SM (2001) Medico-geographical analysis of the territories: mapping, assessment, forecast. Nauchny Mir, Moscow, p 239 (in Russian)

Malkhazova SM (ed) (2007) Medical demographics atlas of Kaliningrad Oblast. Moscow-Kaliningrad, p 85 (in Russian)

Malkhazova SM (ed) (2015) Medico-geographical Atlas of Russia "Natural Focal Diseases. Faculty of Geography, Lomonosov Moscow State University, Moscow, p 208 (in Russian)

Malkhazova SM (ed) (2017) Medico-geographical atlas of Russia "Natural Focal Diseases", 2nd revised edn. Faculty of Geography, Lomonosov Moscow State University, Moscow, p 216 (in Russian)

Malkhazova SM, Gurov AN (eds) (2007) Medico-demographical atlas of Moscow oblast. Moscow State University, Moscow, p 110 (in Russian)

Meade MS, Emch M (2010) Medical geography, 3rd edn. The Guilford Press, NY, London, p 498

Medical Geography (1964) Results and prospects. Irkutsk (in Russian)

Medico-ecological Atlas of Voronezh oblast. Voronezh: Istoki; 2010, p 167 (in Russian)

Medico-geographical Atlas of Khabarovsk Krai and the Jewish Autonomous Oblast. Khabarovsk; 2005, p 112 (in Russian)

Medico-geographical Atlas of Primorsky Krai. Vladivostok; 1969, p 40 (in Russian)

Morse SS (1996) Factors in the emergence of infectious diseases. Emerg Inf Dis 1(1):7–15

Murray Feshbach (ed) (1995) Atlas "The environment and health of Russia's population". Moscow: PAIMS, p 448 (in Russian)

Nafeev AA, Samatyia GB, Korobeinikova AS, Fefelova EF (eds) (2009) Atlas of natural focal infections in Ulyanovsk Oblast of the Russian Federation. Ulyanovsk, p 104 (in Russian)

Nefedov PF (ed) (2003) Medico-ecological atlas of Krasnodar Krai. GUP "Pechatniy dvor Kubani", Krasnodar, p 120 (in Russian)

Nikiforov LP (1968) Natural focus and its structure. In: The questions of the epidemiology of tick-borne encephalitis and its biological characteristics in a natural focus. Moscow: Medicine, pp 227–231 (in Russian)

Olsufiev NG, Dunaeva TN (1970) Natural focality, epidemiology and prevention of Tularemia. Moscow: Medicine, p 270 (in Russian)

On the status of sanitary and epidemiological wellbeing of the population in the Russian Federation in 2013: State report. Moscow: The Federal service for supervision of consumer rights protection and human well-being; 2014, p 191 (in Russian)

Pavlovsky EN (1946) The basis of the doctrine of natural nidality of vector-borne human diseases. Zhurn Obshch Biol 7(1):3–33 (in Russian)

Pavlovsky EN (1964) Natural Nidality of Transmissible Diseases. Nauka, Moscow, p 211 (in Russian)

Pavlovsky EN (1966) Natural nidality of transmissible diseases: with special reference to the landscape epidemiology of Zooanthroponoses. University of Illinois Press, p 257

Pospelov VP (1970) Medico-geographical Atlas of Krasnoyarsk Krai. Krasnoyarskoye kn. izd-vo, Krasnoyarsk, p 41 (in Russian)

Pospelov VP (1987) The landscape epidemiology atlas of the European Part of RSFSR, Ural and Crimean oblast of USSR. Moscow, p 162 (in Russian)

Primorsky Krai Atlas. Vladivostok: Primorsky center of geodesy and cartography; 1998, p 49 (in Russian)

Rodenwaldt E (ed) (1952–1957) Welt Seuchen Atlas, vol. 1–3. Hamburg

Semenko OV (ed) (2014) Atlas of the distribution of natural focal infections in the Resort City of Sochi. Stavropol, p 18 (in Russian)

Semenov VY, Gurov AN (eds) (2004) Medico-ecological atlas of Moscow Oblast. Moscow, p 40 (in Russian)

Shoigu S (ed) (2005) Atlas of natural and technogenic hazards and risks of emergency situations in the Russian Federation. Moscow: CPI “Design. Information. Cartography”, p 271 (in Russian)

Spira WM, Huq A, Ahmed OS, Saeed YA (1981) Uptake of *Vibrio cholerae* biotype El Tor from contaminated water by water Hyacinth (*Eichorniacrassipes*). Appl Environ Microbiol 42:550–553

Stepanov NM (ed) (1992) Atlas of natural focal diseases of Sakhalin oblast. Khabarovsk: VTU GSH, p 41 (in Russian)

The National Atlas of Russia. vol 2. Nature. Ecology. Roskartografiya, Moscow; 2007, p 495 (in Russian)

The National Atlas of Russia. vol 3. Population. Economy. Roskartografiya, Moscow; 2008, p 496 (in Russian)

Tikunov VS (1997) Modeling in cartography. Moscow State University, Moscow, p 405 (in Russian)

Transbaikal Atlas. Moscow-Irkutsk: GUGK; 1967, p 176 (in Russian)

Turyshev AG, Sherstnev VM, Sherstneva MB, Skachkov MV, Korneev AG, Petrisheva GV (eds) (2007) Medico-biological atlas of natural focal infections of Orenburg Oblast. Orenburg, p 55 (in Russian)

Vatlina TV (2012) Medico-demographic Atlas of Smolensk Oblast. Publishing House of Smolensk University, Smolensk, p 105 (in Russian)

Vershinsky BV (1964) Mapping of natural focal diseases in connection with the study of their geography in the USSR. In: Medical geography. Results. Prospects. Irkutsk, pp 62–98 (in Russian)

Visualization of Data with Ring Maps. Arc Rev 2012 1(60):24 (in Russian)

Vorobyov VV, Ignatiev EI (eds) (1968) Principles and methods of medico-geographical mapping. SB RAS, Irkutsk, p 222 (in Russian)

Voronov AG (1981) Medical geography, vol 1. General questions. Moscow State University, Moscow, p 161 (in Russian)

Voronov GA, Oborin MS, Malkhazova SM, Gavrilova IN (2014) Human ecology with the basics of medical geography: a textbook. Perm State University, Perm, p 329 (in Russian)

Zakrutkin VE (ed) (2000) Ecological atlas of the Rostov Oblast. Izd-vo SKNTS VSH, Rostov-na-Donu, p 120 (in Russian)

2 Natural and Socio-economic Conditions

Abstract

This chapter examines the contemporary understanding of natural preconditions for disease as unique features of the environment that determine the emergence and function of host–parasite systems and the spread of natural foci.

This chapter examines the contemporary understanding of natural preconditions for disease as unique features of the environment that determine the emergence and function of host–parasite systems and the spread of natural foci. A brief history of the Soviet medical geographers, epidemiologists and parasitologists that laid the groundwork for this understanding beginning in the 1960s is also examined. A brief characterization of the features and current condition of the natural environment in which host–parasite systems (pathogen–vector–warm-blooded host) form and function in Russia is also provided. The chapter includes a series of maps that reflect some of the natural and socio-economic conditions of Russia that influence the geography and dynamics of natural focal diseases. Special attention is devoted to the tendencies of population distribution across Russia and population demographics (population density and ratio of urban-to-rural populations), which allows for the elucidation of territorial features of the population structure as one of the links that contribute to the epidemiological danger posed by a territory.

2.1 Natural and Socio-economic Preconditions for Natural Focal Diseases

2.1.1 Historical Background

The study of factors that lead to the spread of infectious diseases has a long history. As evidenced by historical sources, humans have been suffering from infectious diseases since ancient times and have been forced to look for their causes long before the discovery of pathogens and parasites. Writings from the second and third centuries BCE contain information about the links between certain diseases and the environment. 'Miasma' theory—the notion that invisible clouds of mysterious particles caused disease in certain locations—first gained popularity in ancient times (Avtsyn 1972).

In *On Airs, Waters, and Places*, Hippocrates (Fig. 2.1) classified environmental and natural conditions and demonstrated how certain areas influence the spread of disease (MIT Classics Archive, trans. Francis James http://classics.mit.edu/Hippocrates/airwatpl.mb.txt):

> When one comes into a city to which he is a stranger, he ought to consider its situation, how it lies as to the winds and the rising of the sun; for its influence is not the same whether it lies to the north or the south, to the rising or to the setting sun. […] and concerning the waters which the inhabitants use, whether they be marshy and soft, or hard, and running from elevated and rocky situations, and then if saltish and unfit for cooking; and the ground, whether it be naked and deficient in water, or wooded and well watered, and whether it lies in a hollow, confined situation, or is elevated and cold.

Hippocrates was one of the first to notice how weather conditions in different years influence fevers, and even proposed making prediction based on annual weather conditions.

In *The Canon of Medicine*, Avicenna (Fig. 2.2) devoted a great deal of attention to the role of the environment in the emergence of various diseases, noting the relationship between health and geographical conditions, especially climate, and wrote about the influence of weather and seasons on disease (Fig. 2.3): 'if the summer be hot and dry, illnesses are quickly broken up; whereas if it be wet, the humours becomes glutinous in character, the stages of the diseases are prolonged, and recovery is delayed. […] Dropsy and lienteric diarrhoea, and looseness of the bowels are liable to occur. This is all because of the flow of redundant humours downwards from the upper parts to the lower. […] Diseases

S. Malkhazova et al., *Mapping Russia's Natural Focal Diseases*, Global Perspectives on Health Geography, https://doi.org/10.1007/978-3-319-89605-2_2

Fig. 2.1 Hippocrates. Etching by Peter Paul Rubens, 1638. *Credit* https://en.wikipedia.org/wiki/Hippocrates#/media/File:Hippocrates_rubens.jpg **Public Domain**

Fig. 2.2 Avicenna. Postal stamp commemorating the 1000-year anniversary of his birth. Image by E. E. Sokolova based on a drawing by M. M. Gerasimov. *Credit* http://markimira.ru/catalog_stamps_ussr/series_ussr/detail.php?ID=18767; https://commons.wikimedia.org/wiki/File:Title_page_of_the_Canon_of_Avicenna_Wellcome_M0013102.jpg#/media/File:Title_page_of_the_Canon_of_Avicenna_Wellcome_M0013102.jpg **Public Domain**

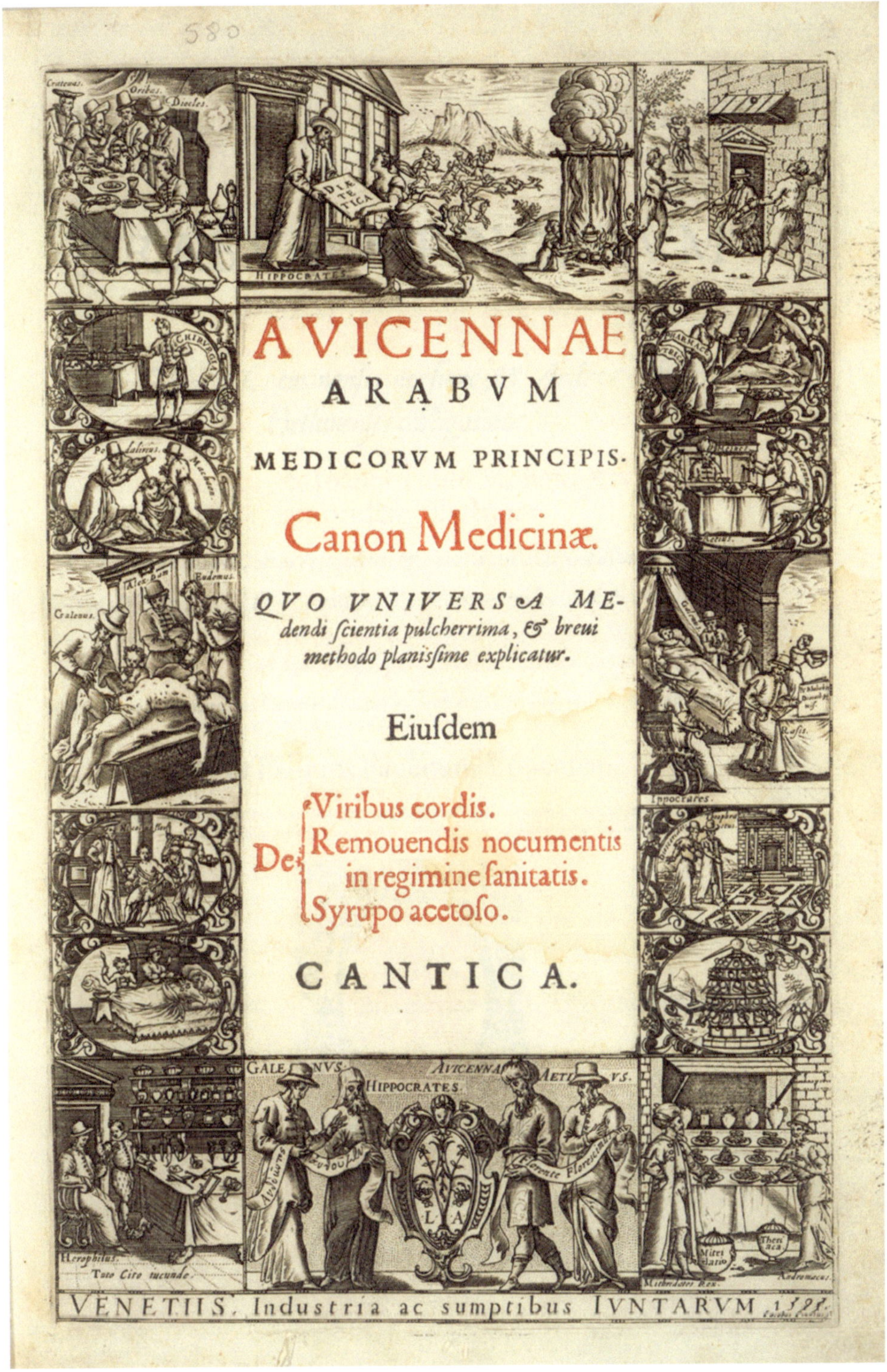

AVICENNAE
ARABVM
MEDICORVM PRINCIPIS.
Canon Medicinæ.
QVO VNIVERSA ME-
dendi scientia pulcherrima, & breui
methodo planissime explicatur.
Eiusdem
De Viribus cordis.
Remouendis nocumentis in regimine sanitatis.
Syrupo acetoso.
CANTICA.
VENETIIS, Industria ac sumptibus IVNTARVM 1595.

Fig. 2.3 Title page of Avicenna's *Canon of Medical Science*, printed in Venice in Latin in 1544. *Credit* https://commons.wikimedia.org/wiki/Category:The_Canon_of_Medicine#/media/File:Title_page_of_the_Canon_of_Avicenna_Wellcome_M0013102.jpg **CC BY 4.0**

specially associated with the hot season. If very hot: tertian, continued, and burning fevers' (189–190, Avicenna, The Canon of Medicine. AMS Edition, 1973). The subheadings of this chapter (Extracorporeal [Causes of Disease]) are 'Climate', 'Altitude', 'Winds', etc. An entire subsection of Avicenna's book is devoted to a detailed discussion of the influence of location, including geographical factors that have an effect on human health: The choice of a place of residence depends on (1) the soil (see 307f); (2) position, whether high or low lying (atmospheric pressure, see 307b); (3) whether exposed or sheltered, bare or covered with trees or woodlands or forests (forests harbour moisture and foster the decay of vegetation); (4) the water supply—its quality, whether the water is covered in (artificially) or exposed to

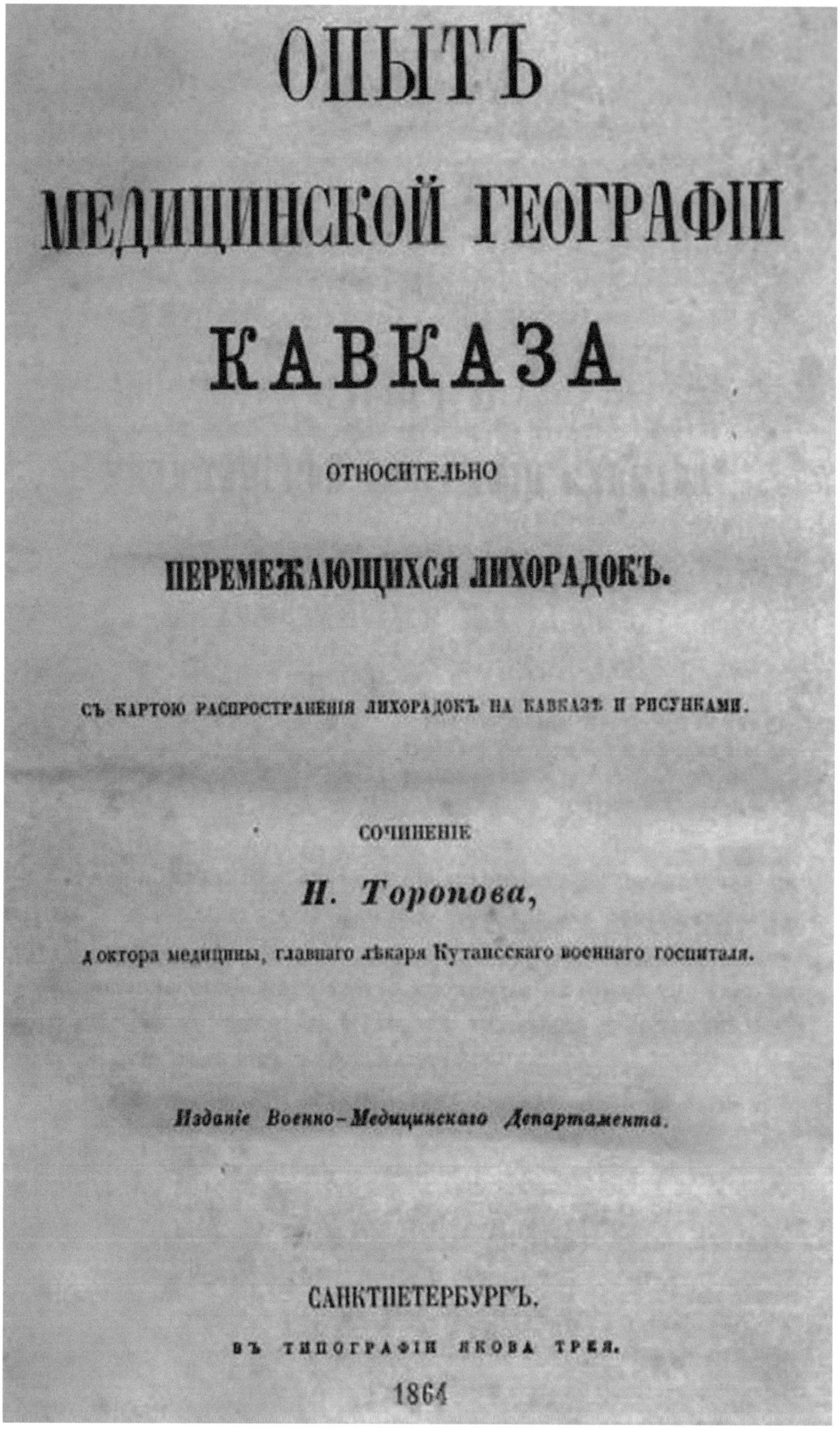

ОПЫТЪ

МЕДИЦИНСКОЙ ГЕОГРАФІИ

КАВКАЗА

ОТНОСИТЕЛЬНО

ПЕРЕМЕЖАЮЩИХСЯ ЛИХОРАДОКЪ.

СЪ КАРТОЮ РАСПРОСТРАНЕНІЯ ЛИХОРАДОКЪ НА КАВКАЗѢ И РИСУНКАМИ.

СОЧИНЕНІЕ

Н. Торопова,

доктора медицины, главнаго лѣкаря Кутаисскаго военнаго госпиталя.

Изданіе Военно-Медицинскаго Департамента.

САНКТПЕТЕРБУРГЪ.

ВЪ ТИПОГРАФІИ ЯКОВА ТРЕЯ.

1864

Fig. 2.4 Title page of N. I. Toporov's Medico-geography of the Caucasus, related to intermittent fevers (1864)

the air, whether concealed or deep, marshy, whether thick or limpid, whether flowing from a height or running over stones and whether salty or 'crude'; (5) the prevailing winds —whether exposed to or sheltered from the sun, whether salubrious or not, fresh (cold) and bracing or dry and sultry (having blown over wide tracts of land), or moist, and whether cold and healthy; (6) the neighbouring country (maritime, marshy, presence of lakes; mountainous or flat; rich in minerals or not; forests, jungles, etc.); (7) whether the ground air is pure and healthy, impure and unhealthy, making the natives prone to illness; what sort of illnesses prevail; [...] (9) the construction of house: whether with large, roomy or with narrow entrances; good ventilation, wide chimneys. Do the doors and windows face east and north? One must be especially careful to arrange to have the easterly winds able to enter the house and see that the rays of the sun can enter all day because the sun's rays render the air pure. One should be close to plenty of sweet running water, open to the sky, frozen in winter (209).

Although in ancient times the direct cause of diseases, which is now known to be live pathogens, was unknown, scholars still noted the influence of environmental factors on their spread and knew that some of these factors had an effect on human health.

In 1746, Friedrich Hoffmann, a British scholar, gave one of the first definitions of endemic disease in his 'Dissertation on Endemic Diseases, or Those Disorders which Arise from Particular Climates, Situations, and Methods of Living' (Avtsyn 1972).

Nikolay Toporov, a military doctor, began the discussion on endemic disease in Russia in 1864. In Medico-geography of the Caucasus, related to intermittent fevers (Toropov 1864) (Fig. 2.4), he defined endemic diseases as 'those that are found only in certain locations'. Using malaria as an example, he demonstrated the influence of landscape, hydrographic conditions and meteorological factors on the spread of infection and even compiled a map of intermittent fevers.

In Okhotin (1890), a navy doctor devoted much of his book, entitled *Introduction to the Study of Exotic Diseases (an Experience of Sharing the foundations of climate pathology and ethnopathology)*, to an analysis of the spread of various diseases and came to the conclusion that, like plants, some of which are ubiquitous and some of which are endemic and found only in certain areas, many diseases also have characteristic territories. Okhotin emphasized the influence of climate conditions on the spread of disease, 'For instance, it is known that malarial diseases [...] follow the same rule of reduction [of warmth] from the equator to the poles as plants do, and disappear altogether at certain latitudes' (9).

Therefore, by the late nineteenth century, it was clear that the spread of a number of diseases is linked to specific biotic and abiotic environmental factors. This understanding served as the groundwork for the theory of preconditions for human disease that was conceived by Russian scientists in the mid-twentieth century.

2.1.2 Disease Precondition Theory

According to the theory developed by Ignatiev (1964), Voronov (1965, 1981), Prokhorov (1968) and Vershinsky (1976), geographic preconditions for the disease can be defined as environmental characteristics that allow for the emergence and function of host–parasite systems. In addition, as noted by Isachenko (1980), although one natural component is usually the direct cause for the spread of a particular disease, the conditions and preconditions of disease are actually determined by the nature and structure of the overall geosystem. This can be clearly observed while studying natural focal diseases. One factor (such as meteorological conditions) is always seen as playing the leading role, but the disease's spatial distribution is connected to a number of other environmental characteristics.

These factors act differently depending on levels of spatial organization. Latitude zoning, which is primarily defined by climate, plays the main role on the highest level. The number of commonly found natural focal diseases in a region increases significantly from cold zones to hot ones, which can be easily traced throughout the territory of the Old World (see Sect. 1.1).

Aside from the relationship between latitude, environment and natural focal diseases, certain regions have unique medico-geographical conditions, which manifest in varying combinations of characteristic diseases. One of the reasons for these unique characteristics is the influence of flora and fauna on the formation of parasitic systems. Territories with the same biomes have the greatest similarity between combinations of natural endemic diseases (Malkhazova and Alexeev 2010).

Regional characteristics can also have an effect on the existence of disease foci, such as orographic conditions, i.e. elevation above sea level. Altitude data provides an understanding of whether or not suitable host and vector habitats exist in a certain area. Diseases transmitted by mosquitoes require suitable breeding grounds. For a number of climate-dependent diseases, the altitude can serve as the upper barrier for the existence of a natural focus. It can change, especially in warm years. The natural focus can shift up due to temperature anomalies, setting conditions for an outbreak.

A physical zoning diagram demonstrates the regional characteristics of Russia's territory. Natural geographic borders can sometimes also serve as the borders of focal territories, as geographic conditions often determine the viability of parasitic systems. Both landscapes alone and their individual components can be seen as preconditions to disease.

This chapter presents maps of the distribution of the main environmental factors that influence the spread of natural focal diseases.

The function of climate (Maps 2.1, 2.2, 2.3, 2.4, 2.5) manifests in parameters, such as average monthly temperature, precipitation levels, snow cover height, duration of the frost-free period, etc. These markers define the heat supply of a region, which has a significant influence on pathogen hosts and vectors. Climate characteristics are especially crucial for poikilothermic animals like arthropods and for those pathogens that spend part of their life cycles inside them.

Atmospheric temperature has an effect on the pathogen's survival in vectors and in the outside environment. Many natural focal diseases begin developing once they have crossed a certain average daily temperature threshold. West Nile fever can develop in mosquitoes of the *Culex* genus at temperatures beginning at 14.3 °C, and a full cycle requires a sum temperature of 109 °C (Reisen et al. 2006).

Atmospheric temperature determines the existence of disease vector populations. Heat supply is important for tick survival and thus for the infections that they carry. The taiga tick (*Ixodes persulcatus*), whose choice of habitat plays a defining role in the existence of tick-borne encephalitis in Europe, a sum temperature of 1600–1900 °C is a weak heat supply, 2000–2300 °C is moderate and over 2300 °C is good (Korenberg 2004). Overall, the northern border of tick-borne encephalitis distribution is defined by the effective sum isotherm of 1400 °C (in a period when the temperature is stable above 10 °C) (Korenberg 1979). Mosquitoes have a thermal optimum and threshold temperatures above and below which they cannot survive. Moreover, higher atmospheric temperatures speed up egg development in mosquitoes and pathogen development within them. This leads to a higher intensity of contact between mosquitoes and their nutrient sources, as well as an increase in mosquito populations. Rapid virus development reduces the time between host infection and new vector infection from the host. This ultimately leads to intensifying epizootic and epidemic processes (Hartley et al. 2012).

The duration of the frost-free period is an important climate marker as it defines the viability of pathogens, hosts and vectors during unfavourable times of year. The function of climate factors also influences the seasonality of human disease. Nearly, all natural focal diseases have some form of seasonality in terms of human morbidity, which is usually defined by periods of maximum vector ability and thus their ecological needs.

Precipitation levels determine the availability and condition of bodies of water, which is where vectors breed (primarily mosquitoes, horseflies, etc.). Bodies of water also serve as habitats for hydrophilic species that carry natural focal infections like tularemia or leptospirosis. Other meteorological indices (relative humidity, minimum temperature, maximum temperature, etc.) can play a significant role, especially on the local level.

Global climate change was a background process for the entire twentieth century and became especially apparent beginning in the 1970s. Questions of climate prognosis, environmental responses to climate change and its ultimate influence on human activity are major issues for scientists today. A great deal of attention has been continuously given to the possible influence of climate change on the existence of disease natural foci (Be'er and Elpiner 2004; Malkhazova 2006).

The hydrological factor can be measured using water temperature, chemical content and pollution as markers. For instance, organically polluted bodies of water do not have favourable conditions for the breeding of *Anopheles* mosquitoes but are suitable for *Culex* mosquitoes. The type of body of water, its size and physical characteristics (like current velocity) play a major role. All these parameters determine the possibility of pathogen, host and vector existence. The body of water's type characterizes the ability of aquatic ecosystems (including artificial reservoirs and other man-made bodies of water) to support disease foci and mollusc populations, which serve as intermediate hosts for various helminths (such as opisthorchis). The quality of surface waters is especially important for the distribution of diseases linked to aquatic or semi-aquatic ecosystems, like tularemia, leptospirosis, opisthorchiasis and diphyllobothriasis.

Soils have a significant effect on diseases whose pathogens or hosts spend part of their lives there. The soil's mechanical content, frost depth, etc. have an effect on the existence of mammal species that dig holes in the ground. This primarily concerns rodents, as well as larger animals, like foxes and arctic foxes. Soil humidity determines the possibility of breeding for certain moisture-loving insects, like the *Aedes communis* mosquito, some tick species, etc. Soil temperature influences the survival of parasite cysts and stages that spend part of their life cycle underground (toxocara roundworms, human whipworms, etc.). Edaphic conditions have a large effect on sapronoses—natural focal diseases whose pathogens can live in the soil for often long periods of time. These pathogens rely on soil temperature, acidity, mechanical conditions (at times) and pollution, including anthropogenic pollution.

Vegetation characteristics define the types of host and vector animal habitats that are available. Taiga and castor bean ticks prefer forests, while the ornate cow tick (*Dermacentor reticulatus*) inhabits open spaces: forest meadows, clearings, fields and pastures. Different forest types provide varied nutrient sources for animals that carry natural focal disease pathogens. It has been established that bank vole populations increase in linden and linden-evergreen forests in years with high linden yields. This, in turn, correlates with

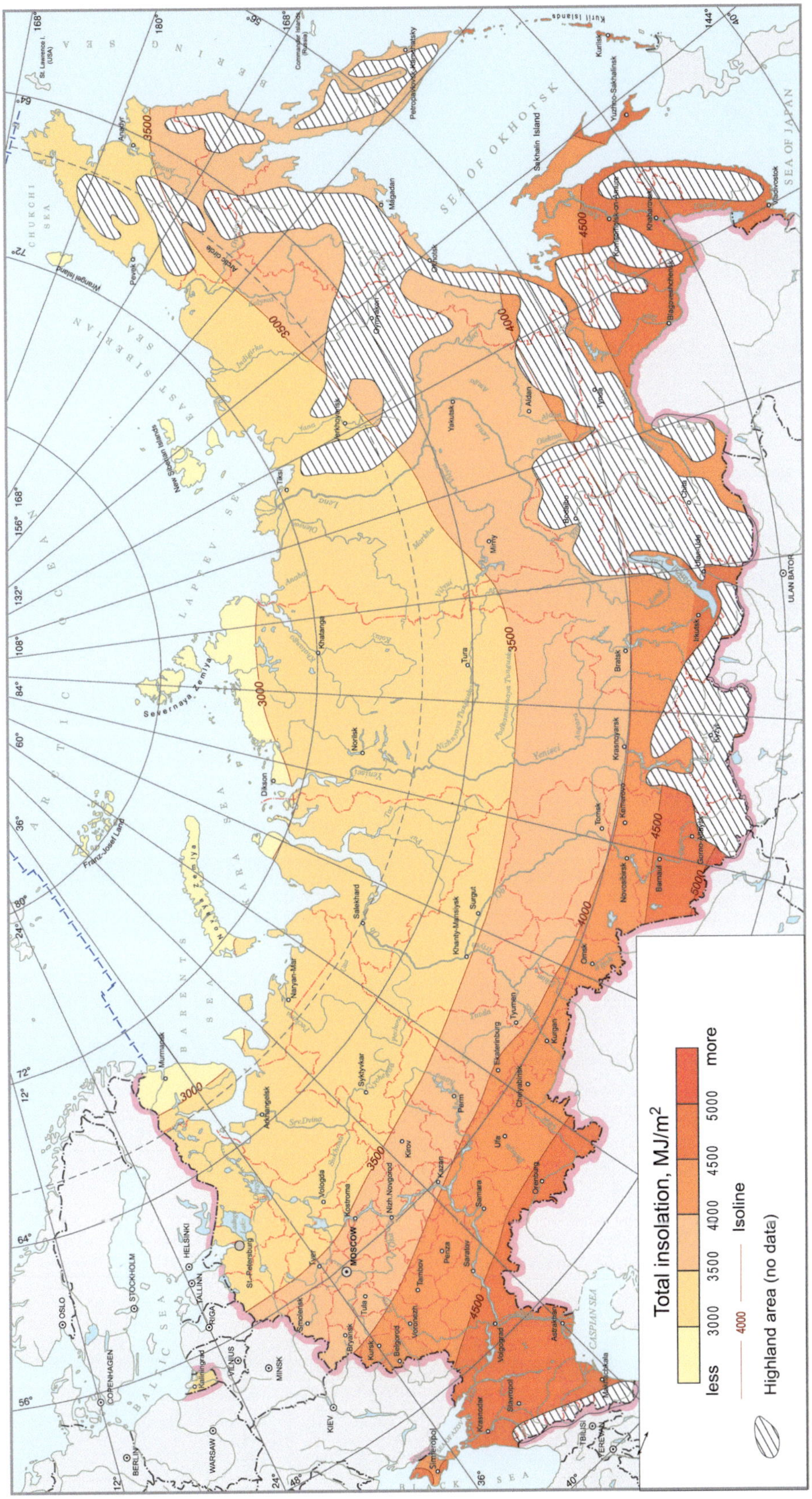

Map 2.1 Annual sunlight

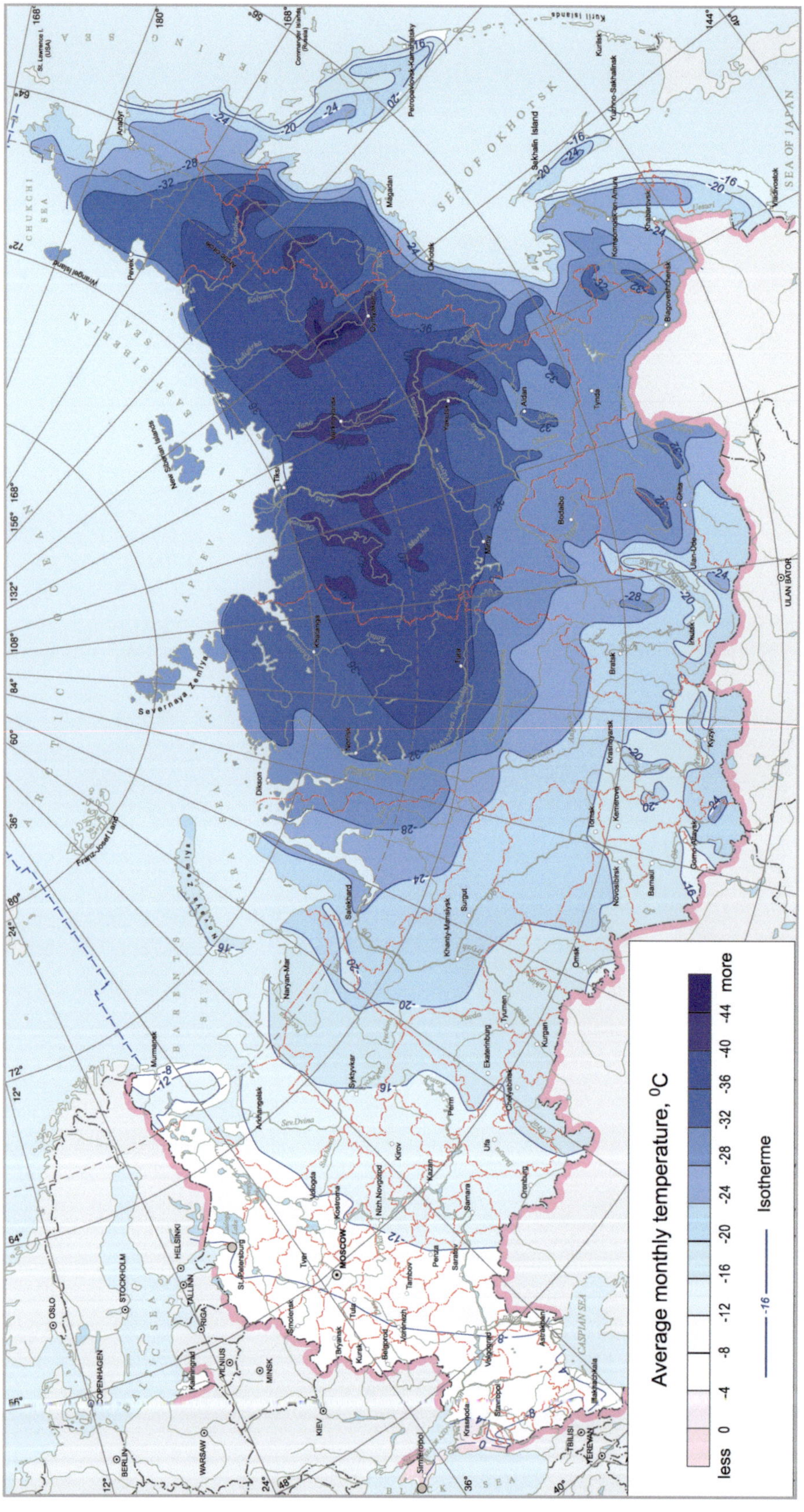

Map 2.2 Atmospheric temperature, January

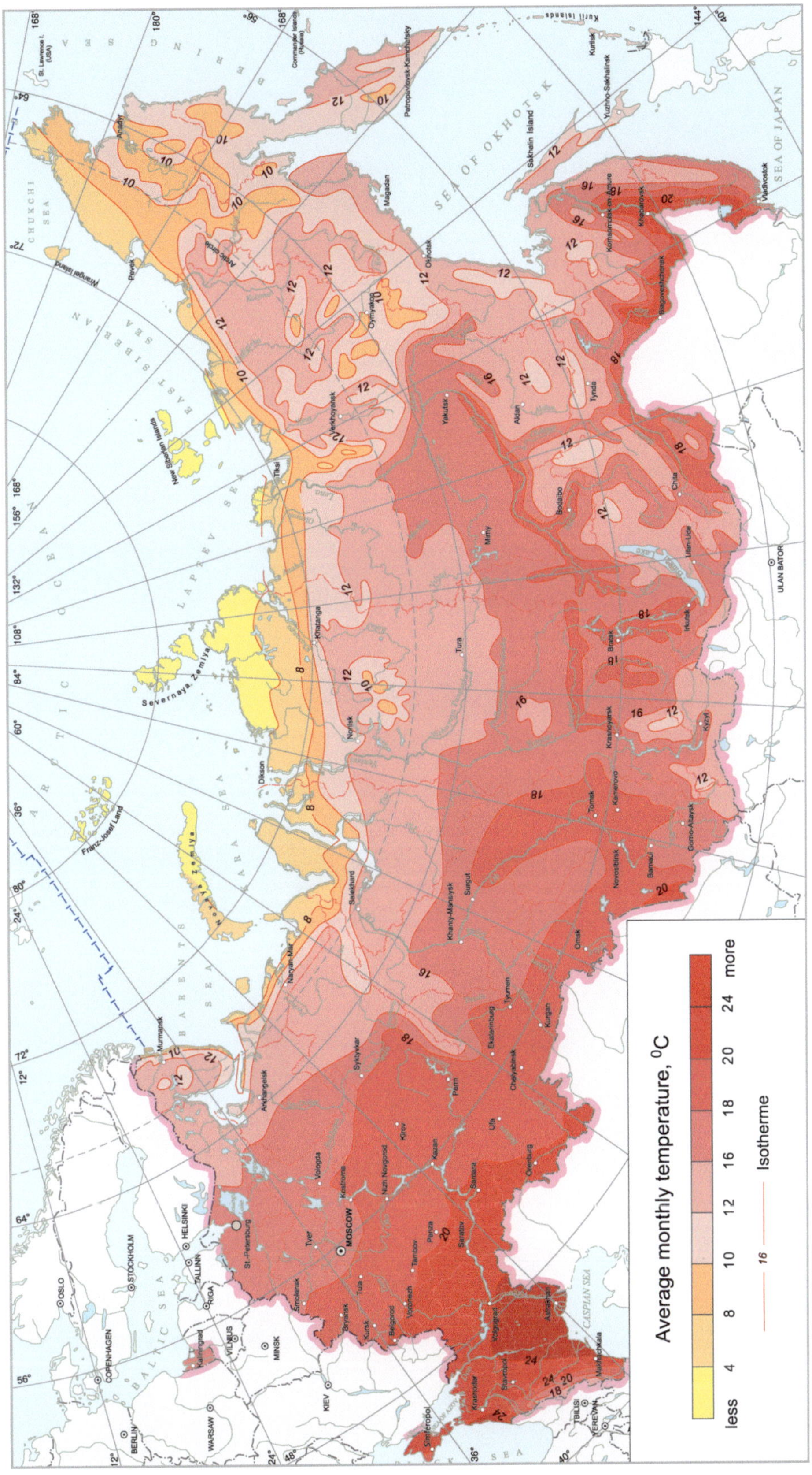

Map 2.3 Atmospheric temperature, July

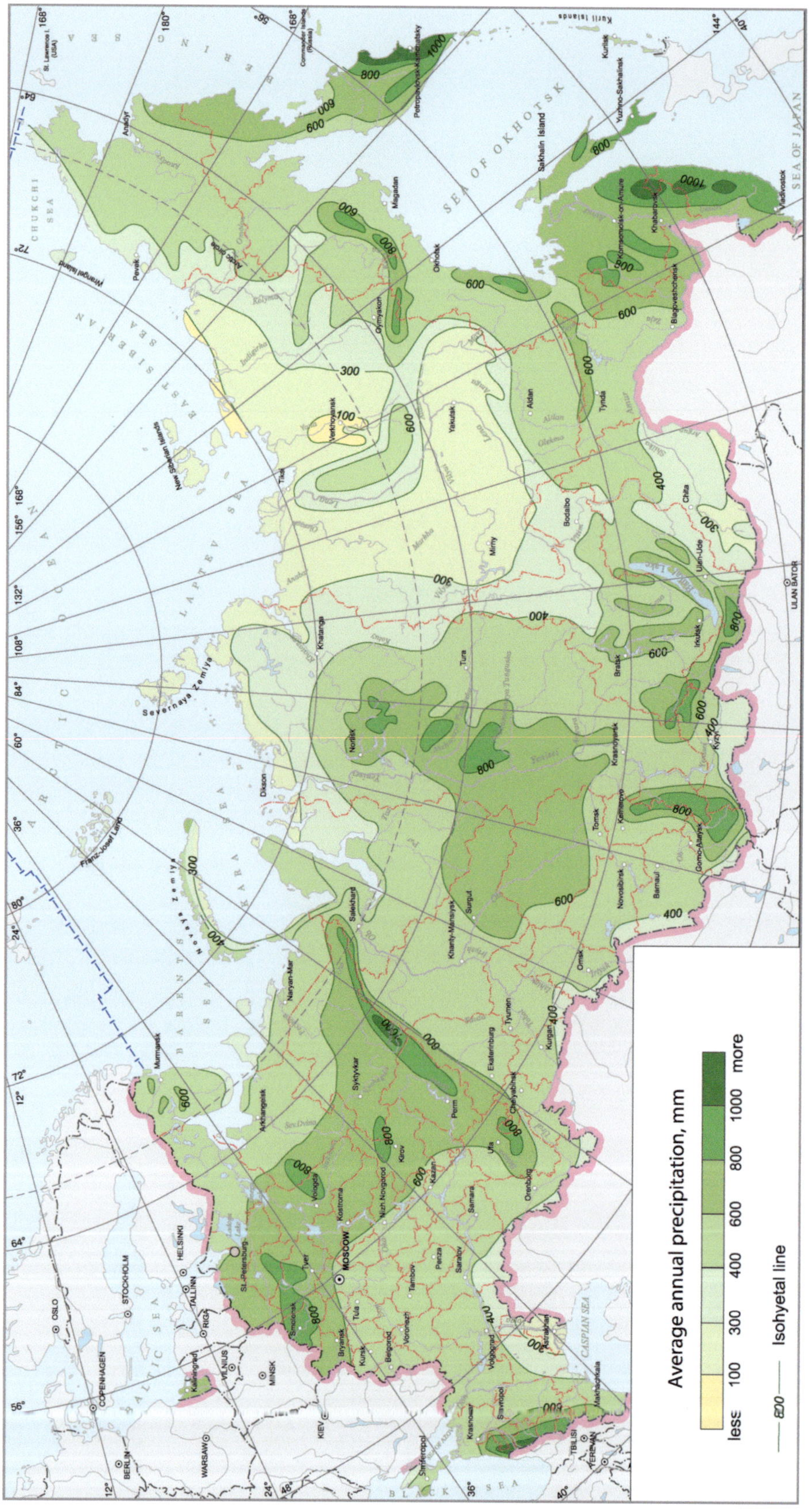

Map 2.4 Annual precipitation

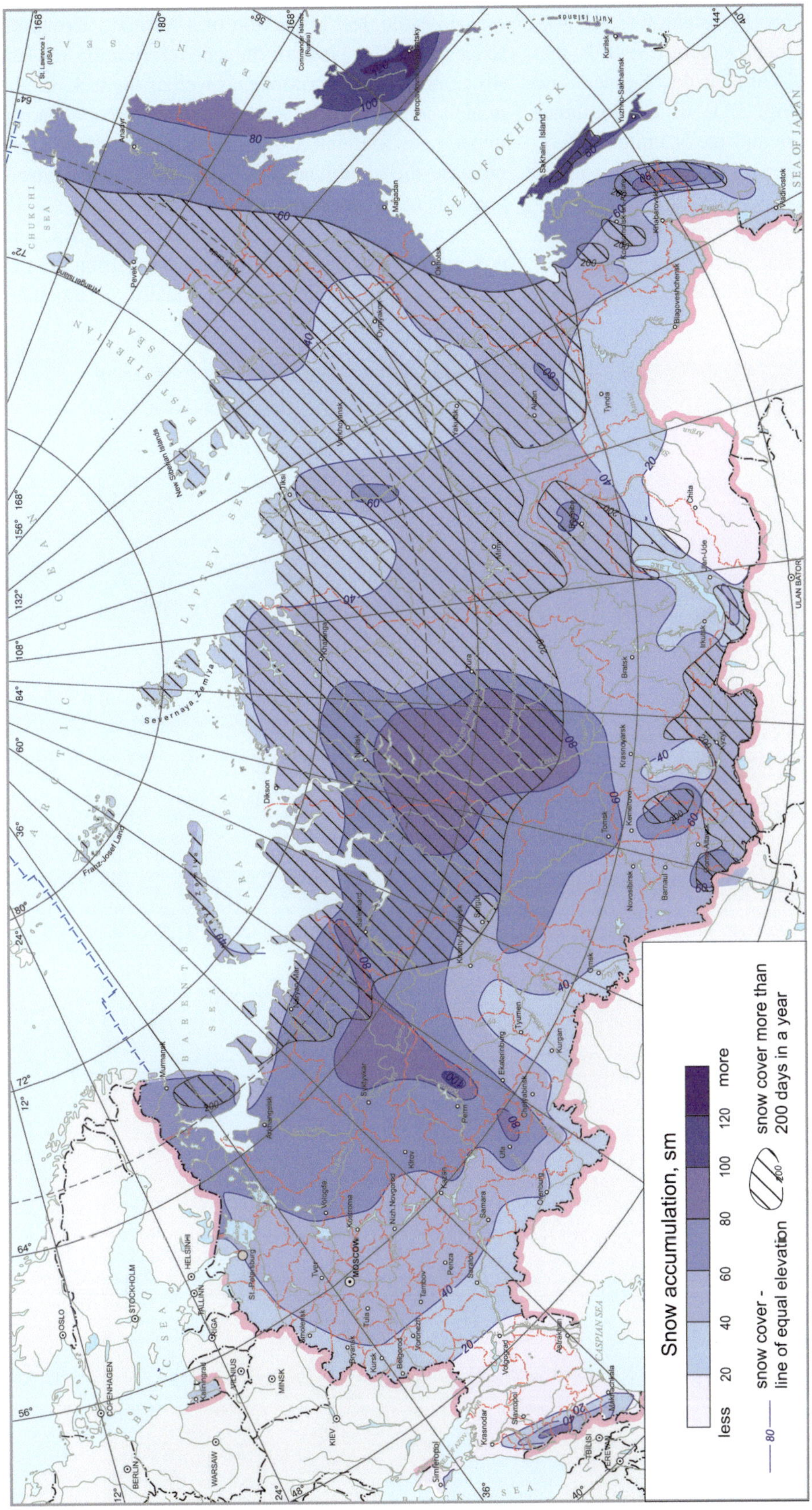

Map 2.5 Height of snow cover

higher hemorrhagic fever with renal syndrome morbidity rates among the human population (Bernstein et al. 2004).

Detailed analysis of natural focus structure reveals that the landscape plays a leading role in their development and maintenance. The landscape (Fig. 2.5), which combines every listed environmental factor and has a unified effect, determines two important conditions. First of all, this is the overall capacity of the ecosystem to support pathogen circulation. Second, this is the ecosystem's ability to support a pathogen introduced by an external source. An understanding of landscape characteristics that determine the existence of disease foci is a crucial component of the medico-ecological inspection of a territory. However, landscape variability and insufficient information on landscape structure in many territories, as well as the lack of developed models of landscape zoning and landscape cartography, impede an in-depth examination of this factor and thus increase the significance of each individual environmental component in landscape epidemiology studies.

Human activity, especially land use, urbanization and pollution, has a major influence on the distribution of natural foci (Fig. 2.6). Examined individually or together, these

Fig. 2.5 Taiga landscape. Tick-borne encephalitis and Lyme borreliosis are natural focal diseases characteristic of the taiga zone. *Credit* N. G. Kadetov (permission obtained)

Fig. 2.6 Human activity facilitates the transformation of natural foci. *Credit* N. G. Kadetov (permission obtained)

factors can have a serious effect on the distribution and characteristics of natural foci.

Forest resource conditions (Map 2.6) and the structure and ecological conditions of agricultural and commercial lands (Map 2.7) are reflected in the distribution of natural foci. Land use characteristics are important for diseases linked to rodents, so knowing the ratio between ploughed and unploughed fields is important for an understanding of the possibility of natural focus appearance. Even in 1855, A. N. Severtsov explored the question of how ploughing influences rodent activity. He came to the conclusion that ploughing previously untouched land and including it in crop rotation is generally deadly for steppe animals, as they cannot adjust to living in an agricultural ecosystem (Oparin et al. 2006). Ploughing leads to a fundamental reorganization of habitats, which is especially evident in steppe biocenosis. This factor usually operates in conjunction with other natural and anthropogenic processes.

Aside from ploughing, forest fires should also be mentioned, as they can completely transform natural ecosystems. The ratio between forests and meadows is important as it defines the species makeup of ticks and thus the potential set of natural focal disease pathogens in a specific location. Finally, the structure of game preserves, as well as the condition of forest resources and wetland ecosystems can determine the population sizes and species diversity of mammals and birds, who serve as pathogen hosts. This factor also increases human contact (hunters, park rangers and hikers) with natural foci.

Livestock population structure (Map 2.8) is a determining factor in the spread of diseases, such as taeniasis, trichinellosis, echinococcosis, etc. Pathogen exchange between wild and domesticated animals facilitates the transformation of natural foci into anthropurgic ones.

The existence of preconditions on a specific territory does not necessarily mean that natural foci will exist on it—rather, these preconditions should signal that there is a risk of natural focus development. Socio-economic conditions have a key effect on the actualization of natural preconditions.

The most important socio-economic indices are levels of urbanization, population demographics, migration levels, lifestyle, the cultural environment, traditions, religious beliefs, material well-being, etc. The population's understanding of sanitary and hygienic norms, as well as the distribution of healthcare and veterinary institutions and infrastructure conditions, should also be taken into account. Like that of natural preconditions, the influence of socio-economic conditions on the spread of natural focal diseases has varying effects and manifestations, depending on levels of territorial differentiation.

On a national level, the socio-economic development of Russia's regions plays a defining role. It depends on numerous factors like the type of natural resources in the region and the focus of its economy. Direction and speed of labour migration influence the introduction of infections from one region into another. Moreover, international and national tourism also have an effect on population health.

On a regional level, the agricultural priorities of a region and the construction of large facilities (which may attract labourers from other regions, etc.) play an important role. Large-scale projects that may change natural ecosystems,

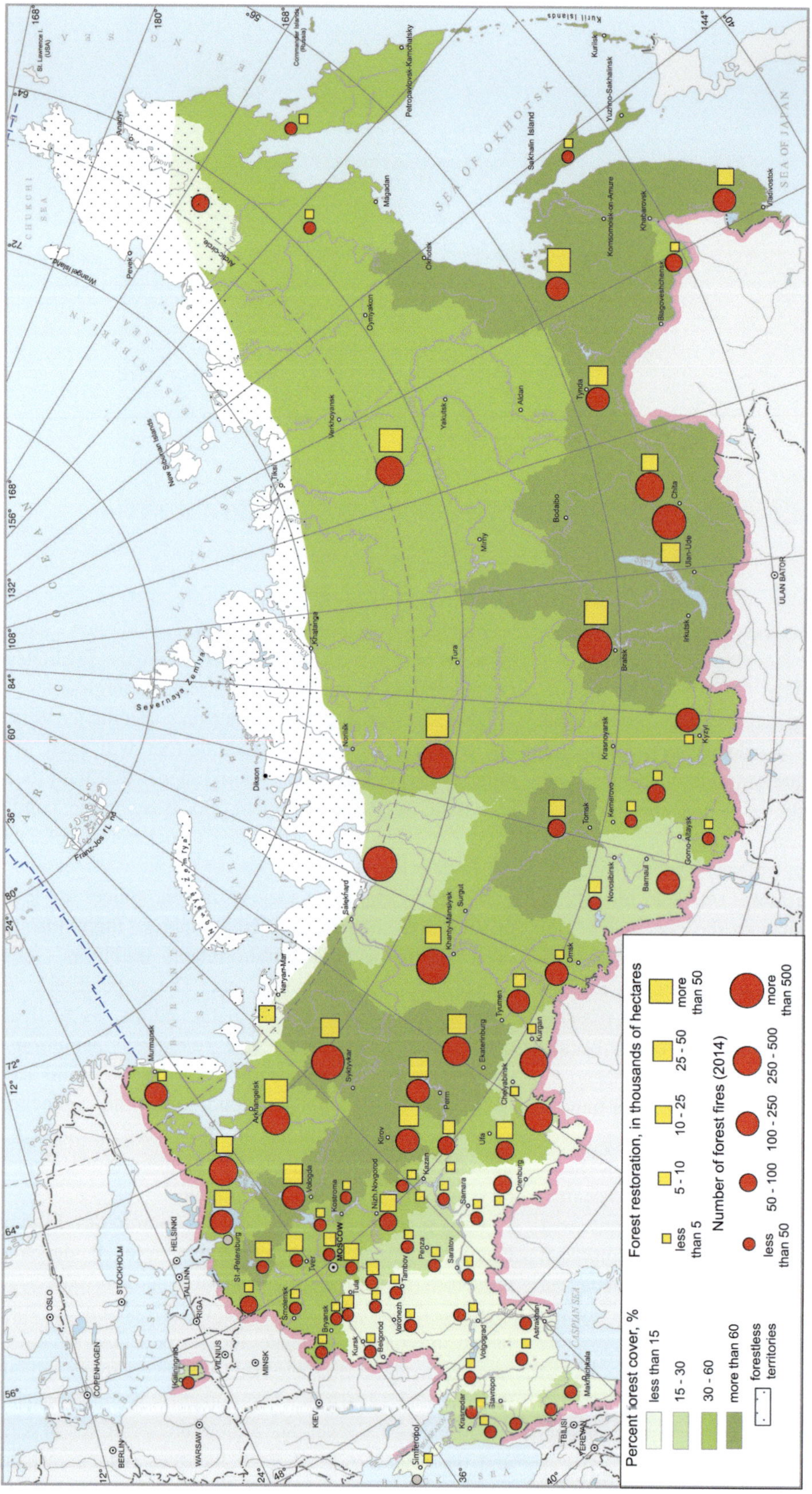

Map 2.6 Forest resources

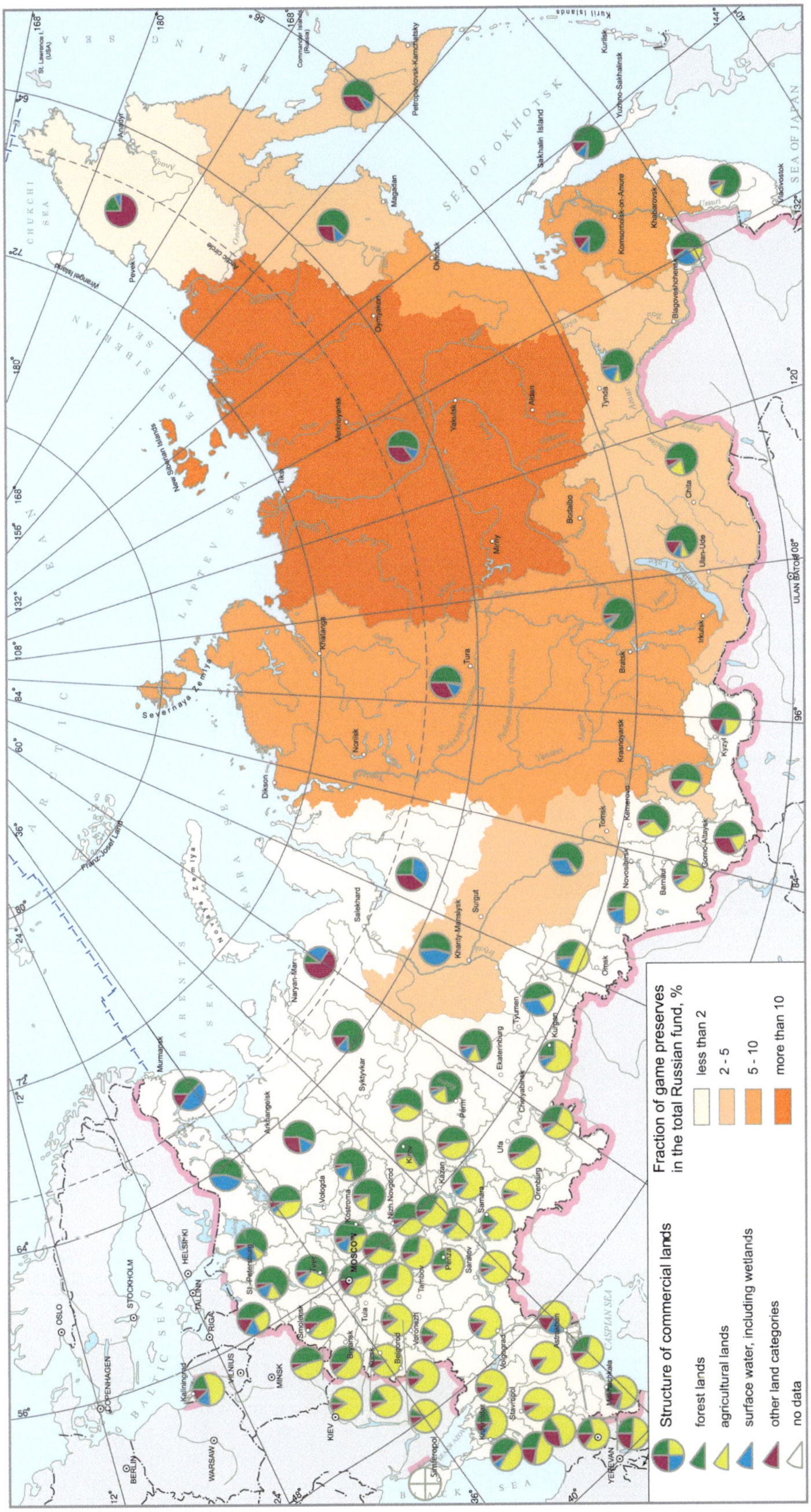

Map 2.7 Agricultural and commercial land structure

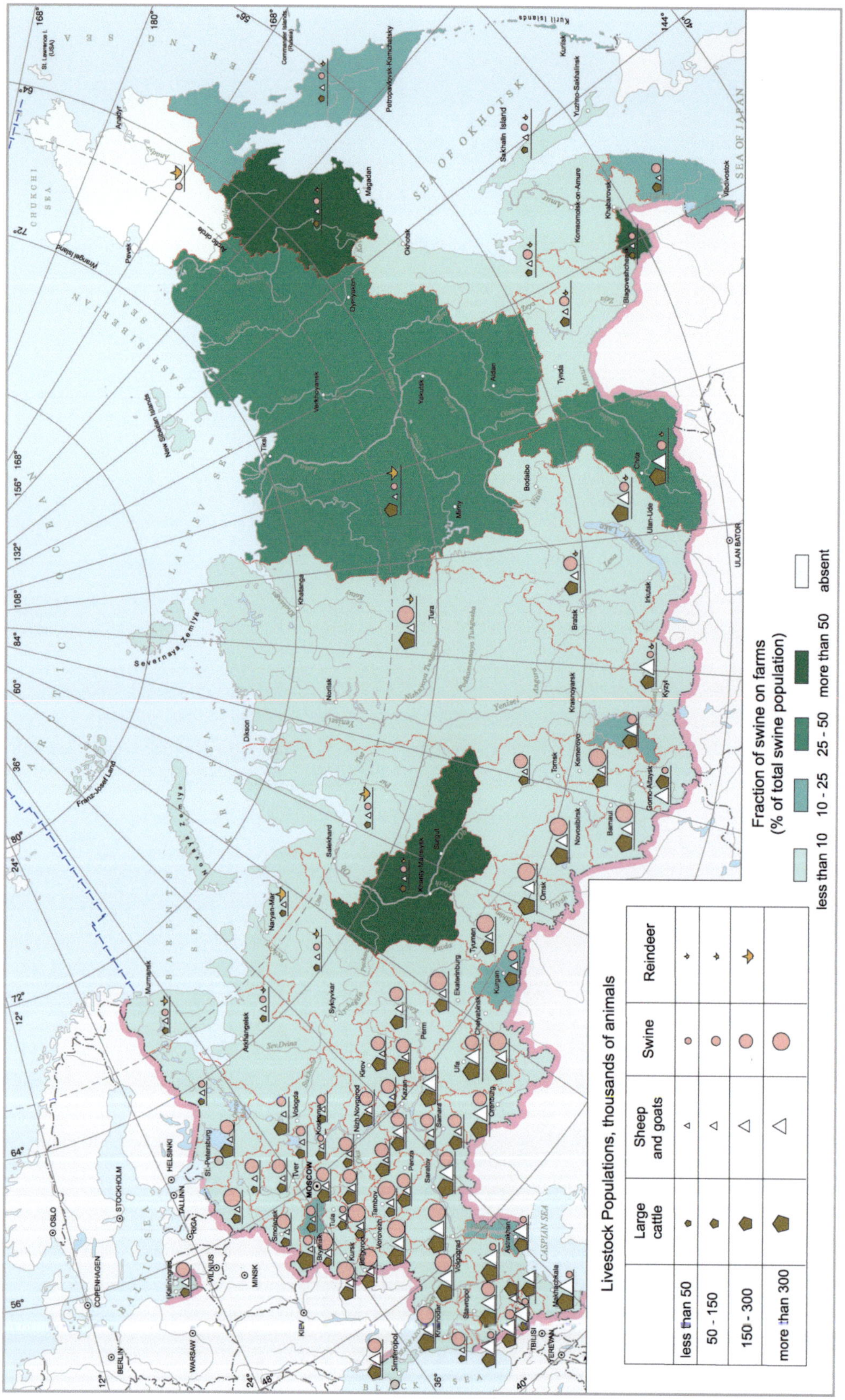

Map 2.8 Livestock population structure

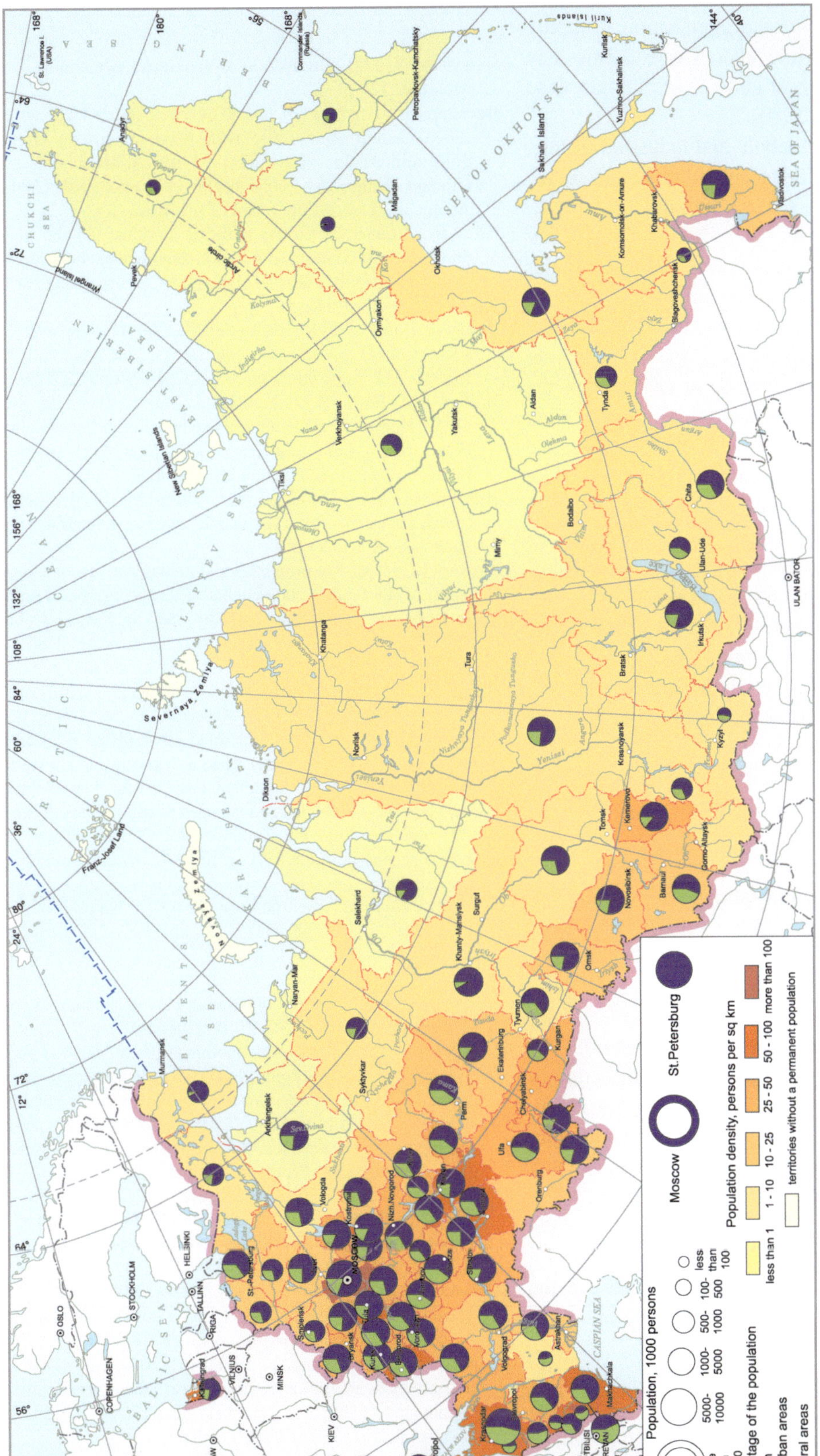

Map 2.9 Population density Urban-to-rural population ratio

ike highway or dam construction can also influence the function of natural foci. Workforce aggregation primarily determines disease spread, while altered ecosystems may lead to the formation of new host and vector biotopes (The Nature of the Middle Region 1980).

On a local level, social, household and industrial factors, such as population income, living conditions, food source quality, potable water quality, sanitary and hygienic conditions, professional occupations, etc., have a significant influence on natural foci. Industrial pollution plays a major role on both the local and regional levels.

The condition of healthcare institutions, the accessibility of medical help, vaccination rates, medication availability, etc. are a defining factor for all types of disease at all levels of the national hierarchy. Local lifestyles and traditions also often allow for infection with certain diseases. For instance, the culinary traditions of some of Russia's northern ethnicities allow for the high levels of opisthorchiasis and taeniasis due to *T. saginata* among indigenous populations, who consume foods made out of raw frozen fish (stroganina) or raw reindeer brains.

Perhaps, population density and the urban-to-rural population ratio play the leading role among all demographic factors (Map 2.9). These factors, in particular, determine the level of contact between humans and disease natural foci and thus lay the conditions for disease morbidity.

Recently, booming types of tourism, such as ecotourism, which involves visits to rural and remote areas, ethnographic tourism, which leads to contact with indigenous populations and gastro-tourism, which involves consumption of local foods, raise the risk of infection and the morbidity rate among individuals who visit wild landscapes for recreational purposes (Cascio et al. 2011; Malkhazova et al. 2014; 2015). This is also facilitated by lack of information among tourists and tourist firms on potential health risks (Van Herck et al. 2004; Rack et al. 2005; Namikawa et al. 2010).

Overall, a number of natural and socio-economic factors have a combined influence on the spread and distribution of natural foci and natural focal disease morbidity rates. Epizootic processes in a natural focus, as well as their rapidity and manifestations, can be seen as the result of interaction between pathogen, vector and/or host populations. The epidemiological situation in a specific focus depends, on the one hand, on the intensity of pathogen circulation, and, on the other hand, from the frequency of contact between the human population and the ecosystem (Korenberg and Yurkova 1983; Korenberg et al. 2013).

The search for new connections between parasitic systems and a combination of geographic factors and research on existing ones are a key part of any medico-geographical evaluation of the risk of infection with natural focal diseases.

References

Avtsyn AP (1972) Introduction to geographical pathology. Medicine, Moscow, 328 p

Bernstein AD, Apekina NS, Korotkov YS, Demina VT, Hvorenkov AV (2004) hemorrhagic fever with renal syndrome: environmental preconditions for activization of the european forest foci. In: Climate change and public health in Russia in the 20th century. Adamant, Moscow, pp 105–113 (in Russian)

Be'er SL, Elpiner LI (2004) Parasitic systems and the incidence of parasitoses in connection with climate change. In: Climate change and public health in Russia in the 20th Century. Adamant, Moscow, p 157–166 (in Russian)

Cascio A, Bosilkovski M, Rodriguez-Morales AJ, Pappas G (2011) The socio-ecology of zoonotic infections. Clin Microbiol Infect 17 (3):336–342. https://doi.org/10.1111/j.1469-0691.2010.03451.x

Hartley DM, Barker CM, Menach A, Niu T, Gaff HD, Reisen WK (2012) Effects of temperature on emergence and seasonality of West Nile virus in California. Am J Trop Med Hyg 1.86(5):884–894. https://doi.org/10.4269/ajtmh.2012.11-0342

Hippocrates (2017) On airs, waters, and places. MIT classics archive (trans: James F). http://classics.mit.edu/Hippocrates/airwatpl.mb.txt. Accessed on 15 Oct 2017

Ignatiev EI (1964) The principles and methods of medico-geographical study of the natural components of the geographical environment. In: Medical geography: results and prospects. Collection of scientific articles. Siberian Branch of the AS USSR, Irkutsk, pp 20–42 (in Russian)

Isachenko AG (1980) The methods of applied landscape studies. Nauka, Leningrag, 222 p (in Russian)

Korenberg EI (1979) Biochorological structure of a species (by the example of the Taiga Tick). Nauka, Moscow, 171 p (in Russian)

Korenberg EI (2004) Environmental causes for possible relationship between climate change and changes of natural foci of diseases, and their epidemiologic consequences. In: Climate change and public health in Russia in the XXI century. Adamant, Moscow, pp 54–67 (in Russian)

Korenberg EI, Pomelova VG, Osin NS (2013) Natural focal infections transmitted by Ixodid Ticks. Kommentariy, Moscow, 463 p (in Russian)

Korenberg EI, Yurkova EV (1983) The problem of forecasting the epidemic manifestation of natural foci of human diseases. Med Parasitol Parasitic Dis 3:3–10 (in Russian)

Malkhazova SM (2006) The modification of medico-geographical global picture. In: Current global changes of natural environment, vol 2. Scientific World, Moscow, pp 558–576 (in Russian)

Malkhazova SM, Alexeev BA (2010) Diversity in the geography of natural endemic diseases. In: Akhtar R, Izhar N (eds) Global medical geography. Rawat Publications, Jaipur, pp 184–198

Malkhazova SM, Mironova VA, Shartova NV, Pestina PV, Orlov DS (2015) Health risks facing travelers to Russia with special reference to natural-focal diseases. Travel Med Inf Dis 13(6):490–498. https://doi.org/10.1016/j.tmaid.2015.09.007

Malkhazova SM, Shartova NV, Kotova TV, Pestina PV (2014) Medico-geographical aspects of tourism development in Russia. Vestnik NAT 2(30), April–June 13–18 (in Russian)

Namikawa K, Iida T, Ouchi K, Kimura M (2010) Knowledge, attitudes, and practices of Japanese travelers on infectious disease risks and immunization uptake. J Travel (1) Med 17(3):171–175. https://doi.org/10.1111/j.1708-8305.2010.00405.x

Okhotin IA (1890) An introduction to the study of exotic diseases (experience of presenting the basics of climatic pathology and ethnopathology). Saint Petersburg, 169 p (in Russian)

Oparin ML, Oparina MS, Tsvetkova AA (2006) The influence of ploughing services on small steppe mammals. Povolzhskiy Ecol J 1:32–39 (in Russian)

Prokhorov BB (1968) The principles and methods of comprehensive medico-geographical assessment of the territories mapping. In: Principles and methods of medico-geographical mapping: a collection of scientific articles. Irkutsk, pp 154–184 (in Russian)

Rack J, Wichmann O, Kamara B, Gunther M, Cramer J, Schonfeld C, Henning T, Schwarz U, Muhlen M, Weitzel T, Friedrich-Janicke B, Foroutan B, Jelinek T (2005) Risk and spectrum of diseases in travelers to popular tourist destinations. J Travel Med 12(5):248–253

Reisen WK, Fang Y, Lothrop HD, Martinez VM, Wilson J, O'Connor P, Carney R, Cahoon-Young B, Shafii M, Brault A (2006) Overwintering of West Nile virus in Southern California. J Med Entomol 43:344–355

The Canon of Medicine of Avicenna (1973). AMS Press, New York

The Nature of the Middle Region of the USSR (1980) In: Mikhailov NI, Voronov AG, Timashev IE (eds) In connection with the problem of Northern River Reversal. Moscow State University, Moscow, 278 p (in Russian)

Toropov NI (1864) Medico-geography of the Caucasus, related to intermittent fevers. SPb, 416 p (in Russian)

Van Herck K, Van Damme P, Castelli F, Zuckerman J, Nothdurft H, Dahlgren AL, Gisler S, Steffen R, Gargalianos P, Lopez-Velez R, Overbosch D, Caumes E, Walker E (2004) Knowledge, attitudes and practices in travel-related infectious diseases: the European airport survey. J Travel Med 11(1):3–8

Vershinsky BV (1976) Geographical epidemiology (origins, current status, prospects of the development). In: Microbiology and hygiene works, vol 45. Pasteur Institute of Epidemiology, pp 7–28 (in Russian)

Voronov AG (1981) Medical geography, vol 1. General questions. Moscow State University, Moscow, 161 p (in Russian)

Voronov AG (1965) The experience of the classification of diseases according to the degree and nature of their dependence on features of the natural environment. In: Methods of medico-geographical research: Collection of scientific articles. MB GS USSR, Moscow, pp 32–47 (in Russian)

Natural Focal Disease Hosts and Vectors

3

Vadim Rumyantsev and Mikhail Soldatov

Abstract
This chapter provides a brief characterization of the groups of animal hosts and vectors that carry natural focal disease from the perspective of their roles in the formation and function of natural foci. The distribution of the most important epidemiologically significant mammals, birds and fish, as well as bloodsucking insects and ticks is examined and illustrated with a series of maps. The maps were compiled based on materials collected by the authors and are the result of long-term studies in various regions of Russia, as well as the analysis of a substantial amount of literary data and archival material. The animals' ranges are given by individual taxonomic unite (class, order, etc.) and are represented in accordance with traditional methods of small-scale zoogeographical cartography. The textual commentary illuminated the role of each species in the formation and function of natural foci and their association with specific landscapes.

3.1 Mammals

Wild, synanthropic and domesticated mammals (Mammalia) serve as hosts for many natural focal diseases.

Cattle is the primary carrier among domesticated animals. Diseases such as leptospirosis, Q fever, brucellosis, anthrax and a number of helminthiases, especially taeniasis and trichinosis are associated with cattle. Domesticated cattle can also serve as a nutrient source for adult ticks, who are vectors for tick-borne encephalitis and Lyme borreliosis. Among other domesticated animals, dogs are the most important carriers, serving as vectors for rabies, leptospirosis, Q fever, toxocariasis and echinococcosis. The formation of anthropurgic foci of natural focal diseases is linked to domesticated animals in particular.

Rodents are the most significant carriers among synanthropic mammals, especially grey (*Rattus norvegicus*) and black (*R. rattus*) rats, as well as house mice (*Mus musculus*). These organisms are carriers for leptospirosis, hemorrhagic fever with renal syndrome and Far East scarlet-like fever. Synanthropic animals are extremely prevalent in Russia, and their ranges often span its entire territory.

Two groups can be highlighted among wild mammals that serve as carriers for natural focal diseases: relatively 'narrowly specialized carriers', i.e. linked to a single disease or a limited number of disease, and 'universal carriers' which are involved in the circulation of pathogens in foci of a number of diseases.[1] In other words, the pathogens themselves can have a limited number of mammal carrier species or a wide range of hosts in a number of taxonomic groups (Tarasevich et al. 2003). Thus, about 30 rodent species are carriers of rickettsia. An even wider range of mammals serves as hosts for the pathogen of tularemia, to varying degrees (Olsufiev and Dunaeva 1970; Mescheryakova 2003).

Some insectivorous mammals of the Insectivora (Eulipotyphla or Lipotyphla) order play a significant role in the foci of a relatively small number of diseases. Members of the Soricidae family, especially of the *Sorex* genus, are one of the primary nutrient sources of tick nymphs, which are vectors tick-borne encephalitis and Lyme borreliosis (Korenberg and Kovalevsky 2000; Korenberg 2003a, b). *Sorex* shrews, *Crocidura* shrews, *Neomys* shrews, as well as animals from other families in this order, like hedgehogs (Erinaceidae), moles (Talpidae), and Russian desmans (Desmanidae), can participate in the maintenance of natural foci for tularemia and a few other diseases (Olsufiev and Dunaeva 1970; Mescheryakova 2003). Ranges of the most epidemiologically important Insectivora species are shown on Maps 3.1, 3.2, 3.3 and 3.4.

[1]The Latin names of mammal taxa are generally listed according to the following source: Dinets and Rothschild (1996). Geographic range maps have been compiled based on the following database: "Populations of Terrestrial Vertebrates in Russia," which was developed in the Department of Biogeography at the Faculty of Geography of Lomonosov Moscow State University (Rumiantsev and Danilenko 2012) with the use of additional sources.

S. Malkhazova et al., *Mapping Russia's Natural Focal Diseases*, Global Perspectives on Health Geography, https://doi.org/10.1007/978-3-319-89605-2_3

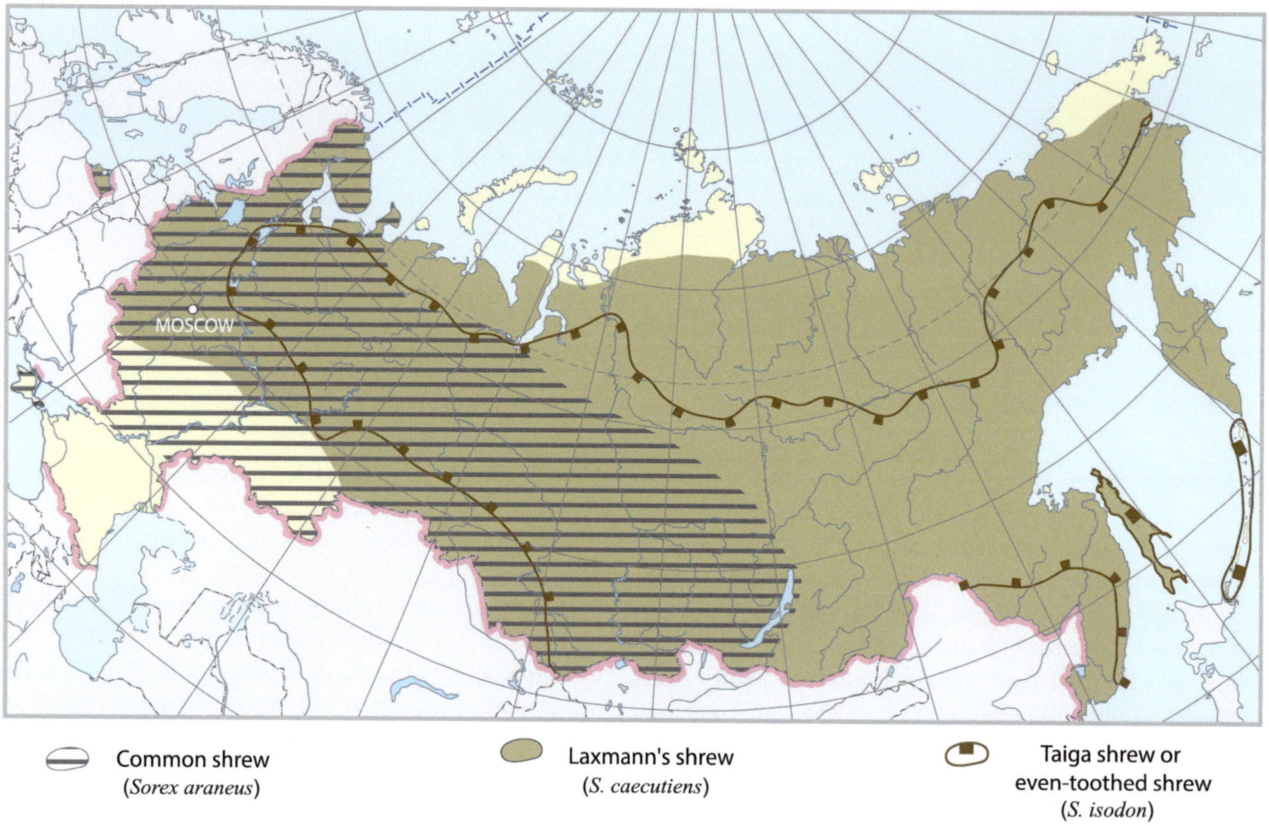

Map 3.1 Shrewmice—*Sorex*

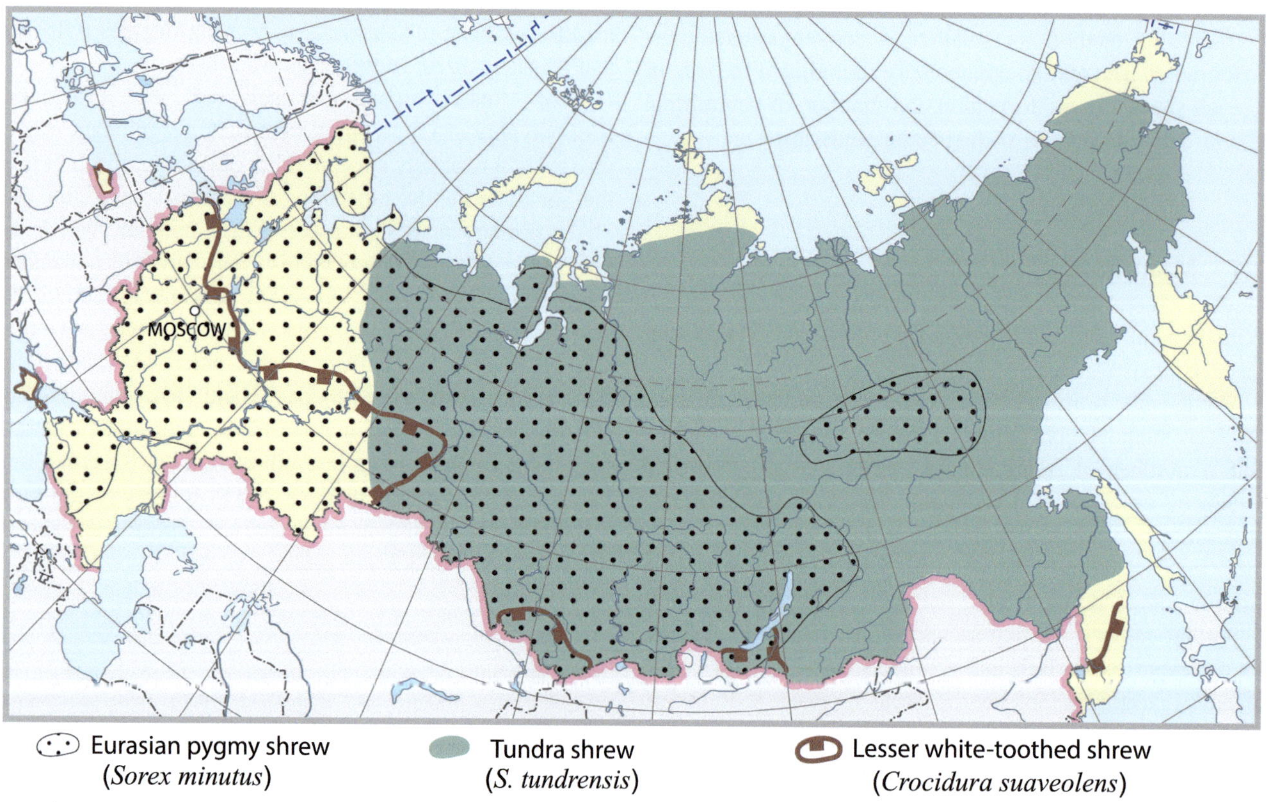

Map 3.2 Shrewmice—*Sorex, Crocidura*

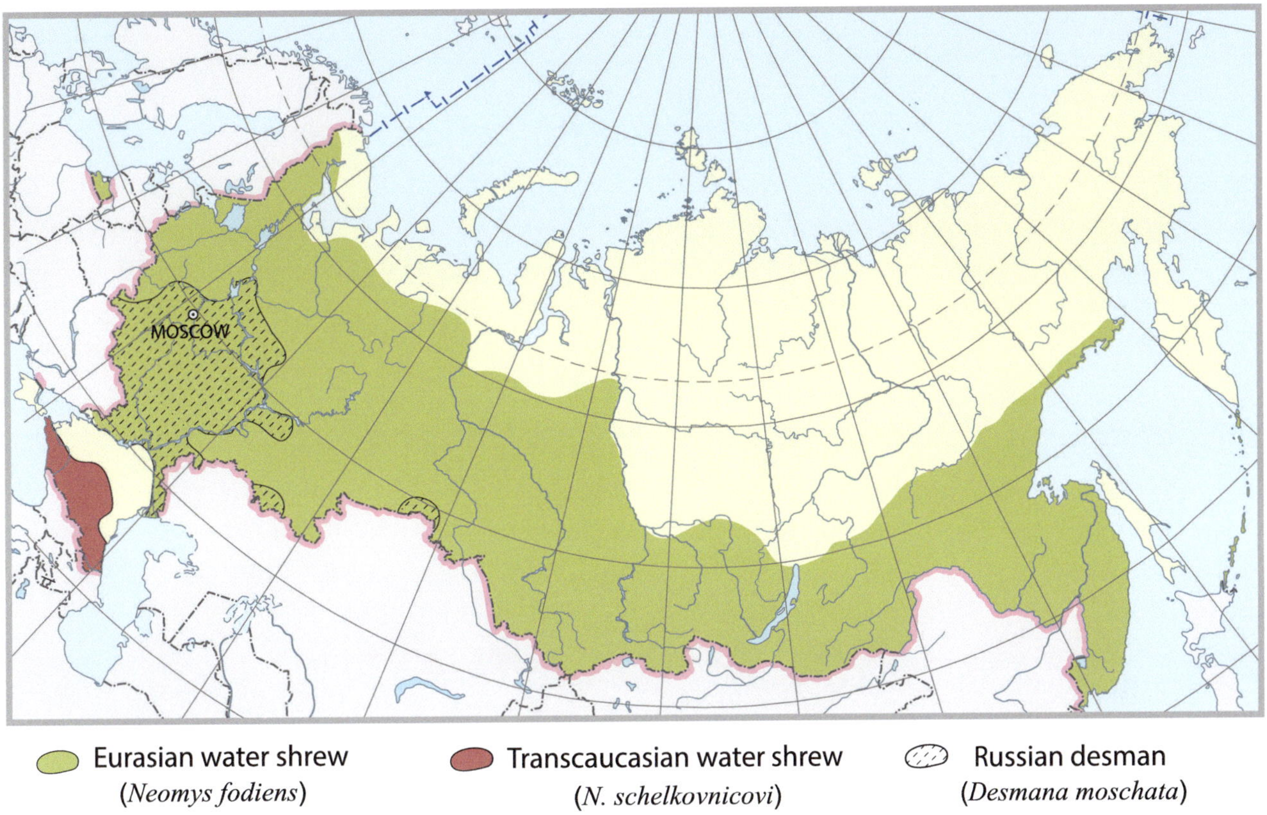

Map 3.3 Water shrews—*Neomys* and Desman—*Desmana*

Members of the Canidae family of the Carnivora order are the main carriers of rabies. The red fox (*Vulpes vulpes*) and the raccoon dog (*Nyctereutes procyonoides*) are especially important carriers. These two species have wide ranges (the red fox is found across Russia's entire territory), and are often ubiquitous and abundant, making the likelihood of their contact with humans or domesticated dogs very high. The wolf (*Canis lupus*), the golden jackal (*C. aureus*) and the corsac fox (*V. corsac*) are slightly less important carriers. The arctic fox (*Alopex lagopus*) is the main carrier of rabies in the tundra. All of these species can also participate in the transmission of trichinosis, toxocariasis and echinococcosis. The fox and the raccoon dog, who are facultative piscivores, can participate in the spread of opisthorchiasis and diphyllobothriasis (Sidorov 1983; Lysenko et al. 2002; Stepanova 2002; Semenova et al. 2011) (Map 3.5).

Members of the Ursidae family—the brown bear (*Ursus arctos*) and the polar bear (*U. maritimus*) are involved in the spread of trichinosis.[2] The brown bear sometimes participates in the spread of tick-borne encephalitis and Lyme borreliosis (as a nutrient source for all stages of tick development) and it may also participate in the spread of North Asian tick-borne rickettsiosis. As the bear is a facultative piscivore, it plays a role in the natural foci of opisthorchiasis and diphyllobothriasis, especially in Russia's eastern regions (Sidorov 1983) (Map 3.6).

Certain species in the Mustelidae family have relatively low epidemiological significance. To a large degree, their role as carriers is related to fish consumption, which leads to the spread of opisthorchiasis and diphyllobothriasis. For the majority of species in the *Mustela* genus—the least weasel, weasels, and the European mink—fish is likely an unintentional food source, but for some, like the Eurasian otter (*Lutra lutra*) fish is the primary nutrient source. However, because the Eurasian otter is a rare species, its epidemiological significance is probably not that high. Many members of the Chiroptera order carry rabies, but their role in natural foci is quite small due to the peculiarities of their ecological function and way of life, which prevents these animals from contact with other mammals and humans (Maps 3.7 and 3.8).

Members of the Leporidae family of the Lagomorpha order (Map 3.9) are one of the primary nutrient sources for ticks at all stages of development. These ticks serve as vectors for tick-borne encephalitis and Lyme borreliosis (Korenberg and Kovalevsky 2000; Korenberg 2003a, b), and

[2]The simplest classification system is in use—all bears are categorized as being in the *Ursus* genus (Dinets and Rothschild 1996). In other sources, the polar bear may be placed in a different genus.

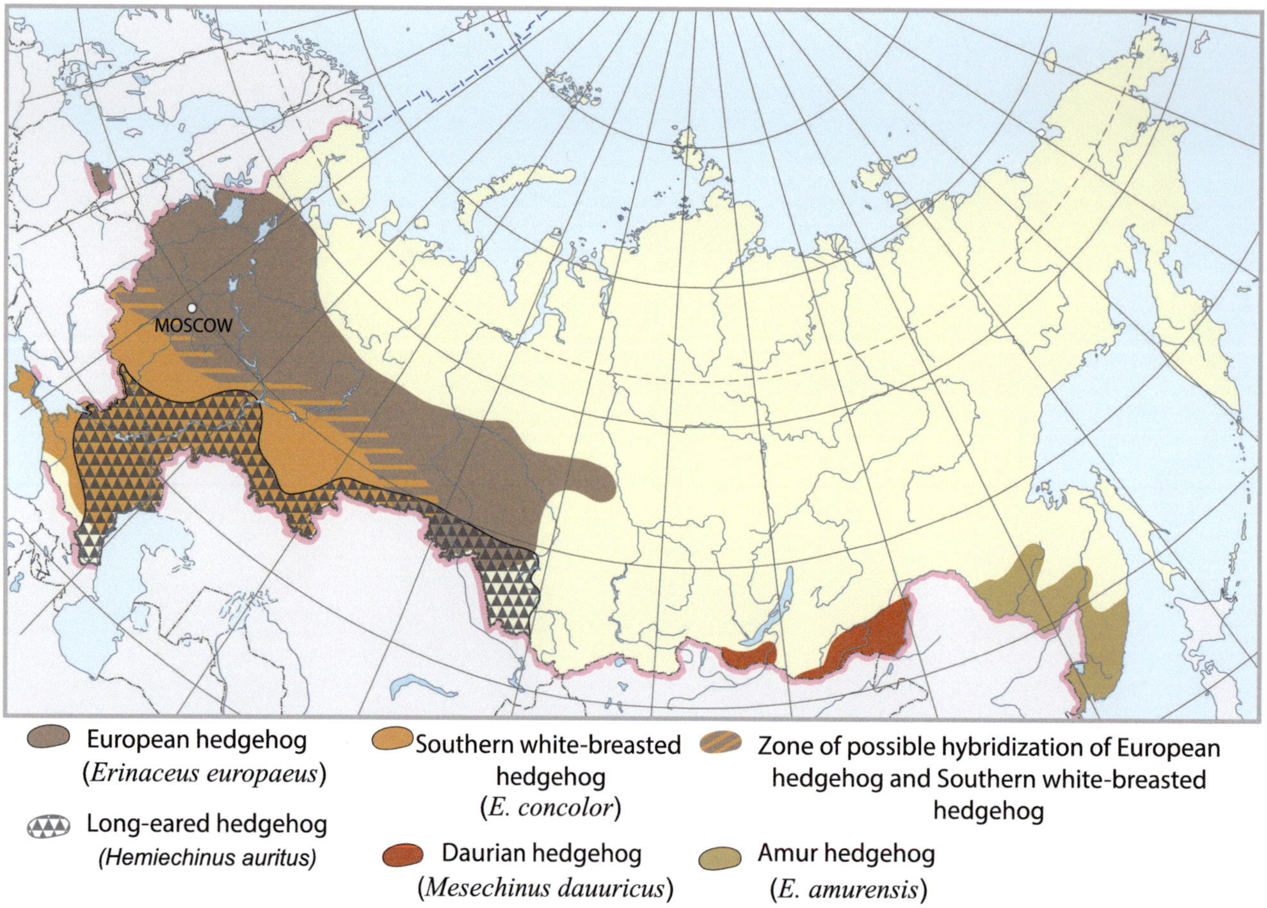

Map 3.4 Hedgehogs—*Erinaceus, Hemiechinus, Mesechinus*

potentially for North Asian tick typhus. Hares and pikas (in the Ochotonidae or Lagomyidae families) may also play a certain role in natural foci of tularemia (Olsufiev and Dunaeva 1970; Mescheryakova 2003).

Members of the Artyodactyla order participate in the spread of many diseases. Almost all of them serve as nutrient sources for ticks, who are vectors of tick-borne encephalitis, Lyme borreliosis and potentially North Asian tick typhus. Wild boars (*Sus scrofa*) participate in the spread of taeniasis and trichinosis, as well as opisthorchiasis and diphyllobothriasis, as they are facultative piscivores. Species in the Cervidae family—the elk (*Alces alces*), red deer (*Cervus elaphus*), sika deer (*C. nippon*), reindeer (*Rangifer tarandus*), roe deer (*Capreolus capreolus*), Siberian roe deer (*C. pygargus*) and the Siberian musk deer (*Moschus moschiferus*) can serve as carriers for the pathogens of echinococcal infections, anthrax and tetanus. Reindeer also carries taeniasis due to *T. saginata* and brucellosis (Maps 3.10, 3.11 and 3.12).

Members of the Rodentia order are the most epidemiologically significant wild mammals. A large number of species in this order play crucial roles in the natural foci of various diseases.

Species in the Sciuridae family—marmots (*Marmota* genus) and ground squirrels (*Spermophilus* genus) are very important carriers in natural foci of the plague (Neronov et al. 1991). Among marmots living in Russia, the grey marmot (*M. baibacina*) and the Tarbagan marmot (*M. sibirica*) (Map 3.13) are the most common carriers of infectious disease. As for ground squirrels, in Russia the little ground squirrel (*S. pygmaeus*) and the Daurian ground squirrel (*S. dauricus*) are the most significant carriers (Map 3.14). Ground squirrels can also act as carriers for North Asian tick typhus and tularemia (Tarasevich et al. 2003; Olsufiev and Dunaeva 1970; Mescheryakova 2003). Some other types of squirrels periodically participate in the natural foci of these diseases, especially the squirrel (*Sciuris vulgaris*) and the chipmunk (*Tamias sibiricus*) (Olsufiev and Dunaeva 1970; Mescheryakova 2003; Tarasevich et al. 2003).

Gerbils (Gerbillidae) have traditionally been classified as a subfamily of the Cricetidae family, but today they are

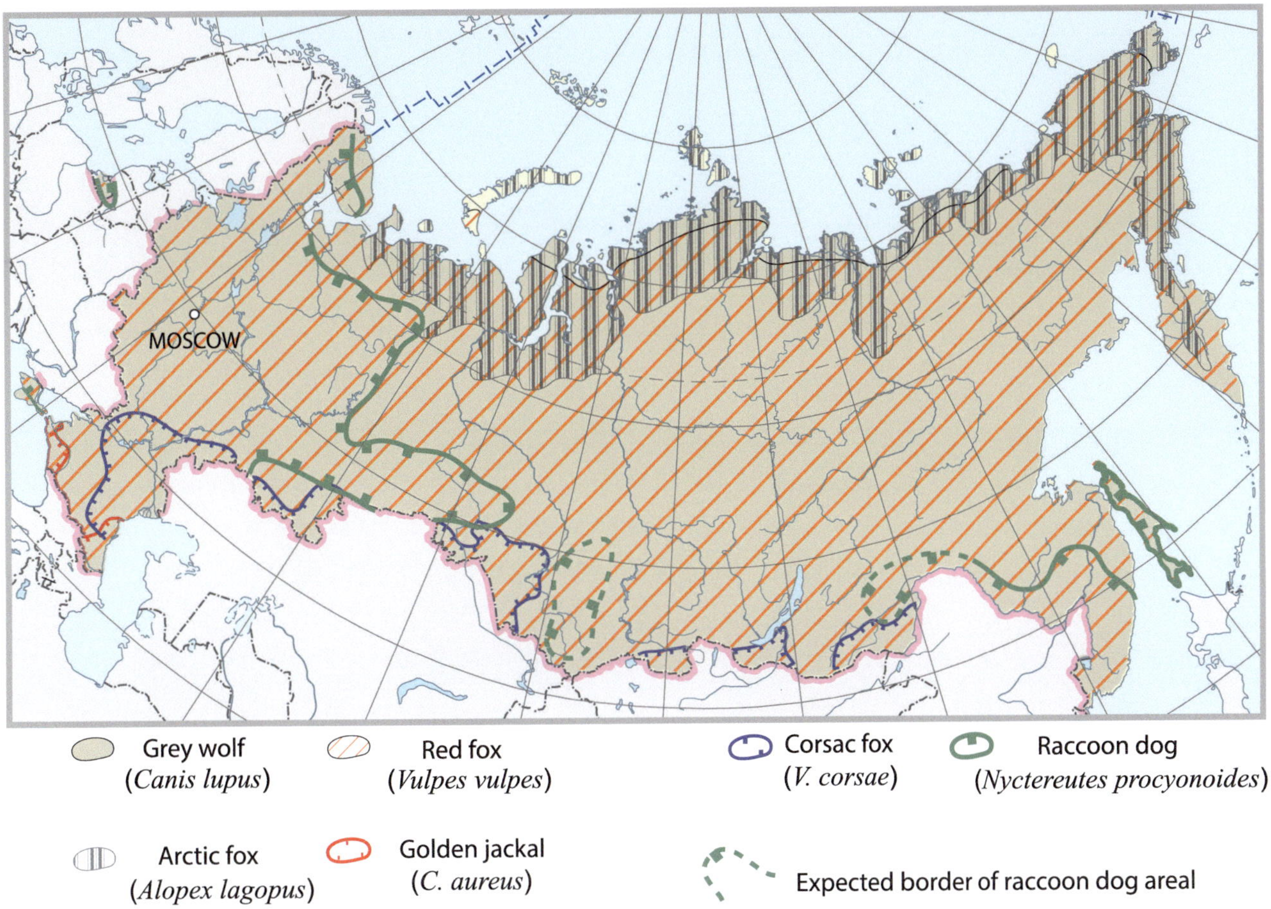

Map 3.5 Canids—Canidae

believed to make up an independent family. Along with marmots and ground squirrels, they are extremely important carriers of the plague (Neronov et al. 1991). On Russia's territory, most of these animals are in the *Meriones* genus: the tamarisk gerbil (*M. tamariscinus*), the midday gerbil (*M. meridianus*) and the mongolian gerbil (*M. unguiculatus*) (Map 3.15). The only species in the *Rhombus*'s genus that serves as a carrier is the great gerbil (*R. opimus*), which is the main carrier of the plague in larger territories of Kazakhstan and Central Asia, as well as beyond the borders of the former USSR. This gerbil is not found in Russia, although sometimes it briefly appears in the Northern Caspian Depression.

Species in the Muridae family play a significant role in the foci of many diseases, such as North Asian tick typhus, carried by a number of mice (the *Micromys, Mus*, and *Apodemus* genera), hemorrhagic fever with renal syndrome, carried by field mice of the *Apodemus* genus, tularemia, carried by the striped field mouse (*A. agrarius*) and rats (*Rattus* genus), as well as pseudotuberculosis and leptospirosis, which are carried by a number of mouse species in the *Micromys* and *Apodemus* genera (Olsufiev and Dunaeva 1970; Mescheryakova 2003; Tarasevich et al. 2003) (Maps 3.16 and 3.17). Synanthropic rats and the house mouse also serve as carriers for the aforementioned diseases.

Microtidae voles (Microtidae), like gerbils (see above), were traditionally classified as a subfamily of the Cricetidae family, but today they are often designated as a separate family. They often serve as important carriers of a whole variety of diseases. Members of the *Clethrionomys* or *Myodes* genus are one of the main nutrient sources for nymph ticks, which transmit tick-borne encephalitis and North Asian tick typhus (Korenberg and Kovalevsky 2000; Korenberg 2003a, b). The bank vole (*Cl. glareolus*) in the European part of Russia and the grey red-backed vole (*Cl.

Map 3.6 Bears—Ursidae

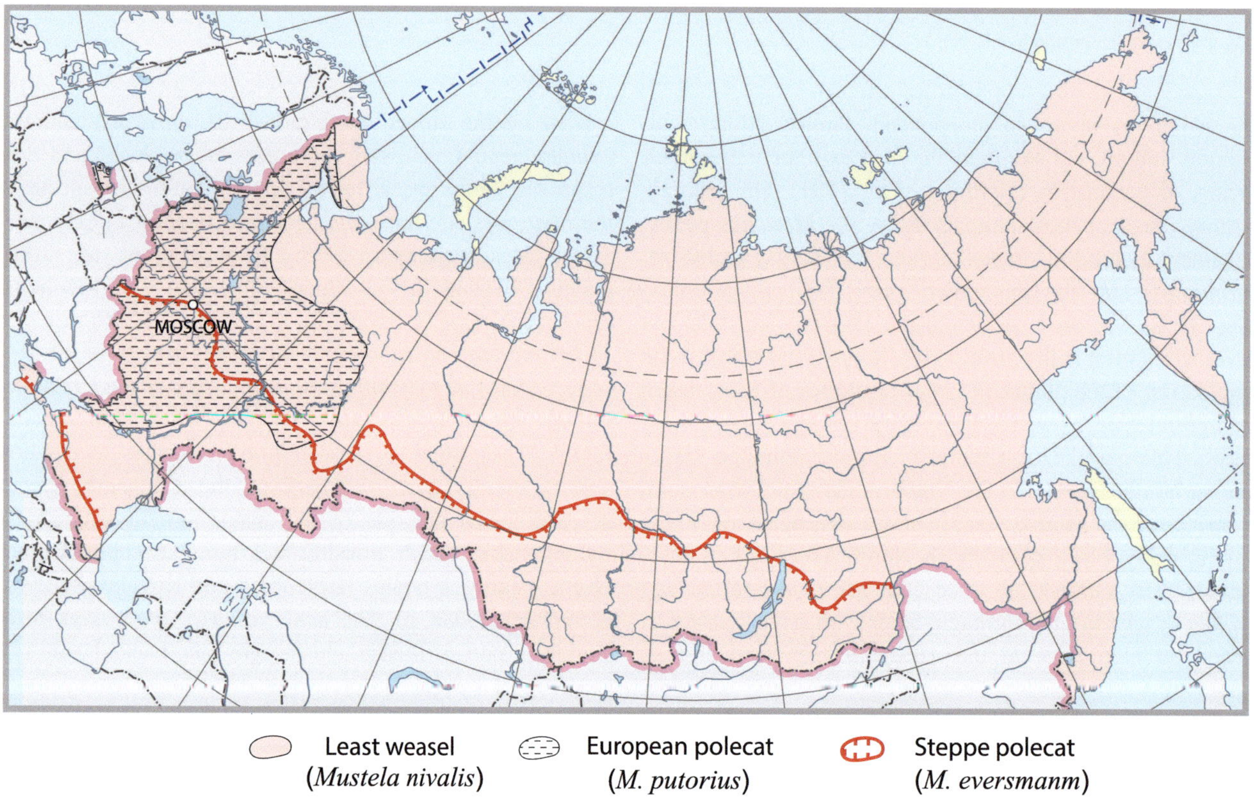

Map 3.7 Weasels and Polecats—*Mustela*

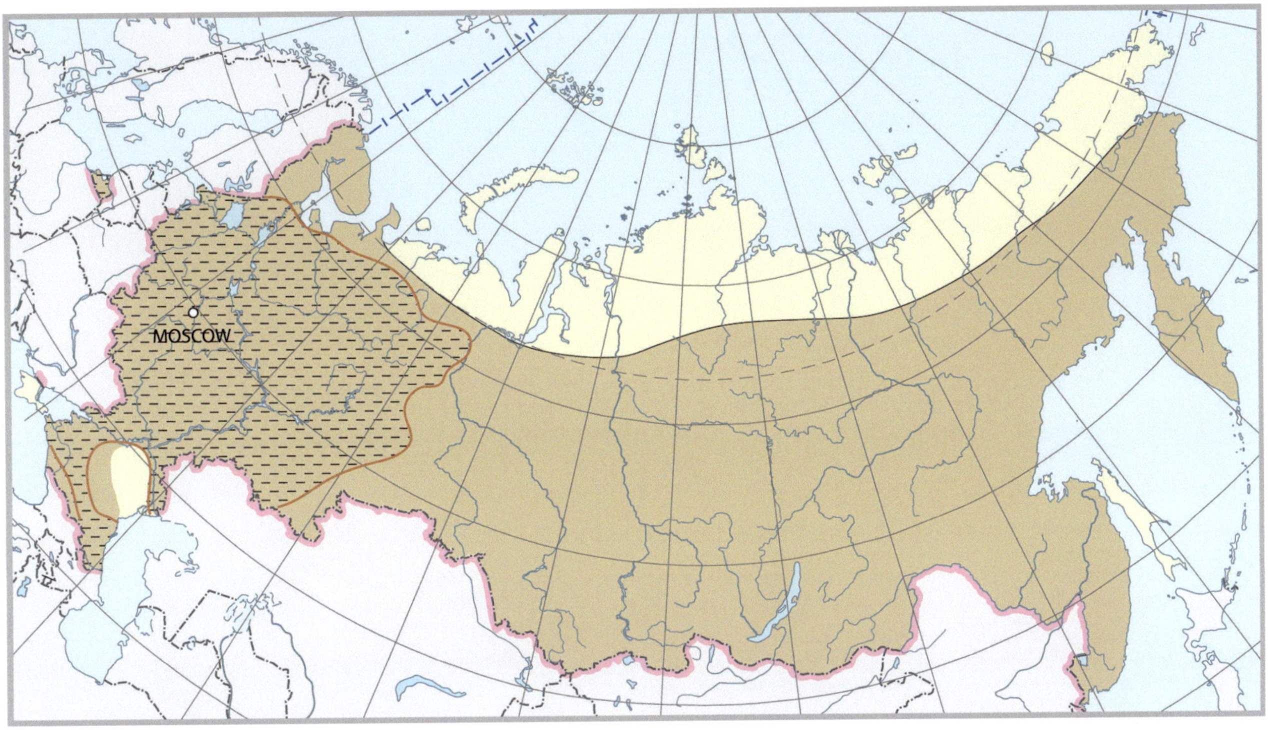

American mink (*Mustela vison*) European mink (*M. lutreola*)

Map 3.8 Minks—*Mustela*

Mountain hare (*Lepus timidus*) European hare (*L. europaeus*) Tolai hare (*L. tolai*)

Map 3.9 Hares—*Lepus*

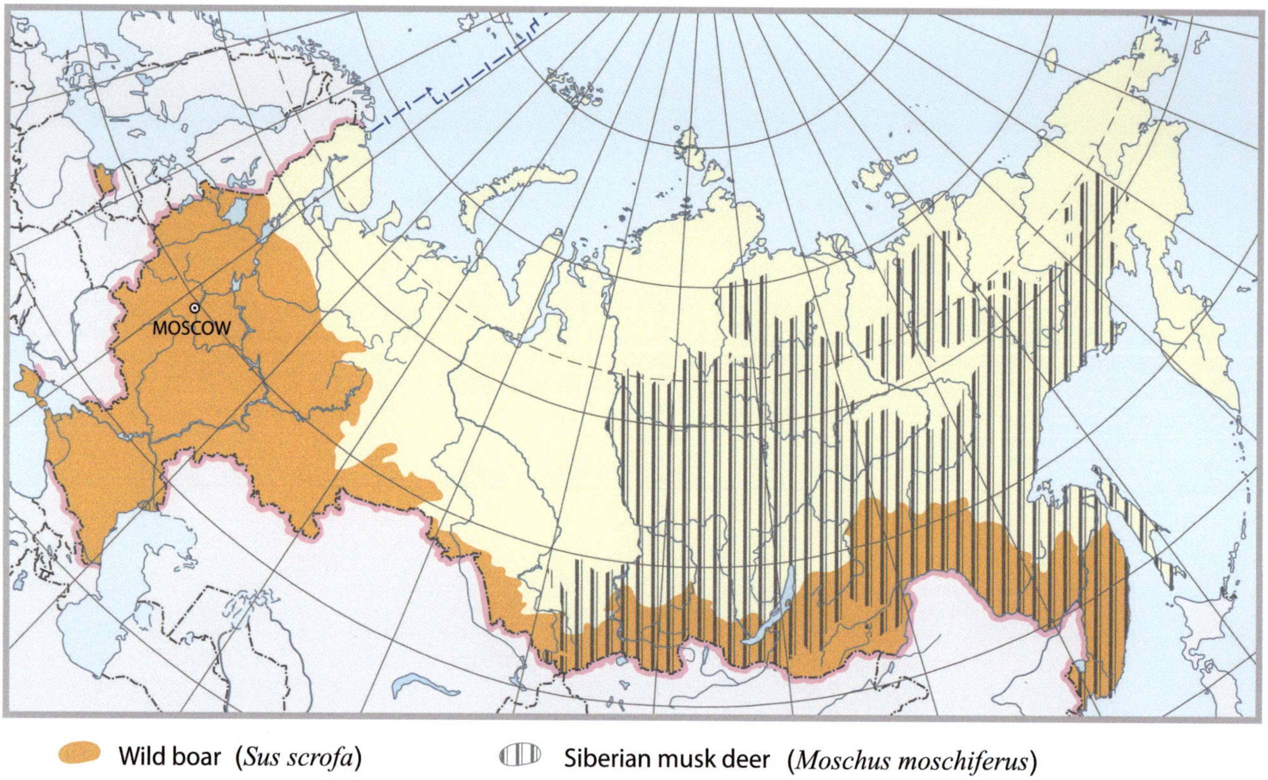

Map 3.10 Boar—*Sus*, Musk Deer—*Moschus*

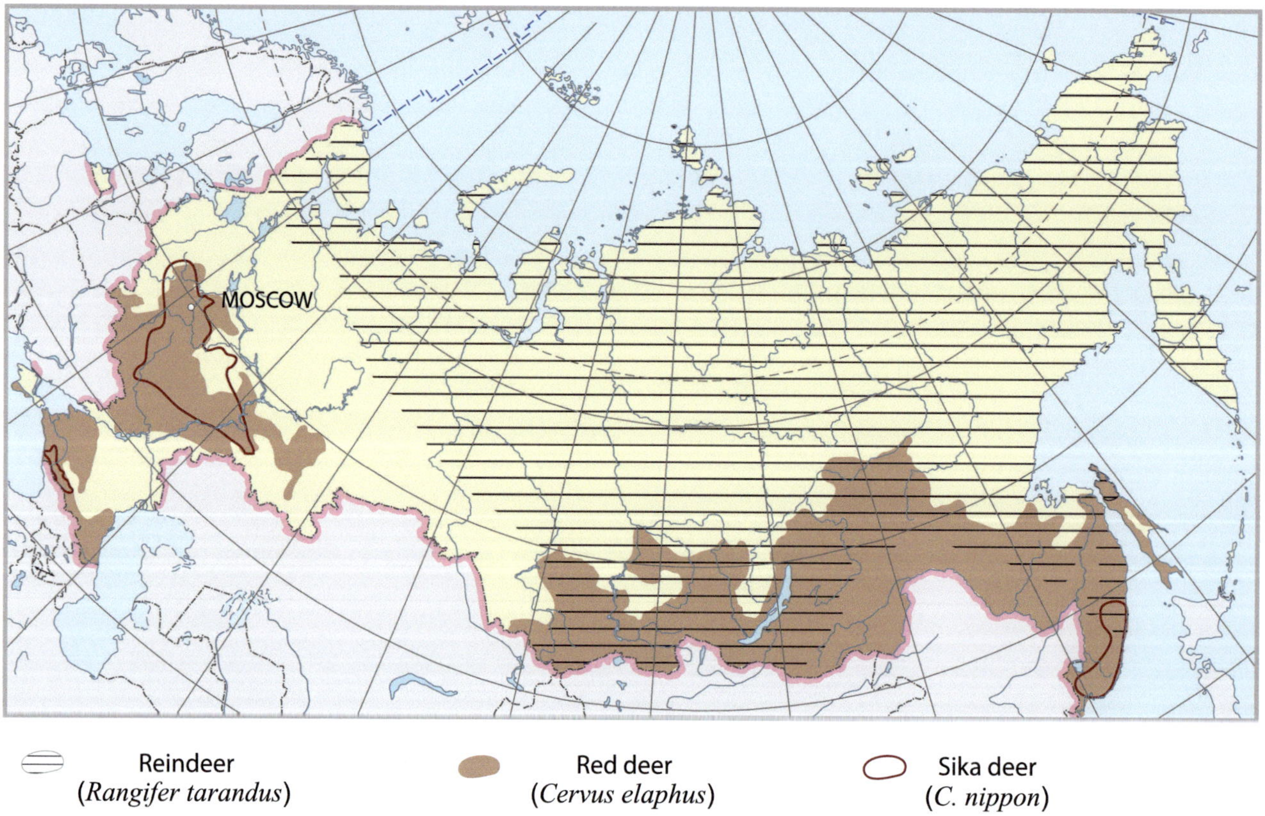

Map 3.11 Deers—Cervidae

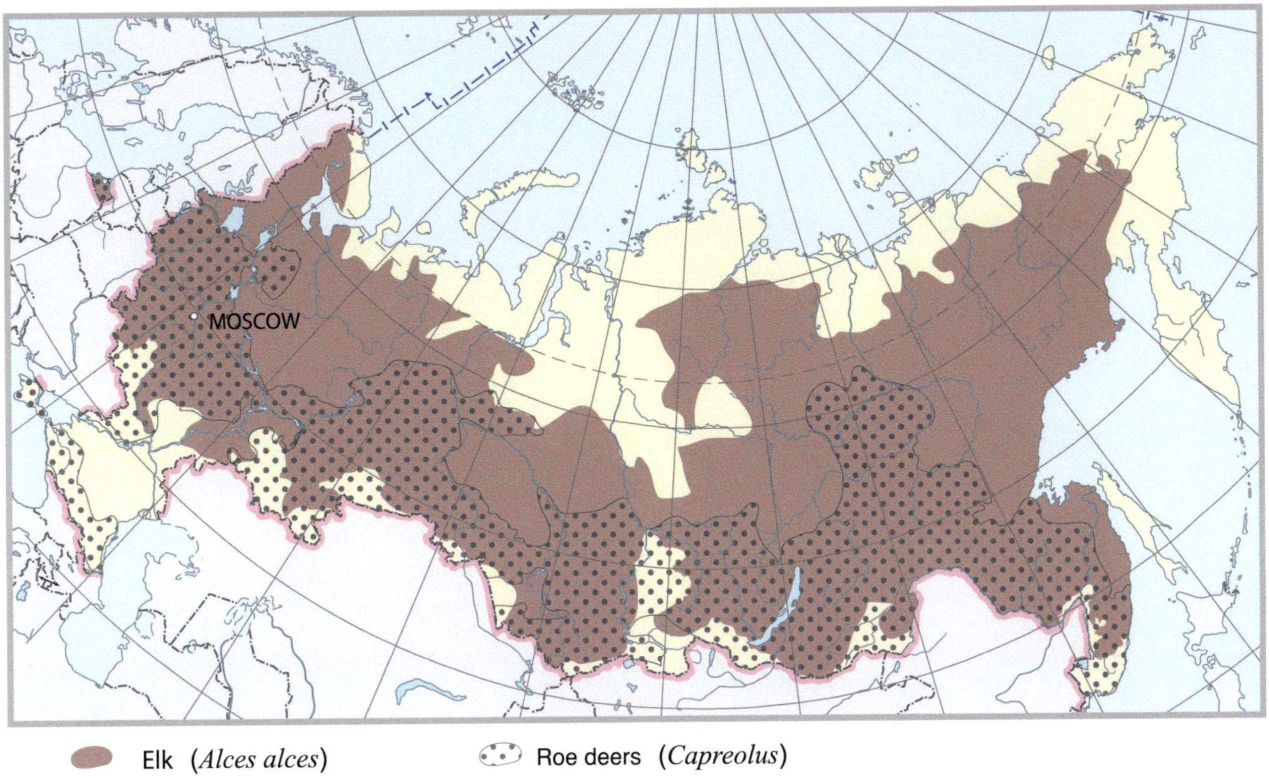

Map 3.12 Elk and Roe deers—Cervidae

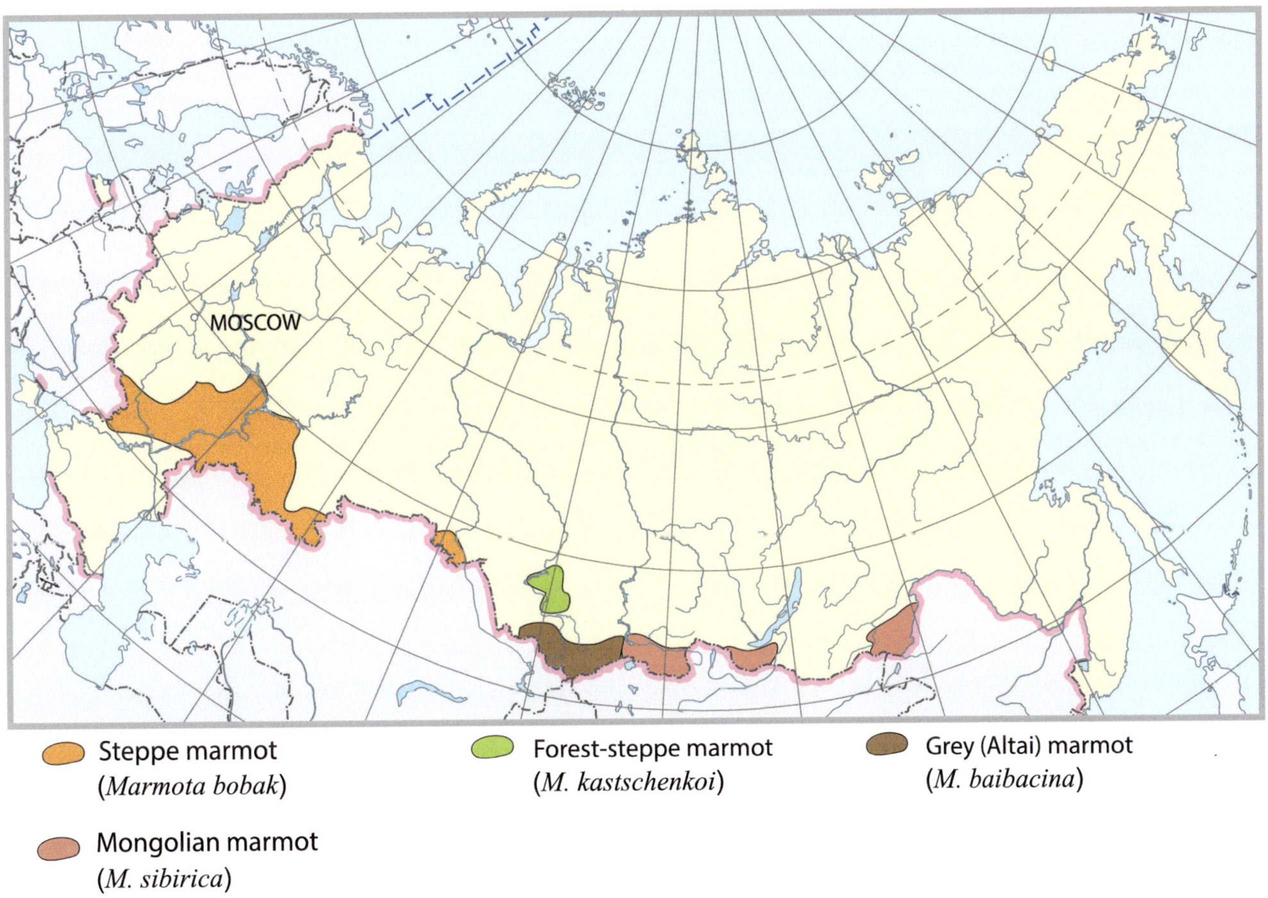

Map 3.13 Marmots—*Marmota*

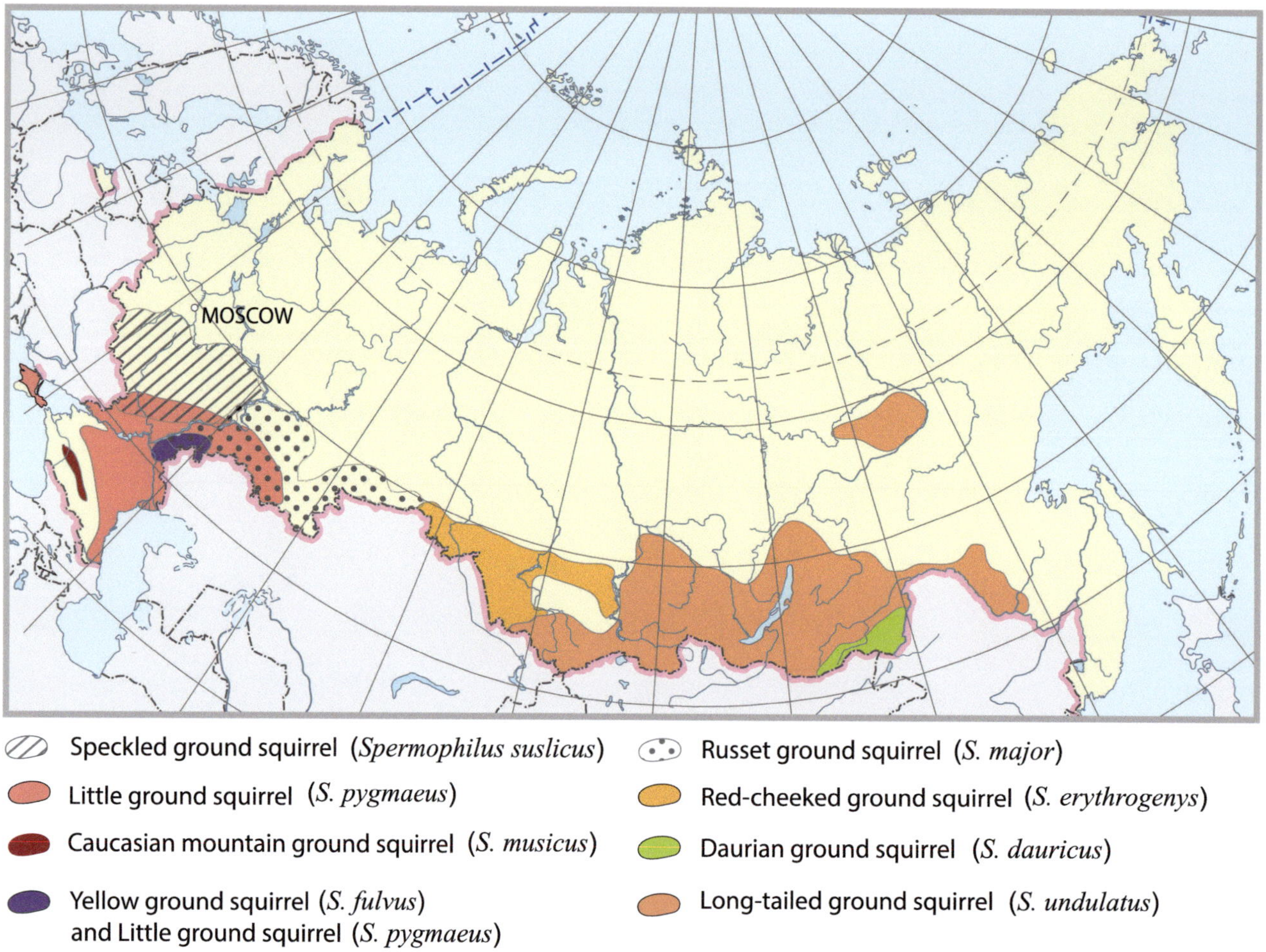

Map 3.14 Ground squirrels—*Spermophilus*

rufocanus) in Siberia and the Russian Far East are the main carriers of hemorrhagic fever with renal syndrome (Map 3.18). Voles in the *Microtus* genus are the main carriers of leptospirosis [the tundra vole (*M. oeconomus*) and the common vole (*M. arvalis*)] (Ananyina 2003) (Map 3.19).

The European water vole (*Arvicola terrestris*) and the muskrat (*Ondatra zibethicus*) are also important carriers of leptospirosis, tularemia and certain other diseases, especially those connected to aquatic and semiaquatic ecosystems (Map 3.20). Water voles also participate in the natural foci of other diseases, like North Asian tick typhus, pseudotuberculosis, etc. The muskrat is the main reservoir for the pathogen of hemorrhagic fever with renal syndrome.

To conclude, it is important to note that recently the question of the role of certain water and wetland mammals (beavers—*Castor canadensis* and *C. fiber*, otters—*Myocastor coypus*—and muskrats) in the spread of giardiasis (Stepanova 2001; Dunlap and Thies 2002) has been raised.

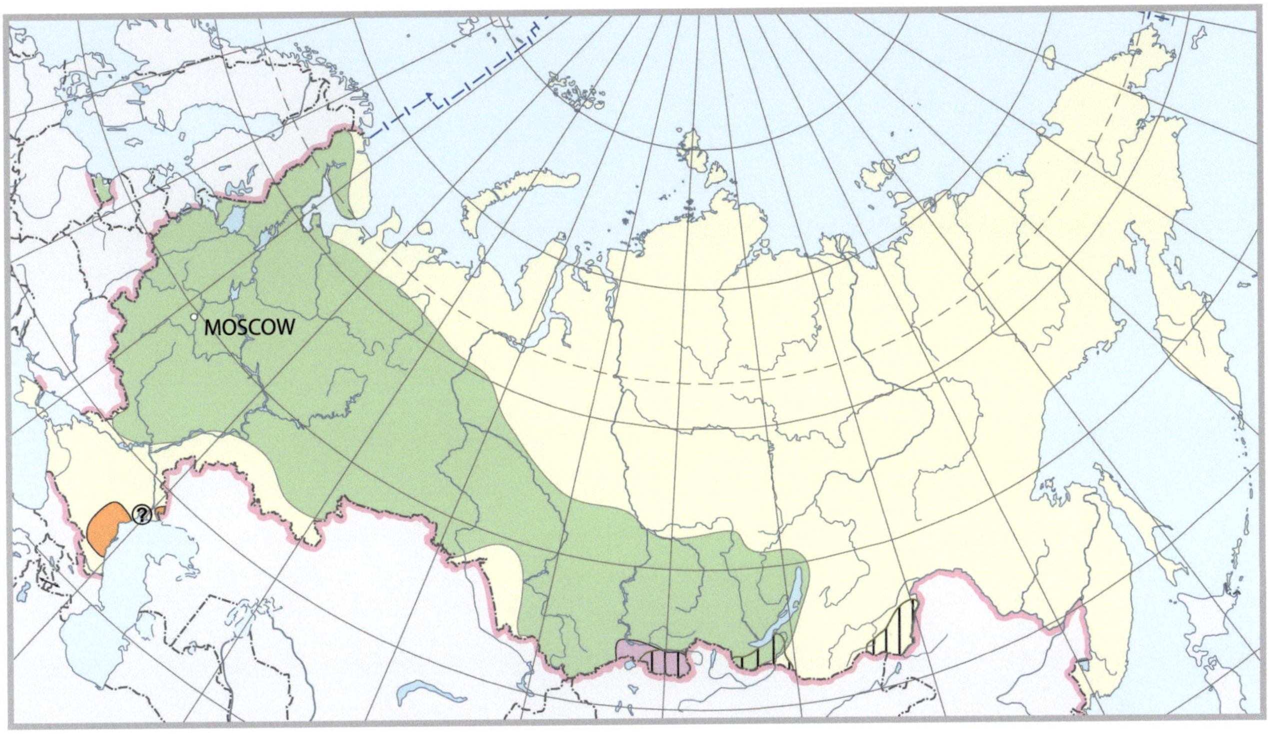

Northern birch mouse (*Sicista betulina*)

Tamarisk gerbil (*Meriones tamariscinus*) and Midday gerbil (*M. meridianus*)

Midday gerbil (*M. meridianus*)

Mongolian gerbil (*M. unguiculatus*)

Map 3.15 Bush Mice—*Sicista* and Gerbils—*Meriones*

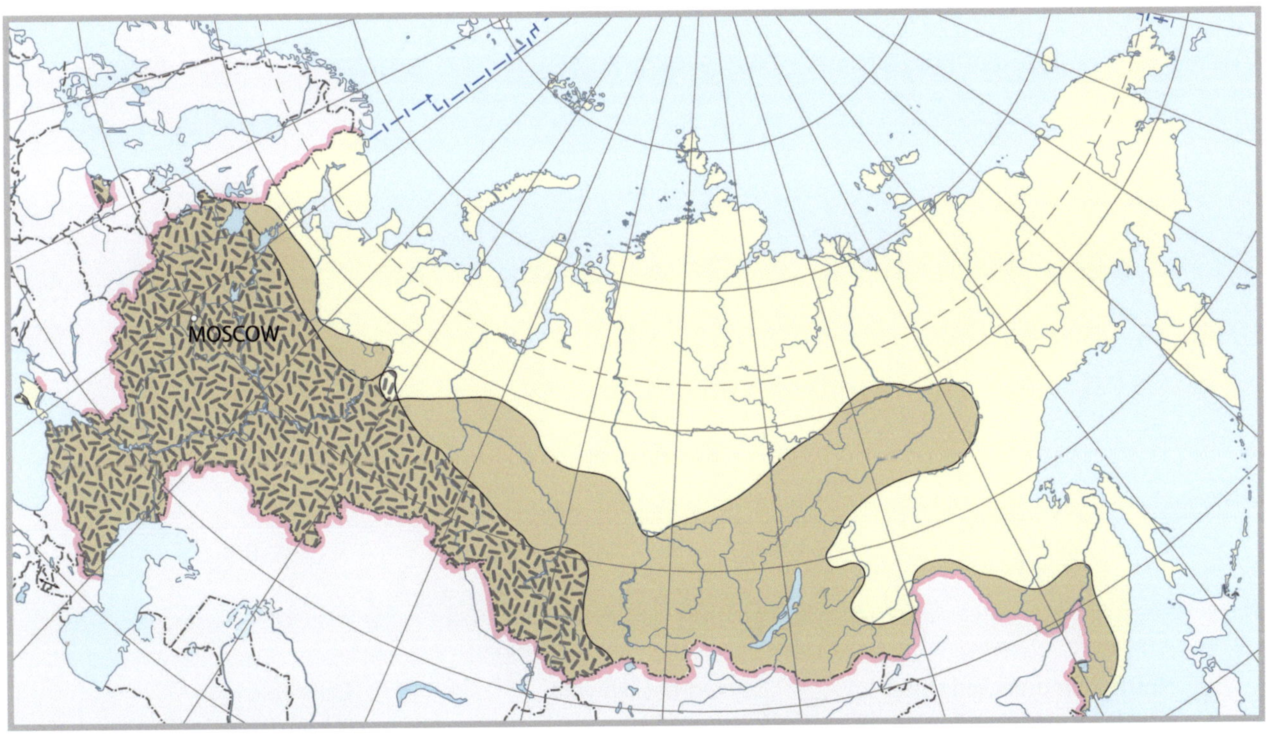

Eurasian harvest mouse (*Micromys minutus*)

Ural field mouse (*Apodemus uralensis*)

Map 3.16 Mice—*Micromys, Apodemus*

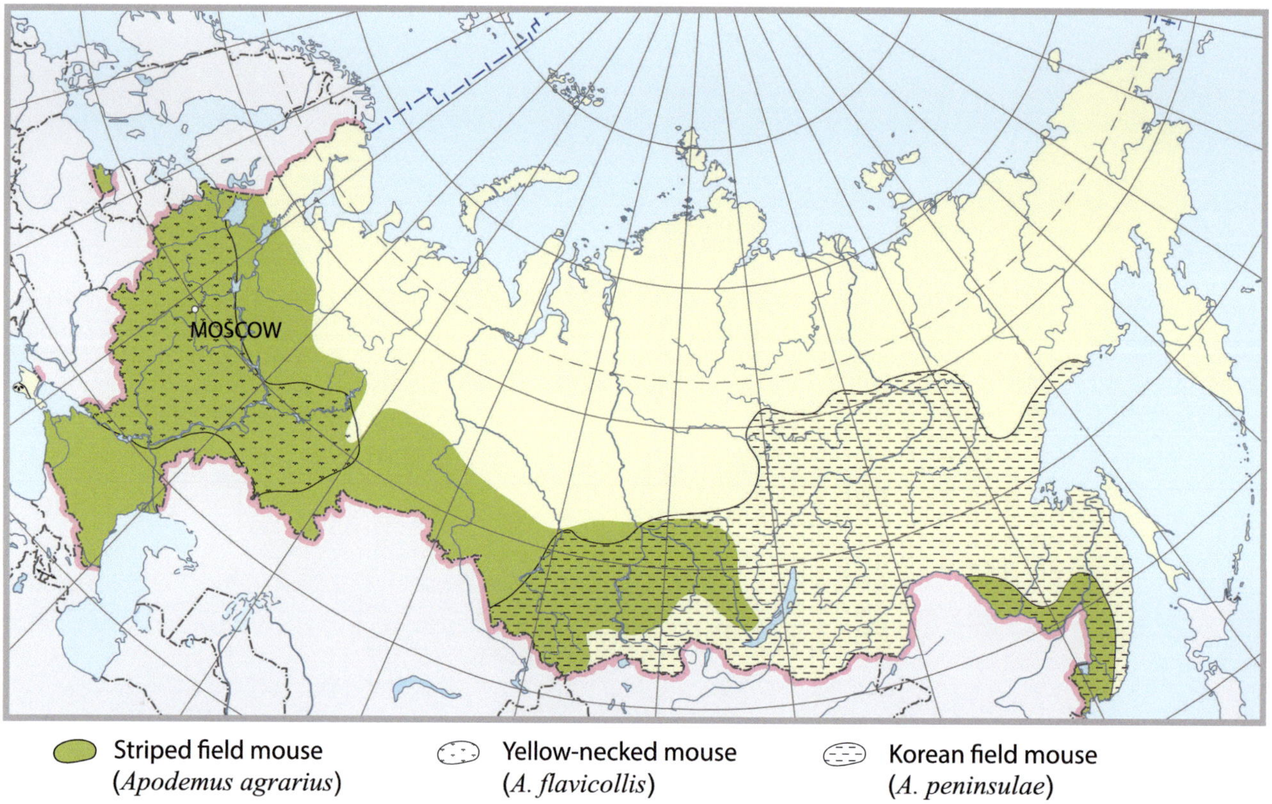

Map 3.17 Mice—*Apodemus*

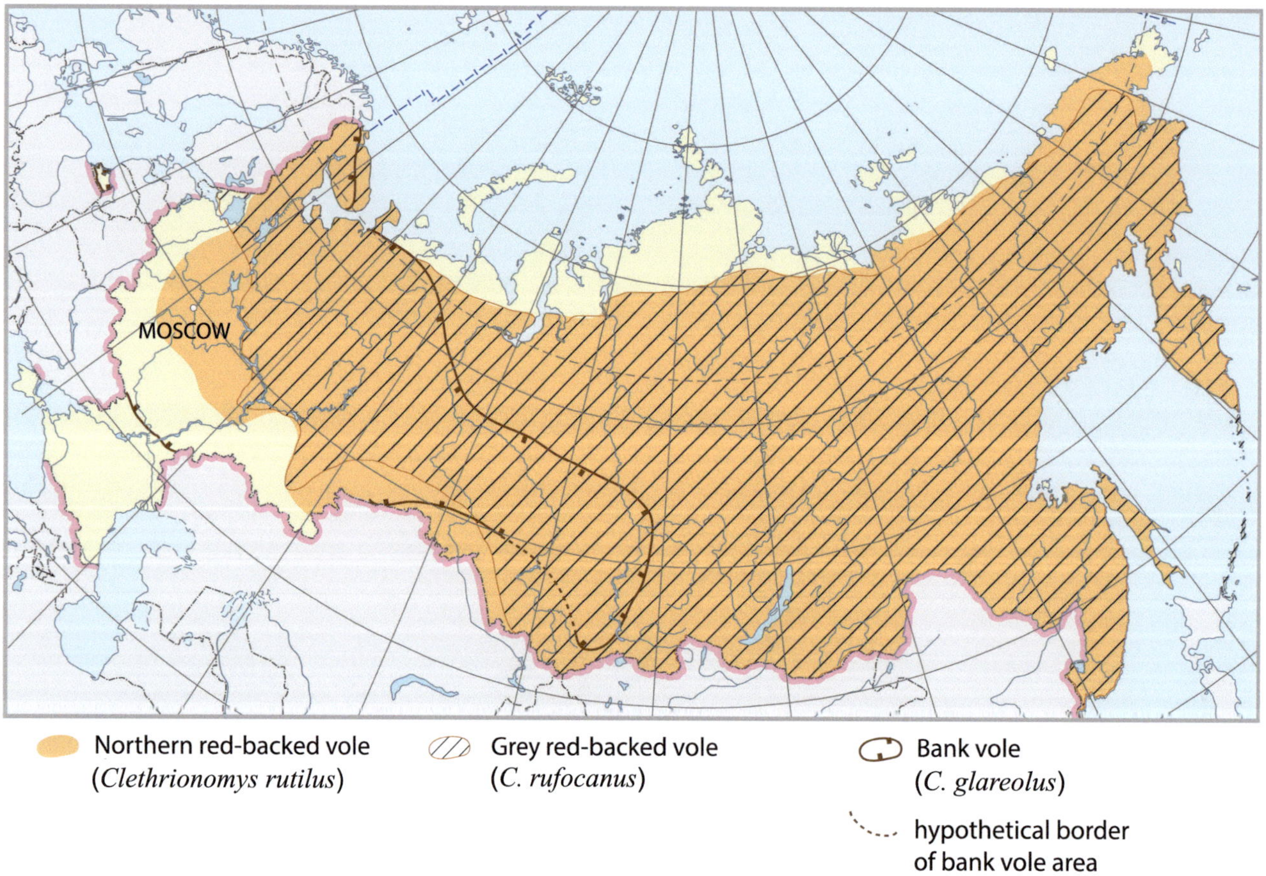

Map 3.18 Red-Backed Voles—*Clethrionomys*

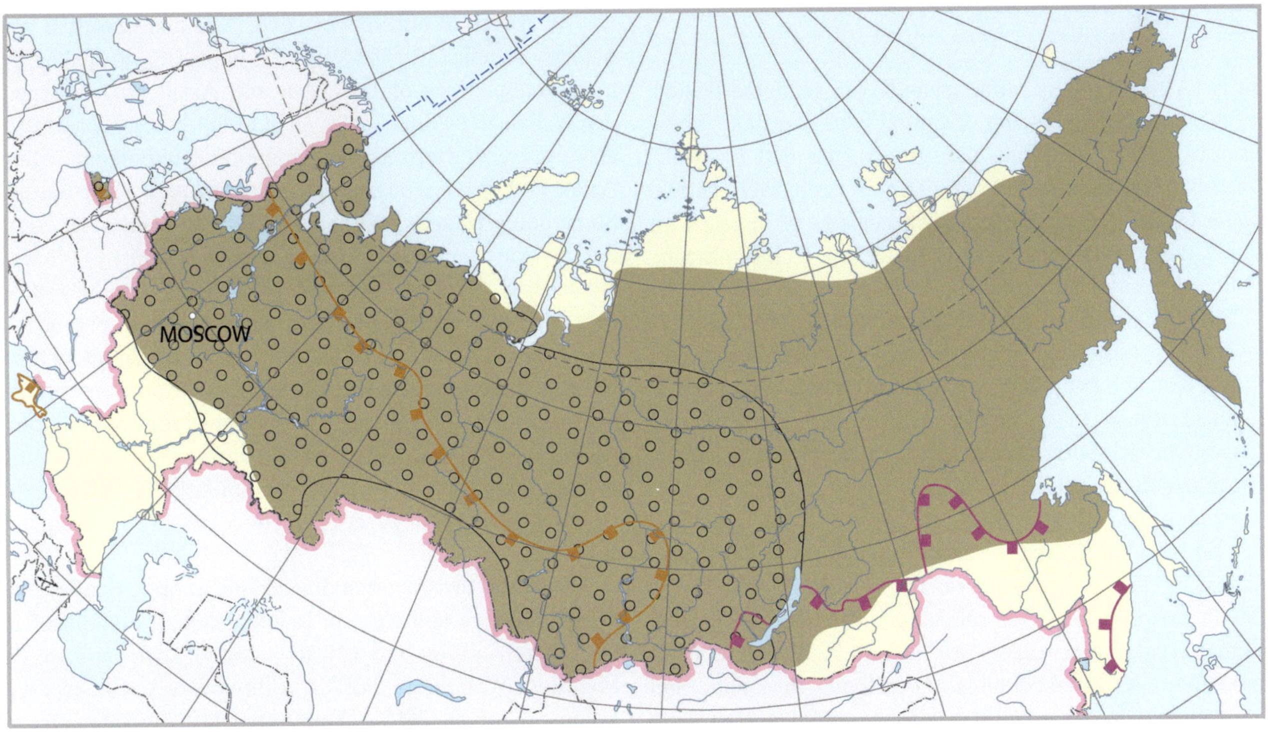

Tundra vole (*Microtus oeconomus*)

Field vole (*M. agrestis*)

Common vole (*M. arvalis*)

Reed vole (*M. fortis*)

Map 3.19 Meadow Voles—*Microtus*

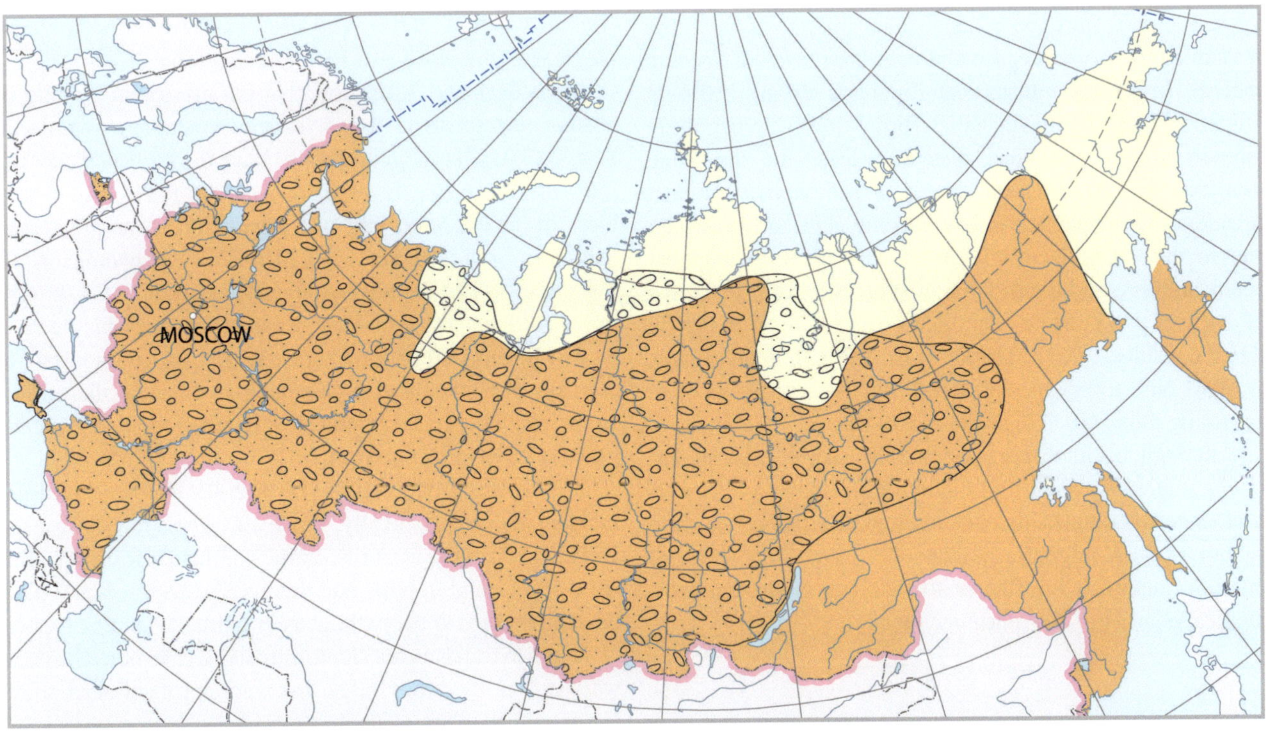

Muskrat (*Ondatra zibethicus*)

European water vole (*Arvicola terrestris*)

Map 3.20 Muskrat—*Ondatra* and Water vole—*Arvicola*

3.2 Birds

Birds (Aves), be they wild, synanthropic or domesticated, are the carriers of a number of natural focal diseases.[3] These diseases can be split into two groups, the first of which consists of diseases that are primarily or exclusively carried by birds, and the second of which consists of diseases for which the role of bird hosts is incidental or insignificant, while other animals (mammals) serve as the primary hosts.

The first group includes historically widespread diseases (psittacosis), diseases that are new to Russian territory (West Nile fever), and extremely rare diseases: arboviruses such as Karelian fever (also known as Pogosta disease) and Kemerovo tick-borne viral fever (Kurkela et al. 2004; Encyclopedic… 2001).

Among synanthropic birds, the pigeon (*Columbia livia*) and birds in the Corvidae family pose the greatest threat. According to Rospotrebnadzor, up to 50% of pigeons in Russia are infected with psittacosis. Corvidae are active participants in the so-called 'urban' form of West Nile fever circulation. As for wild birds, those living in aquatic and wetland ecosystems are most significant, and they contribute to the spread of psittacosis and the rural form of West Nile fever circulation.

The psittacosis pathogen has been found in more than 140 bird species. The corresponding exact data for West Nile fever are unavailable, however, based on indirect information, the list of such species is no smaller at the very least. Special studies must be conducted in order to analyse the epidemiological significance and map the ranges of specific species. Infectious transmission by birds during seasonal migration is a significant issue. Birds' ability to regularly migrate across enormous distances leads to the wide (transcontinental) and rapid spread of pathogens. In recent decades, this phenomenon has become especially relevant due to the categorization of West Nile fever as a global healthcare risk, and the addition of West Nile fever to the list of diseases that must be controlled on a state level in Russia (On the results… 2012; The Forecast… 2013). With regards to West Nile fever, special attention must be paid to birds migrating along the Africa-Eurasia flyway.

Although ornithologists around the world have extensively studied seasonal bird migration, we have a very cursory understanding of its epidemiological significance. The pattern of such migration is generally clear (Fig. 3.1), but this information is insufficient for analysis of the epidemiological aspects of this phenomenon.

The system of bird flyways over Asia and Europe is particularly complex. By and large, it is impossible to analyse the migration patterns of European and Asian birds independently. Palearctic Eurasia (i.e. all of Europe and the majority of Asia) are characterized by a common ornithofauna. Many Asian bird species migrate to and from Africa via Europe. Some long-distance migrants spend the winter in North Africa, many—in tropical Africa, but the majority of birds go all the way to South Africa. Overall, over a quarter of all Palearctic bird species (approximately 2 million organisms) spend the winter south of the Sahara Desert. As most birds migrate to the southwest in autumn, flyways are concentrated in Europe, especially given that most birds' final destination is south of Europe, in Africa. However, many birds migrate directly to the southwest from Asia to Africa, either not flying over Europe at all or only passing its periphery (Fig. 3.2).

The study of aquatic and semiaquatic birds that migrate between Africa and Eurasia is internationally regulated by the Agreement on the Conservation of African-Eurasian Migratory Waterbirds (AEWA) (http://www.unep-aewa.org; Belousova et al. 2007). The Agreement's documents are available on its official website (http://www.unep-aewa.org). AEWA's full list includes more than 230 species and subspecies of birds and more than 500 populations. All of these species are under conservation within the AEWA. Over 250 populations in AEWA are found in Russian territory, and a complete list of these populations has been published (Belousova et al. 2007). The study of these birds' epidemiological significance is a very relevant topic.

Maps 3.21 and 3.22 show that the maximum number of nesting waterbirds is found in certain southern regions in Russia (3.21). These regions also have the highest morbidity rates for West Nile fever. However, the fraction of waterbirds in the entire nesting bird population is the greatest in Russia's arctic regions (3.22). This phenomenon is insignificant with regards to West Nile fever, as these territories are located beyond the possible range of this disease. However, this population characteristic may be important in the case of psittacosis, in part because many of these birds are intensively hunted by the local population, increasing the probability of direct contact.

The second group consists of diseases that birds transmit incidentally or episodically. There are quite a few diseases of this kind. The viruses of tick-borne and Japanese encephalitis have been found in birds and their ectoparasites. Many birds have antibodies to the pathogens of North Asian tick typhus, Q fever, toxoplasmosis, leptospirosis, Lyme borreliosis, as well as to a few arboviruses that were not mentioned above (Lvov and Ilyichev 1979; Lvov et al. 1989).

The role of wild birds in these processes has not been sufficiently studied (Lvov and Ilyichev 1979, etc.). But there is a number of synanthropic birds among those who

[3]Domesticated and pet birds will not be discussed in this volume.

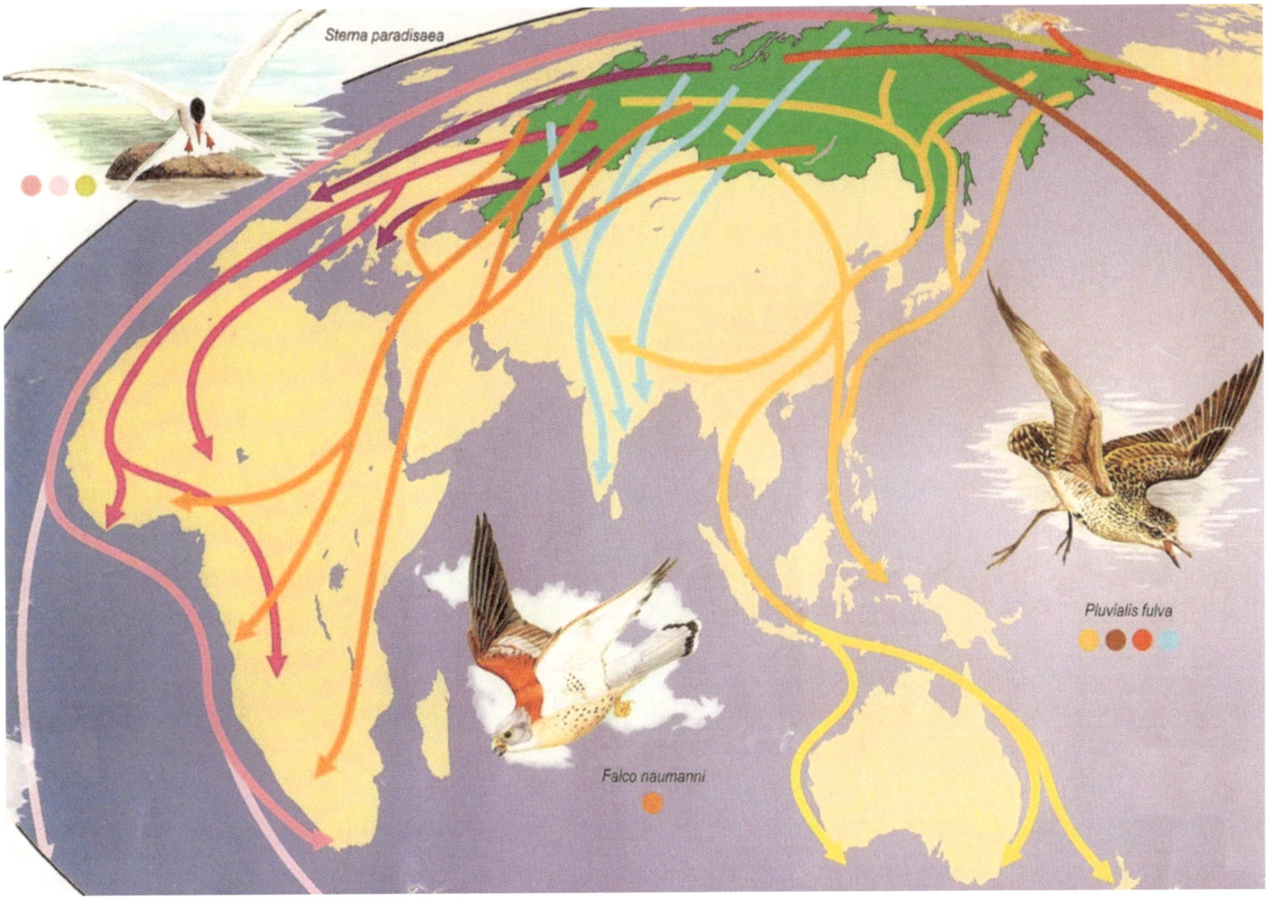

Fig. 3.1 General pattern of the migration of birds inhabiting Russia (Russia is highlighted in green). Edited poster fragment *Credit* Migratory… 2001, amended

participate in the spread of the aforementioned diseases. However, these birds rely on human settlements and activity to varying degrees. Based on this, Lipin (1974) divides them into three groups.

- Obligate synanthropes—species that completely rely on human settlements. They nest and feed in them (pigeons, sparrows, swallows, etc.).
- Facultative nesting synanthropes—species that settle and feed both in human settlements and in natural biotopes (hill pigeons, common swifts, hoopoes, lesser spotted woodpeckers, common ravens, western jackdaws, common starlings, Eurasian tree sparrows, white wagtails, great tits, common house martins, etc.).
- Facultative non-nesting synanthropes—species that do not inhabit human settlements but are closely linked to them due to their nutrient sources (Eastern imperial eagles, kites, common ravens, rooks, magpies, Bohemian waxwings, etc.).

Synanthropic groups of the third type serve as middlemen between natural foci and human settlements. These birds in particular significantly influence the infection rates of birds that nest in human settlements. The link between birds in the second group and natural biotopes is also of some significance. For example, some Eurasian tree sparrows nest outside of human settlements, but move closer to humans once they have stopped nesting. All of this provides the opportunity for humans to have indirect contact (via birds in the first group and domesticated animals) with wild bird species (Lipin 1974).

Tens of species of wild birds enter human settlements episodically, but they do not remain there for prolonged time periods, so their role in infection transmission is small. However, it is important to mention direct contact between humans and wild birds, especially with species that are wild game. Direct contact also influences transmission of diseases in the first group (see above), as game fowl in Russia primarily inhabits aquatic and wetland ecosystems.

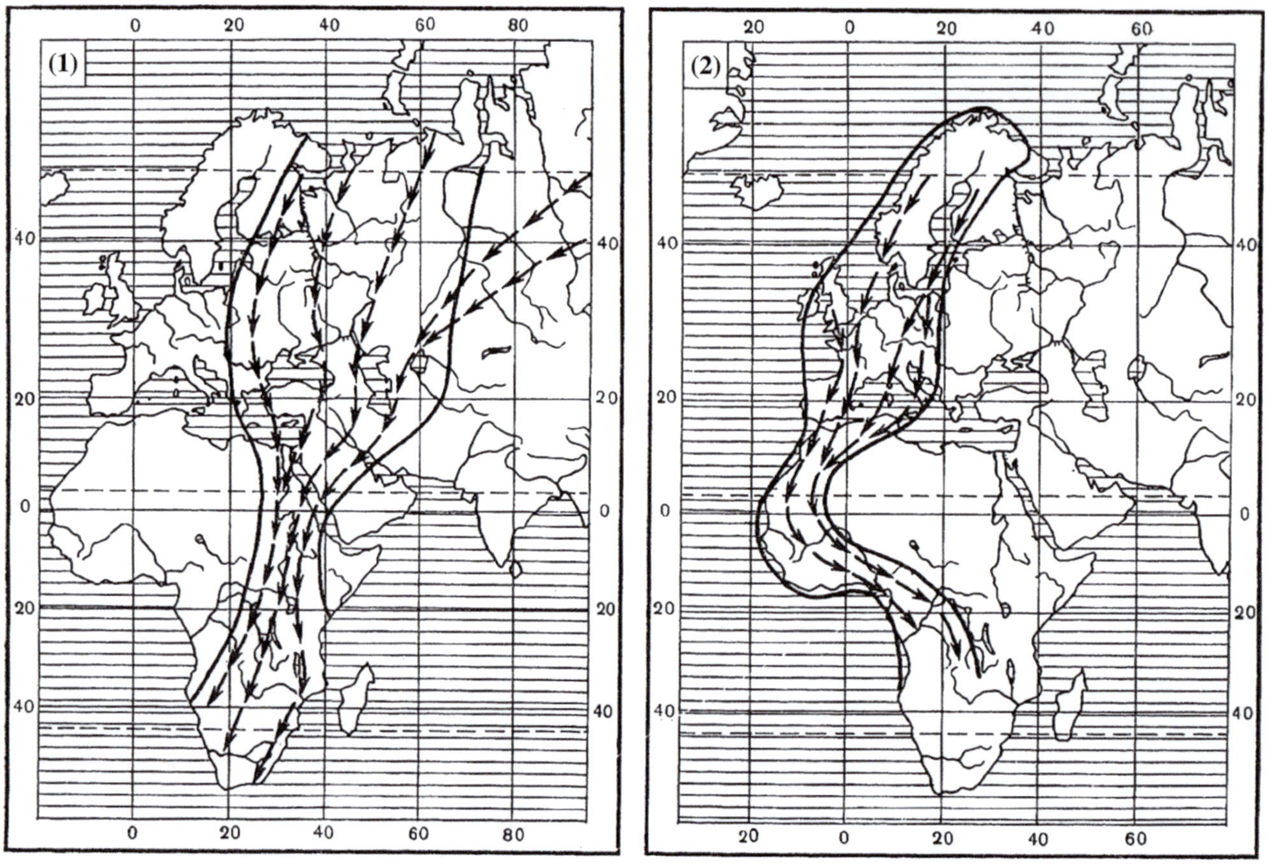

Fig. 3.2 Bird transcontinental migration corridors between Eurasia and Africa: East European (1) and West European (2) *Credit* Lvov and Ilyichev 1979, amended

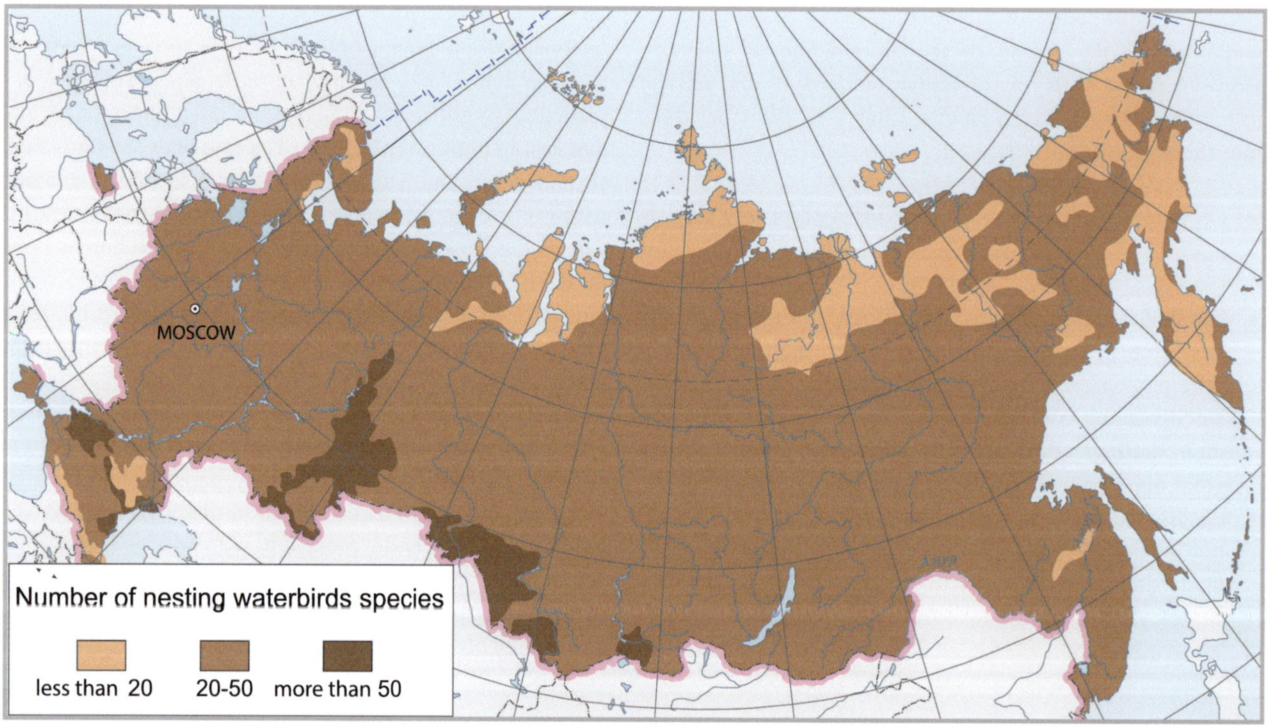

Map 3.21 Nesting waterbirds

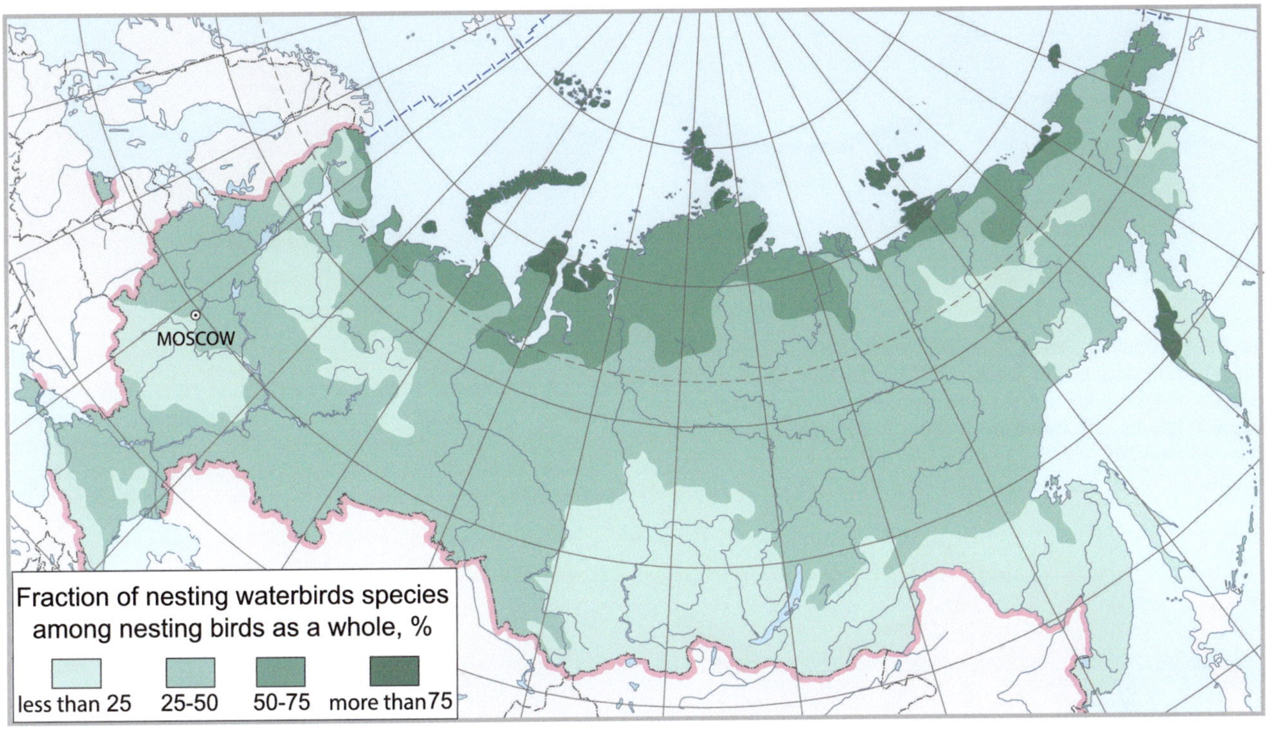

Map 3.22 Fraction of nesting waterbirds among nesting birds as a whole

3.3 Fish

Fish serve as hosts for many helminthiases, among which opisthorchiasis and diphyllobothriasis pose the greatest risk for humans.

Fish in the *Cyprinidae* family, especially so-called limnophilic species, which prefer calm waters and avoid rivers with fast currents, are intermediate hosts for the pathogen of opisthorchiasis. Widely prevalent species include the common roach (*Rutilus rutilus*) (Fig. 3.3),[4] the white bream (*Blicca bjoerkna*), the common bream (*Abramis brama*) (Fig. 3.4), the ide (*Leuciscus idus*) (Fig. 3.5), the European chub (*Leuciscus cephalus*), the common dace (*Leuciscus leuciscus*), the blue bream or zope (*Ballerus ballerus*), the white bream (*Ballerus sapa*), the common nase (*Chondrostoma nasus*), the common rudd (*Scardinius erythrophthalmus*) (Fig. 3.6), the sabrefish (*Pelecus cultratus*), the tench (*Tinea tinea*), the gudgeon (*Gobio gobio*), the common bleak (*Alburnus alburnus*) (Fig. 3.7), the phoxinus (*Phoxinus phoxinus*, *Phoxinus czecanowskii*), the asp (*Aspius aspius*), the common barbel (*Barbus barbus*) and the sunbleak (*Leucaspius delineateus*). Previously, the common carp (*Cyprinus carpio*) and the carassius (*Carassius carassius*, *C. auratus*) were not thought of as carriers, but recently they have been added to a list of potential parasitic hosts (Atlas-determinant… 1957; Be'er 2005). Within the Cyprinidae family, the common roach, the common dace and the ide are most prevalent, and their ranges spread not only across the basins of the major rivers in Russia's European territory, but also further East, reaching the basin of the Lena river. The gudgeon's range is just as wide: it is found in the rivers of Russia's European territory, Siberia and even in the basin of the Amur river, where its population is quite sizeable. Evidently, the phoxinus has the widest range, as it inhabits the waters of European Russia, Siberia and the Russian Far East.

Local species in the Cyprinidae family (over 50 total) inhabit the basins of the Amur and other large rivers in the Russian Far East, and are also potential parasitic carriers.

Limnophilic species in this group are at the greatest risk of being infected by opisthorchiasis. *Abbotina rivularis, Acanthorhodeus asmussi,* the bighead carp (*Aristichthys nobilis*), the grass carp (*Ctenopharyngodon idella*), *Gobio soldatovi*, the sharpbelly (*Hemiculter leucisculus*), *Leuciscus waleckii*, *Phoxinus lagowskii*, the stone moroko (*Pseudorasbora parva*) and a number of other species are members of this group. Not long ago, opisthorchiasis was never reported east of the Ob-Irtysh river basin, but there has been a recent increase in opisthorchiasis morbidity in regions in

[4]The fish images in Figs. 3.3, 3.4, 3.5, 3.6, 3.7, 3.8, 3.9, 3.10, 3.11, 3.12, 3.13, 3.14, 3.15, 3.16, 3.17, 3.18 and 3.19 are adapted from the *Atlas-determinant "Fish resources of the USSR"*. Moscow: VNIRO 1957. 224p. (in Russian).

Fig. 3.3 Cyprinidae Family. Common roach (*Rutilus rutilus*) *Credit* Atlas of freshwater fish of Russia / Ed. by YS Reshetnikov. Vol. 1. Moscow: Nauka; 2003. 379p. (in Russian).Atlas of freshwater fish of Russia / Ed. by YS Reshetnikov. Vol. 2. Moscow: Nauka; 2003. 253p. (in Russian)

Fig. 3.6 Cyprinidae Family. Common rudd (*Scardinius erythrophthalmus*) *Credit* Atlas of freshwater fish of Russia / Ed. by YS Reshetnikov. Vol. 1. Moscow: Nauka; 2003. 379p. (in Russian).Atlas of freshwater fish of Russia / Ed. by YS Reshetnikov. Vol. 2. Moscow: Nauka; 2003. 253p. (in Russian)

Fig. 3.4 Cyprinidae Family. Common bream (*Abramis brama*) *Credit* Atlas of freshwater fish of Russia / Ed. by YS Reshetnikov. Vol. 1. Moscow: Nauka; 2003. 379p. (in Russian).Atlas of freshwater fish of Russia / Ed. by YS Reshetnikov. Vol. 2. Moscow: Nauka; 2003. 253p. (in Russian)

Fig. 3.7 Cyprinidae Family. Common bleak (*Alburnus alburnus*) *Credit* Atlas of freshwater fish of Russia / Ed. by YS Reshetnikov. Vol. 1. Moscow: Nauka; 2003. 379p. (in Russian).Atlas of freshwater fish of Russia / Ed. by YS Reshetnikov. Vol. 2. Moscow: Nauka; 2003. 253p. (in Russian)

Fig. 3.5 Cyprinidae Family. Ide (*Leuciscus idus*) *Credit* Atlas of freshwater fish of Russia / Ed. by YS Reshetnikov. Vol. 1. Moscow: Nauka; 2003. 379p. (in Russian).Atlas of freshwater fish of Russia / Ed. by YS Reshetnikov. Vol. 2. Moscow: Nauka; 2003. 253p. (in Russian)

the Yenisei, Lena and Amur river basin: Krasnoyarsk Krai, Buryatia, Sakha (Yakutia) Republic, Tyva Republic, Primorsky Krai, and Khabarovsk Krai (Stepanova 2002). As a result, study of disease in local fish species is becoming more and more relevant.

The opisthorchiasis epidemiological situation is the most complex in the Ob-Irtysh river basin. The ide, common roach, common dace and phoxinus are the most prevalent host species. The common bream, which previously inhabited only European Russia and the Aral Sea, was let loose in the Novosibirsk Reservoir. The common bream assimilated well and spread along the Ob river. Today, it has settled nearly the entire river basin and has become the main fished species in some regions. Like other species in this family, the common bream can be infected with opisthorchiasis, so its spread has sharply increased the epidemiological significance of opisthorchiasis in Siberia.

Fish are infected during feeding and spawning in inundable shallow bodies of water that connect to the main river during seasonal flooding. These areas have all the necessary conditions for the development of parasite larvae and the infection of intermediate hosts. According to some researchers, the synchronization of the *Opisthorchis* life cycle with seasonal flood timing and the migration of fish at times when molluscs are emitting maximal numbers of *Opisthorchis* larvae can be used to predict the epidemiological risk in a specific territory (Ozherelyev and Potseluev 1987).

The longer the seasonal flooding lasts, the higher the probability that fish are infected. This phenomenon is especially evident in the Ob-Irtysh basin, where flood waters recede and are reinforced by rainfall nearly until the river freezes over. As a result, the rivers spread far and thus link numerous inundable shallow bodies of water, which serve as a habituating for feeding and spawning fish in the Cyprinidae family.

Carpidae have a unique ecology due to their nomadic way of life and active migration, which are caused by cyclical changes in their environment. The fish regularly migrate to inundable waters to spawn and feed after seasonal floods begin. After the flood recedes, they return to the main river corridor, from which they move to waters that do not undergo fish kills, such as the entrance of ground waters from the riverbed or headwaters. For instance, large common breams and ides, having fed in the middle Ob during the summer, swim upstream to rivers in the Tomsk Oblast for the winter. Fertile common breams and ides from the lower Irtysh often swim downstream far north along the current of the Ob in search of oxygen-rich waters. Ides that winter in the estuary of the Ob can migrate to northern rivers like Nadym and Pur via the Ob delta. Small fish, like the common roach, the common dace and the gudgeon, migrate twice across tens of kilometres in warm times of year in search of food or to avoid fish kills (Soldatov 2011). Only non-migratory crucian carps are an exception to this phenomenon as they are extremely resilient in low-oxygen environments and frozen bodies of water.

As such, fish widely spread the pathogens of opisthorchiasis during spring and autumn migration, thus creating significant differences between infection levels in different parts of the river–lake network.

Diphyllobothriasis is another group of parasitic diseases hosted by fish that pose no less danger to humans than opisthorchiasis. Many fish in a number of families serve as hosts for this disease (Diphyllobothriasis… 1999).

The *Esocidae* Family. The nosogenic territory of the fish tapeworm (*Diphyllobothrium latum*) almost completely coincides with the range of the northern pike (*Esox lucius*) (Fig. 3.8).

The pike is a typical ambush predator. In rivers, it constantly inhabits the overgrown zone near the riverbank, but after reaching a reproductive age and a length of 50 cm, it moves to the central part of lakes and reservoirs. Juvenile fish generally feed on zooplankton in the first months of their lives, switching to feeding off other juvenile fish, predominantly in the Carpidae and Percidae families, later on. Given its enormous range, the pike can inhabit all kinds of bodies of water, enabling it to be the primary host of the fish tapeworm (Atlas of freshwater fish… 2002, 2003).

Fig. 3.8 Esocidae Family. Northern pike (*Esox lucius*) *Credit* Atlas of freshwater fish of Russia / Ed. by YS Reshetnikov. Vol. 1. Moscow: Nauka; 2003. 379p. (in Russian).Atlas of freshwater fish of Russia / Ed. by YS Reshetnikov. Vol. 2. Moscow: Nauka; 2003. 253p. (in Russian)

Fig. 3.9 Percidae Family. European perch (*Perca fluviatis*) *Credit* Atlas of freshwater fish of Russia / Ed. by YS Reshetnikov. Vol. 1. Moscow: Nauka; 2003. 379p. (in Russian).Atlas of freshwater fish of Russia / Ed. by YS Reshetnikov. Vol. 2. Moscow: Nauka; 2003. 253p. (in Russian)

Fig. 3.10 Percidae Family. Ruffe (*Gymnocephalus cernuus*) *Credit* Atlas of freshwater fish of Russia / Ed. by YS Reshetnikov. Vol. 1. Moscow: Nauka; 2003. 379p. (in Russian).Atlas of freshwater fish of Russia / Ed. by YS Reshetnikov. Vol. 2. Moscow: Nauka; 2003. 253p. (in Russian)

The *Percidae* Family. Aside from the pike, the fish tapeworm can also parasitize on other predatory fish, such as the European perch (*Perca fluviatis*) (Fig. 3.9), the ruffe (*Gymnocephalus cernuus*) (Fig. 3.10) and the zander (*Sander lucioperca*).

The European perch is a river and lake species. It is an active predator that is accustomed to life near the bank of body of water, where it feeds on zooplankton, benthos and juvenile fish of a variety of species, which replace one another in the perch's diet as it grows. The European perch's nutrition largely depends on the nutrient reserves in the body of water it inhabits. In some bodies of water, it may remain a planktivore or benthivore for its entire life, without becoming a predator. The European perch is one of the most widespread fish in Russia. Its range almost completely coincides with that of the pike. Like pikes, European perches can inhabit a variety of bodies of water, and in some lakes of Western Siberia, they have formed a monoculture.

The ruffe also has a wide range, spreading from Russia's western borders to the Kolyma river. It does not inhabit any rivers in the Russian Far East or the Transbaikal. The northern border of its range lies practically along the coast of the Arctic Sea. Ruffes are benthic fish; however, they can move to any location in the water mass. They feed primarily on benthos, but can also eat a variety of zooplankton, fish roe, especially pike roe and minnows. All of this allows the ruffe to reach large population size in bodies of water.

As such, the ruffe and European perch, given their wide ranges, large populations, and willingness to inhabit different bodies of water, are the primary hosts of the fish tapeworm.

The zander—a predator that inhabits the open portion of large rivers, lakes and reservoirs—can also serve as a host for the fish tapeworm. It is also found in the coastal parts of oceans, where the water has low salinity. It is spread out in bodies of water in European Russia, reaching the Ural Mountains. It was let into the large lakes of Karelia, from where it moved to the Northern Dvina and Mezen rivers. Today, the zander is also found in the Ob-Irtysh river basin, although its abundance there is low, except in the Novosibirsk Reservoir. The mature zander is an active predator and hunts the European perch, the ruffe and fish in the Carpidae family, although juvenile zanders feed on zooplankton and benthos. Like other predators, the zander supports the function of diphyllobothriasis natural foci in Russia's European territory.

The Volga pikeperch or Volga zander (*Stizostedion volgensis*) is a close relative of the zander and inhabits the Volga and Don river basins. It can also carry the fish tapeworm.

The *Lotidae* Family. The burbot (*Lota lota*) (Fig. 3.11) is the only member of the Gadiformes order that inhabits freshwater in the northern regions of the Northern Hemisphere. In Russia, the burbot has a wide range, but it is most abundant in the Arctic ocean's basin.

The burbot prefers cold and clear bodies of water with a rocky floor and spring water sources. It spawns and feeds during the cold season, beginning approximately in November. In the warm season, it hides near the floor, entering a state of dormancy. As such, burbots are usually caught in the autumn or winter. The burbot is a benthal fish. Mature burbots feed exclusively on fish, but juveniles also eat zooplankton, benthos and fish roe. Burbots carry *D. latum,* so during the winter months, the burbot in particular is the cause of diphyllobothriasis infections in native Siberian populations, who consume raw burbot liver.

The *Coregonidae* Family. The fish tapeworm can also parasitize on other fish families, such as the coregonidae. However, fish in this family are usually infected with *D. dendriticum* and *D. ditremum.* The family numbers 14 species total, and they primarily inhabit the Arctic ocean basin (Atlas of freshwater fish… 2003a, b). Practically every one of the 14 species exist as both lake-inhabiting and anadromous forms. The European whitefish (*Coregonus lavaretus*) has one of the widest ranges and has a circumpolar distribution. In Russia, it inhabits nearly all the bodies of water in the Arctic Ocean basin, and is also found in the Baltic sea basin and in Zabaikalsky Krai. In North America, the European whitefish has been found in Alaskan and Canadian waters.

C. lavaretus (Fig. 3.12) is a polymorphous species, and in Russia it has six subspecies, including the Siberian one (*C. lavaretus pidschian*), which is found in the basin of the Arctic ocean. This species is characterized by a large variety of ecological forms, such as migratory, river and lake fish: in large lakes, they are further divided into coastal, deepwater and pelagic fish with different feeding practices, ranging from typical planktivores to typical benthivores. For time to

Fig. 3.11 Lotidae Family. Burbot (*Lota lota*) *Credit* Atlas of freshwater fish of Russia / Ed. by YS Reshetnikov. Vol. 1. Moscow: Nauka; 2003. 379p. (in Russian).Atlas of freshwater fish of Russia / Ed. by YS Reshetnikov. Vol. 2. Moscow: Nauka; 2003. 253p. (in Russian)

Fig. 3.12 Coregonidae Family. Common cisco (*Coregonus lavaretus*) *Credit* Atlas of freshwater fish of Russia / Ed. by YS Reshetnikov. Vol. 1. Moscow: Nauka; 2003. 379p. (in Russian).Atlas of freshwater fish of Russia / Ed. by YS Reshetnikov. Vol. 2. Moscow: Nauka; 2003. 253p. (in Russian)

time, the European whitefish acts as a predator, eating their own roe and that of other species.

In bodies of water in Russia's European territory, *C. lavaretus* is joined by the European cisco (*Coregonus albula*), which lives in the basins of the Baltic, White, and Barents seas, the sardine cisco (*C. sardinella*) (Fig. 3.13), the Arctic cisco (*C. autumnalis autumnalis*), the broad whitefish (*C. nasus*), found in the basin of the Kara Sea, the peled (*C. peled*) (Fig. 3.14), the nelma (*Stenodu leucichthys nelma*) and its subspecies the beloribitsa (*S. l. leucichthys*), which is found in the basin of the Caspian Sea. However, due to the beloribitsa's low abundance, its role in the spread of diphyllobothriasis is insignificant.

In recent years, the ranges of the European cisco and the peled have substantially grown thanks to acclimation, while broodstock have been created in places where these fish do not naturally spawn. Today, these species inhabit not only bother bodies of water, but also the lakes and reservoirs of the Volga basin, increasing the likelihood of diphyllobothriasis natural focus formation with *D. dendriticum* parasitism in these territories.

Fig. 3.13 Coregonidae Family. Sardine cisco (*Coregonus sardinella*) *Credit* Atlas of freshwater fish of Russia / Ed. by YS Reshetnikov. Vol. 1. Moscow: Nauka; 2003. 379p. (in Russian).Atlas of freshwater fish of Russia / Ed. by YS Reshetnikov. Vol. 2. Moscow: Nauka; 2003. 253p. (in Russian)

The greatest Coregonidae biodiversity is found in Siberia, especially in the basins of the great Siberian rivers: the Ob, the Yenisei, and the Lena. Ten fish species in this family inhabit these territories. The majority of these fish have wide ranges that spread from the Urals to the Chukotka peninsula. The muksun (*Coregonus muksun*), the tugun (*C. tugun*), the round whitefish (*Prosopium cylindraceum*) and the pygmy whitefish (*P. coulteri*) are only found in Siberia.

Coregonidae are unique in their ability to broadly migrate within the limits of basins. As a rule, they feed in river deltas or coastal seawater with low salinity. To spawn, Coregonidae travel upstream, often to the riverhead to find rocky areas with spring water. This allows these fish to spread parasites beyond the limits of natural foci, increasing infection rates near basin boundaries.

The Thymallidae Family. Fish in this family can serve as intermediate hosts for *D. dendriticum* and *D. ditremum*. Two species are widespread in Russia: the grayling (*Thymallus thymallus*) and the Arctic grayling (*T. arcticus*) (Fig. 3.15).

The Arctic grayling is found in rivers and lakes of the Arctic Ocean basin, from the Kara Sea to the Chukotka Peninsula and North America. Both graylings are found in the Pechora River basin and a few of its tributaries. It also inhabits Lake Baikal and the Amur River basin, including the Ussuri and Songhua rivers. The Arctic grayling is a typical omnivore, feeding on invertebrates and aquatic vegetation. Juvenile Arctic graylings feed on Branchiopoda and small forms of other invertebrates. As they age, the spectrum of their nutrient source broadens: worms, molluscs, airborne insects, whitefish roe, caddisfly larvae, Gammarus crustaceans and non-biting midges make up 70% of their food sources.

Graylings are coveted catch for amateur fishermen. Hikers and fishermen often consume it in the form of lightly salted raw fish, which often leads to infection with diphyllobothriasis.

Fig. 3.14 Coregonidae Family. Peled (*Coregonus peled*) *Credit* Atlas of freshwater fish of Russia / Ed. by YS Reshetnikov. Vol. 1. Moscow: Nauka; 2003. 379p. (in Russian).Atlas of freshwater fish of Russia / Ed. by YS Reshetnikov. Vol. 2. Moscow: Nauka; 2003. 253p. (in Russian)

Fig. 3.15 Thymallidae Family. Arctic grayling (*Thymallus arcticus*) *Credit* Atlas of freshwater fish of Russia / Ed. by YS Reshetnikov. Vol. 1. Moscow: Nauka; 2003. 379p. (in Russian).Atlas of freshwater fish of Russia / Ed. by YS Reshetnikov. Vol. 2. Moscow: Nauka; 2003. 253p. (in Russian)

The Osmeridae Family. Fish in this family inhabit the basins of the Arctic and Pacific oceans and also carry the pathogens of diphyllobothriasis.

The European smelt (*Osmerus eperlanus*) (Fig. 3.16) is widespread in European Russia. This is a migratory and lake-inhabiting species that inhabits the basins of the Baltic, White, and Barents seas. It is a coastal migratory fish that has isolated lake populations which include premature small forms known as 'snetok'. Snetok is a pelagic schooling fish that stays in the upper and middle layers of the water. It feeds on small plankton and crustaceans; larger specimens may eat roe, larvae and even young fish.

Recently, the smelt has reached the Volga basin from the White Lake in Vologda Oblast, and is now found in the Rybinsk, Gorky, Kuybyshe, and Saratov reservoirs. In some of these, river smelt is fished commercially. Evidently, it is snetok in particular that serves as a primary carrier for *D. dendriticum*, and *D. ditremum* as they spread to the southern regions of European Russia.

The Salmonidae Family. The majority of these fish are anadromous and inhabit the basins of the Pacific, Arctic and Atlantic oceans (Krivolutsky et al. 2013). 30 species in the Salmonidae family are found in Russia. It has been established that salmon in the *Oncorhynchus* genus are carriers of *D. luxi,* which poses a danger to humans. The diphyllobothriasis natural focus in the Russian Far East is found in the zone where Far East salmon release its pathogen to the mainland and islands during their migration to spawn in freshwater. The sockeye salmon (*Oncorhyncus nerka*) (Fig. 3.17), the chum salmon (*O. keta*), the pink salmon (*O. gorbuscha*) (Fig. 3.18) and the Dolly Varden trout (*Salvenilus malma*) (Fig. 3.19) can all serve as hosts for *D. luxi.* Pacific char and trout species can carry *D. dendriticum* and *D. ditremum* as well. This phenomenon has been seen in bodies of water in the Chukotka Peninsula (Pugachev 1983).

Fish can also serve as hosts for other helminths that pose a health risk to humans.

Fig. 3.16 Osmeridae Family. European smelt (*Osmerus eperlanus*) *Credit* Atlas of freshwater fish of Russia / Ed. by YS Reshetnikov. Vol. 1. Moscow: Nauka; 2003. 379p. (in Russian).Atlas of freshwater fish of Russia / Ed. by YS Reshetnikov. Vol. 2. Moscow: Nauka; 2003. 253p. (in Russian)

Fig. 3.17 Salmonidae Family. Sockeye salmon (*Oncorhyncus nerka*) *Credit* Atlas of freshwater fish of Russia / Ed. by YS Reshetnikov. Vol. 1. Moscow: Nauka; 2003. 379p. (in Russian).Atlas of freshwater fish of Russia / Ed. by YS Reshetnikov. Vol. 2. Moscow: Nauka; 2003. 253p. (in Russian)

Fig. 3.18 Salmonidae Family. Pink salmon (*Oncorhynchus gorbuscha*) *Credit* Atlas of freshwater fish of Russia / Ed. by YS Reshetnikov. Vol. 1. Moscow: Nauka; 2003. 379p. (in Russian).Atlas of freshwater fish of Russia / Ed. by YS Reshetnikov. Vol. 2. Moscow: Nauka; 2003. 253p. (in Russian)

Fig. 3.19 Salmonidae Family. Dolly Varden trout (*Salvelinus malma*) *Credit* Atlas of freshwater fish of Russia / Ed. by YS Reshetnikov. Vol. 1. Moscow: Nauka; 2003. 379p. (in Russian).Atlas of freshwater fish of Russia / Ed. by YS Reshetnikov. Vol. 2. Moscow: Nauka; 2003. 253p. (in Russian)

Metagonimiasis is found in the basin of the Caspian Sea, in the Russian Far East and in Krasnodar Krai. Fish in the Carpidae family can serve as intermediate hosts for its pathogen.

The parasite causing nanophyetiasis is carried by fish in the Russian Far East, including freshwater (the taimen and lenok), thymallidae and freshwater whitefish (Coregonidae).

A detailed map with a description reflecting fish distribution in Russia was prepared by M. S. Soldatov and published in Medico-geographical Atlas of Russia "Natural Focal Diseases" (2015, 2017).

3.4 Arthropods

Arthropods, especially insects and ticks, participate in the spread of many natural focal diseases. These two arthropod groups parasitize on both humans and mammals have extraordinary biodiversity and a number of ecological connections to humans and vertebrates. In a number of cases, the vectors of natural focal diseases also serve as reservoirs for the pathogen, allowing the pathogenic microorganisms to survive for prolonged periods of time.

Long-term coevolution has led to the parasitism of arthropods on vertebrates. As a result, a number of lifestyle systems that allow for the life and reproduction of both species have come into existence (Beklemishev 1970).

V. N. Beklemishev distinguished between a few lifestyle systems of parasitic arthropods (Guide to… 1974).

1. Actively Attacking Periodic Bloodsuckers (mosquitoes, blackflies, horseflies, etc.). These are mobile insects that can fly substantial distances in search of food, which they consume in prolonged intervals. They digest blood outside of the host's body, and the amount of blood they consume is limited by their need for maintaining mobility. Free bloodsuckers, like all insects that undergo complete metamorphosis, live in different biotopes depending on their life stage: adults live in the air, and larvae in most species live in water.
2. Ambush Bloodsuckers
   (a) Pasture, non-hiding ambush parasites (ticks in the Ixodidae family, etc.). The members of this group inhabit the same biotopes as their hosts. They are characterized by a slow sucking of a large amount of blood in one attack and an ability for prolonged fasting. As a result, arthropods are not very mobile, are highly fertile and have individual longevity and slow development.
   (b) Nesting, digging (home) ambush parasites (e.g. bedbugs, ticks in the Argasidae family). Live in their hosts' nests or temporary shelters and cannot spend long periods of time in outside environments. Because they feed on their host only when it is in its nest or shelter, the members of this group suck blood very quickly and can consume a large portion of blood in one attack. Besides, they can go prolonged periods without food.
3. Constant ectoparasites (various lice species). These parasites have morphologically adapted so as to remain on their host's body and move through their hair or fur: they have flat wingless bodies, clingy claws and hard shells which prevent them from being crushed. They consume small quantities of blood at short intervals and cannot go without food for long.

Diseases for which the pathogen is passed to the host via an arthropod are referred to as vector-borne diseases. The pathogen, the vector and the vertebrate host all participate in maintaining foci of vector-borne diseases. These diseases can be both anthroponoses, i.e. diseases that cannot be transmitted without humans (malaria, typhus, etc.) and zoonoses, or diseases that can be transmitted without human participation (the plague, tick-borne encephalitis, tularemia, etc.). The majority of vector-borne diseases also have natural foci.

There are infections in this group for which vector-borne transmission is the only possible means of infection in natural conditions (Lyme borreliosis, Japanese encephalitis, West Nile fever): this phenomenon is termed obligate vector-borne transmission. Some diseases can also be transmitted in other ways, and are called facultative vector-borne diseases (tick-borne encephalitis, tularemia, the plague, etc.) (Tarasov 2002).

Vectors can be specific or mechanic. Vector specificity is the result of prolonged coevolution between the pathogen and the vector. It is a complex biological phenomenon.

The pathogen can reproduce and even go through some of its life stages in the organism of a specific vector. As a result, the vector can only transmit the pathogen to a new host after some time, once it has developed and reproduced. The pathogen often remains in the vector for its entire life, from larval stages to the imago stage (transstadial transmission), and sometimes can even be passed on to the vector's descendants (transovarial transmission). For instance, the tick-borne encephalitis virus is passed on by the female tick to her descendants and remains in the arthropod's organism throughout its entire life. This ability is a crucial factor for the maintenance of pathogen circulation in natural foci.

Unlike in specific vectors, the evolutional ties between pathogens and their mechanical vectors are not as strong or ancient. The pathogen cannot develop and usually cannot reproduce within the mechanical vector's body. Transmission can occur immediately after the mechanical vector received the pathogen from an infected host or from the environment. The pathogen cannot stay alive inside the mechanical vector for long.

Some infections in the natural focus are transmitted by multiple arthropod species, one of which may be the primary vector, while the others are secondary vectors. For instance, both ticks and mosquitoes participate in the transmission of some viruses, and the less mobile ticks provide long-term shelter for the virus and maintain a connection between the pathogen and its primary hosts, thus maintaining focal stability.

Mosquitoes, due to their mobility and ability to attack different forms of prey, allow the virus to be spread across a wide area and a variety of hosts, including humans (Tarasov 2002).

As noted above, two groups of arthropod vectors play the greatest role in natural focal diseases transmission in Russia: mosquitoes and Ixodidae ticks. Besides, fleas (Siphonaptera) are also important as they are specific vectors of the plague. Their role will be discussed in depth in 4.16. Other arthropods, such as horseflies, biting midges, blackflies and sandflies, can also carry certain infections, but their roles are significantly smaller compared to those of the groups mentioned above. Biting midges, which are the smallest bloodsucking insects, are largely a veterinary issue, but they can also transmit some arboviruses to humans (Harrup et al. 2015). Horseflies are vectors for tularemia, while sandflies can transmit leishmaniosis.

Mosquitoes are members of the Culicidae family in the Diptera order. They are relatively small insects with a gracefully elongated torso, a small head, and long legs. Each mosquito goes through several stages of development: the egg phase, a larval phase, a pupa phase and an imago (adult) phase. Mosquitoes are heterotopic animals, meaning that they spend different life stages in different environments. Eggs, larvae and pupae generally inhabit the water, while adults are found in the air. Mosquitoes are periodically attacking free bloodsuckers. Only female mosquitoes suck blood, because they need it for egg development, and they can fly great distances in search of prey. Depending on their species, mosquitoes can attack their prey outside or inside buildings. Mosquitoes usually bite in the evening or twilight, right before the sunset, although some species attack in the morning or on cloudy days.

Mosquitoes serve as vectors for a number of arthropod-borne natural focal diseases (see 4.23), some of which pose a serious risk, such as yellow fever, Japanese encephalitis, dengue fever, etc. In Russia, mosquitoes transmit Japanese encephalitis (in a limited area in South Primorsky Krai) and participate in the spread of tularemia

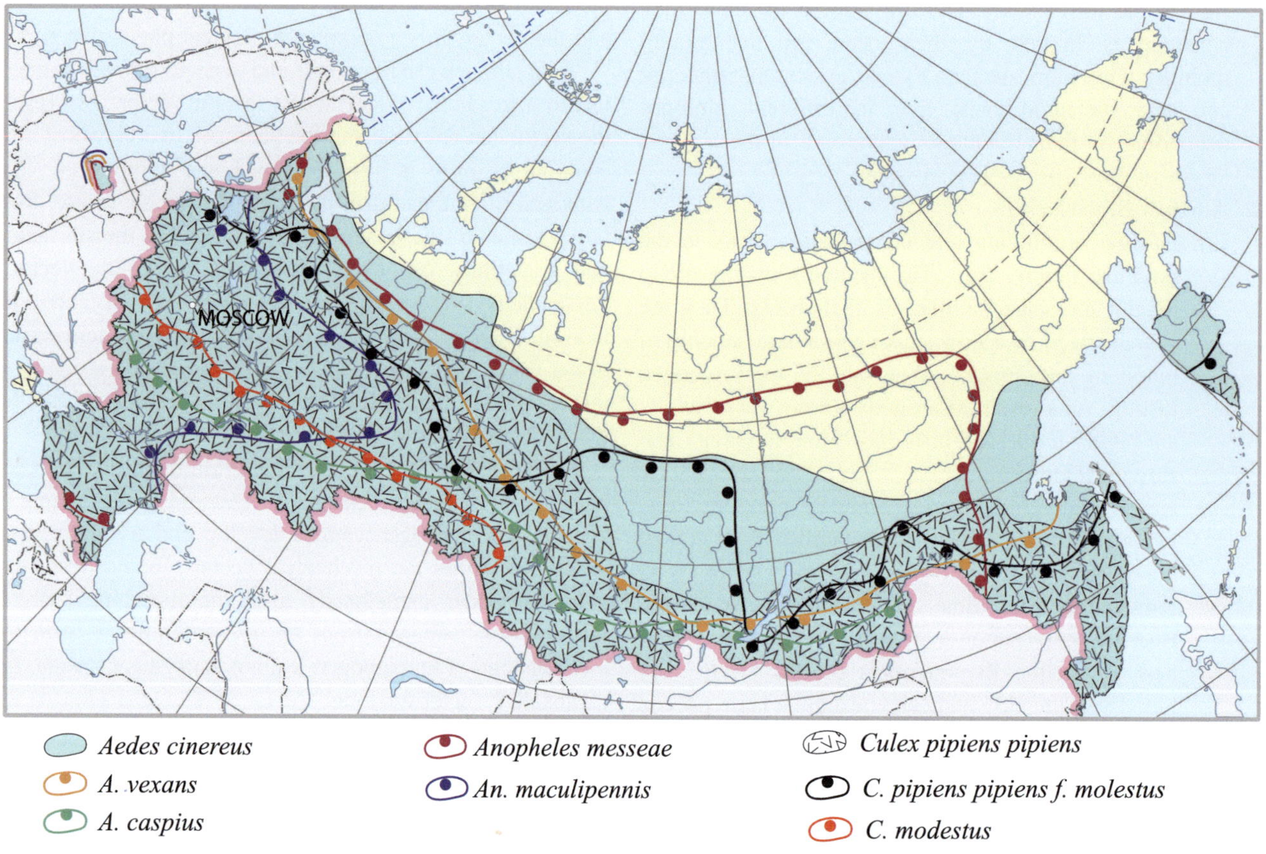

Map 3.23 The spread of primary mosquito species in Russia

and dirofilariasis (vector-borne zoonotic helmintiasis). Today West Nile fever is the most important and common natural focal disease transmitted by mosquitoes.

Map 3.23 shows mosquito range boundaries that are the most epidemiologically significant in Russia. 105 species in the *Culicidae* family are found in Russia, 12 of which are members of the *Anopheles* genus, 63 of which are in the *Aedes* genus and 17 of which are in the *Culex* genus. The rest are members of other genera (Khalin and Gornostaeva 2008).

Ticks (Acari) are arachnids. Approximately 55,000 tick species have been recorded globally (Zhang 2011), but there are likely no fewer than 100,000 tick species total. Ticks are the most diverse and ecologically successful arachnid group. They live in all habitats; many ticks are human and animal parasites, and can cause harm to their host either on their own or by serving as a vector for infectious disease (Westheide and Rieger 2013).

Many ticks' bodies are characterized by a reduction to a small size (on the scale of millimetres), primitive morphology and rapid development. Ticks make use of different nutrient sources at different developmental stages.

Tick classification is a widely discussed topic, and today there are a number of taxonomic divisions. According to the most widely accepted classification, ticks are divided into six orders that are grouped into two superorders: Parasitiformes and Acariformes (A Manual of Acarology 2009). Ticks in the Ixodida order and ticks in the Gamasina cohort of the Monogynaspida suborder of the Mesostigmata order are traditionally seen as the most relevant to human health.

Ixodidae ticks are highly specialized obligate bloodsuckers that parasitize on terrestrial vertebrates at each developmental stage. Ticks also serve as long-term reservoirs for the pathogens of a number of natural focal infections. Due to their individual longevity, polyphagy and capacity for both transovarial and transstadial transmission, ticks can maintain the stable function of natural foci even if no warm-blooded hosts are present. From an epidemiological standpoint, the Ixodidae family is the most important family of ticks in the Ixodida order.

The majority of this family's members are ambush bloodsuckers; only ticks in the *Hyalomma* genus can actively pursue their prey. In the *Ixodes* genus, every active developmental stage has its own set of nutrient sources, which is essential to maintaining the circulation of pathogens in the natural focus. After becoming satiated, the tick leaves its host, sheds its shell and metamorphoses to its next developmental stage, which then must find a new host and feed on it. Therefore, ticks in this genus change hosts as they develop (Korenberg et al. 2013).

Ixodidae ticks feed over a prolonged period of time: larvae take 2–4 days, while nymphs and adult ticks take 4–6 days. Tick development is profoundly seasonal. The majority of ticks manage to go through all stages of development in a season, but some (members of the *Ixodes* and *Haemaphysalis* genera) only go through one or two phases before they begin hibernation.

A crucial characteristic of the spread of Ixodidae ticks is their ability to spread unevenly in one area, which is based on the availability of animals that can serve as nutrient sources, the characteristics of the vegetation, the humidity of the soil, etc.

Ixodidae ticks are extremely widespread. In Russia, *Ixodes persulcatus* and *I. ricinus,* which serve as vectors for Lyme borreliosis, tick-borne encephalitis, and a number of other infections common in the taiga and mixed coniferous forests (see 4.1, Fig. 4.1.1). *Dermacentor reticulatus*, which prefers open areas like meadows and forest edges, is found in European Russia's mixed coniferous and temperate deciduous forests. *D. silvarum, Haemophysalis coincinna*, and other species are common in the forests of Eastern Siberia and the Russian Far East. *D. reticulatus, D. marginatus D. nuttali,* etc. are found in steppes and forest steppes in Eastern Siberia. Tick diversity and abundance is lower in regions with a semi-arid climate, where they often live in rodent holes or riverbeds. Ticks in the *Rhipicephalus* genus are most common in this zone. Ticks in the *Hyalomma* genus, some of which are vectors for hemorrhagic fever, and the *Rhipicephalus* genus are most common in the desert. The most important members of the Ixodida order are shown in Map 3.24.

The medical significance of Gamasid mites as natural focal disease vectors is still unknown. The primary role of this group of ticks in natural foci is providing circulation of pathogens among animal carriers and thus supporting the natural focus (Zemskaya 1973). Gamasid mites can participate in the spread of Q fever and tularemia among small vertebrates (Guide… 1974). The Gamasid mite *Dermanyssus gallinae,* which reproduces in doves and pigeons, can attack people in urban apartments, entering them through vents, causing tick-borne dermatitis and psittacosis. Today, given the growth of synanthropic rodent populations, the rat tick *Ornithonyssus bacoti* has begun to pose a problem, as it can also cause tick-borne dermatitis and transmit the pathogen of murine typhus to humans.

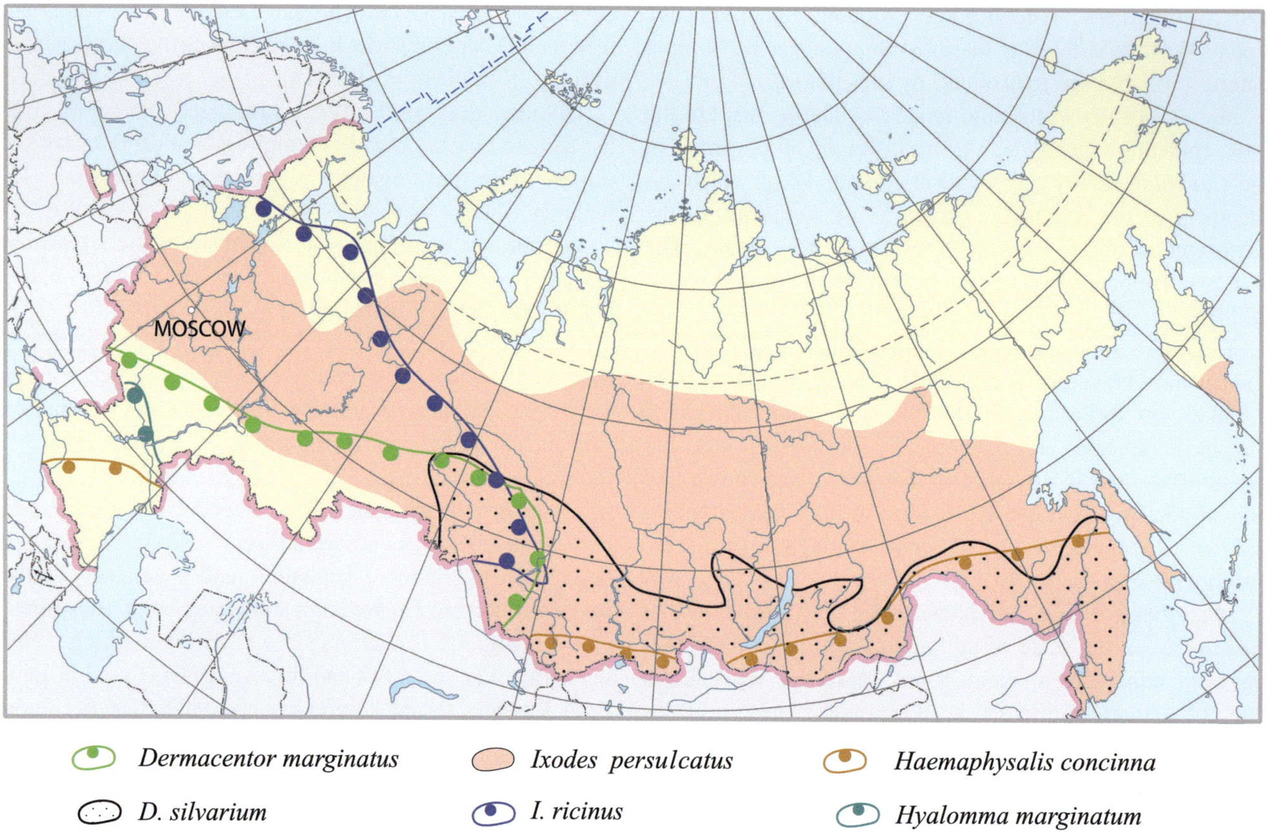

Map 3.24 Ranges of the main species of *Ixodidae* Ticks

References

Agreement on the conservation of African-Eurasian Migratory Waterbirds—AEWA. http://www.unep-aewa.org. Accessed 10 June 2015 (in Russian)

Ananyina YuV (2003) Leptospirosis. In: Natural focality of diseases: the Gamaleya RAMN Institute research. Moscow: Rusaki, pp 122–136 (in Russian)

Atlas-determinant "Fish resources of the USSR". Moscow: VNIRO; 1957, p 224 (in Russian)

Be'er SA (2005) The opisthorchiasis pathogen biology. Moscow: Tov-vo nauch. izd. KMK, p 336 (in Russian)

Beklemishev VN (1970) Biocenological fundamentals of comparative parasitology. Moscow: Nauka, p 410 (in Russian)

Belousova AV, Milyutina ML, Prisyazhnyuk VE, Mezhnev AP, Tomkovich PS, Kharitonov SP, Fefelov IV (2007) Analytical materials on the population of waterbirds, migrating in the African-Eurasian Flyway, existing within the borders of the Russian Federation. Moscow: VNII Okhrany Prirody, p 72 (in Russian)

Derbeneva-Ukhova VP (ed) (1974) Guide to medical entomology. Moscow: Medicine, p 360 (in Russian)

Dinets VL, Rothschild EV (1996) Animals. Encyclopedia of nature of Russia. Moscow: Izd-vo ABF, p 344 (in Russian)

Dunlap BG, Thies ML (2002) Giardia in beaver (*Castor canadensis*) and nutria (*Myocastor coypus*) from east Texas. J Parasitol 88 (6):1254–1258

Harrup LE, Bellis GA, Balenghien T, Garros C (2015) *Culicoides Latreille* (Diptera: Ceratopogonidae) taxonomy: current challenges and future directions. Infect Genet Evol 30:249–266

Khalin AV, Gornostaeva RM (2008) On the taxonomic composition of the hematophagic mosquitoes (Diptera: Culicidae) of the world fauna and the fauna of Russia (critical review). Parasitology 42 (5):360–381 (in Russian)

Korenberg EI (2003a) Tick-borne borreliosis. In: Natural focality of diseases: the Gamaleya RAMN Institute research. Moscow, Rusaki, pp 99–121 (in Russian)

Korenberg EI (2003b) Tick-borne encephalitis. In: Natural focality of diseases: the Gamaleya RAMN Institute research. Moscow, Rusaki, pp 35–63 (in Russian)

Korenberg EI, Kovalevskiy YuV (2000) The main features of the eco-epidemiology of tick-borne encephalitis. In: Problems of tick-borne and parasitic diseases. From the round table of the VI Russian-Italian scientific conf. "Infectious diseases: diagnosis, treatment, prevention" 14–16 Dec 2000, St.-Petersburg, pp 13–20 (in Russian)

Korenberg EI, Pomelova VG, Osin NS (2013) Natural focal infections transmitted by ixodid ticks. Moscow: Kommentariy, p 463 (in Russian)

Krantz GW, Walter DE (eds) (2009) A manual of acarology, 3rd edn. Texas Tech University Press, p 807

Krivolutskiy DA, Matyokin PV, Soldatov MS (2013) An introduction to the Russian sea biogeography. Moscow, Faculty of geography MSU, p 193p (in Russian)

Kurkela S, Manni T, Vaheri A, Vapalahti O (2004) Causative agent of pogosta disease isolated from blood and skin. Emergency Infinity Dis 10(5). http://wwwnc.cdc.gov/eid/article/10/5/03-0689_article

Lipin SI (1974) The importance of synanthropic birds connected with the distribution of natural focal infections and methods of their study in anthropurgic foci. Methodical recommendations. Irkutsk:

IRK. Research Institute of epidemiology and Microbiology. http://lawru.info/dok/1974/10/31/n1187980.htm. Accessed 20 Apr 2015 (in Russian)
Lvov DK, Ilyichev VD (1979) Bird migration and pathogen transfer (ecologico-geographical relations of birds with infectious pathogens). Nauka, Moscow, p 270 (in Russian)
Lvov DK, Klimenko SM, Gaydamovich SYa (1989) Arboviruses and arboviral infections. Moscow: Medicine, p 336 (in Russian)
Lysenko AYa, Vladimova MG, Kondrashin AV, Majori G (2002) Clinical Parasitology. A guide. WHO, Geneva, p 752 (in Russian)
Malkhazova SM (ed) (2015) Medico-geographical Atlas of Russia "Natural Focal Diseases. Faculty of geography, Lomonosov Moscow State University, Moscow, p 208 (in Russian)
Malkhazova SM (ed) (2017) Medico-geographical Atlas of Russia "Natural Focal Diseases", 2nd revised edn. Faculty of Geography, Lomonosov Moscow State University, Moscow, p 216 (in Russian)
Mehlhorn H (ed) (2001) Encyclopedic reference of Parasitology; biology, structure and function, 2nd ed. Springer, Berlin and Heidelberg, p 555
Meshcheryakova IS (2003) Tularemia. In: Natural focality of diseases: the Gamaleya RAMN Institute research. Rusaki, Moscow, pp 137–160 (in Russian)
Migratory birds of Russia: the link between the continents (poster). Russian Bird Conservation Union, Moscow, 2001 (in Russian)
Moiseenko AV (ed) (1999) Diphyllobothriasis. A manual. Tyumen: Tyumen State Medical Academy, p 54 (in Russian)
Neronov VM, Malkhazova SM, Tikunov VS (1991) Regional geography of the plague. In: The results of science and technology. Series Medical geography, vol 17. VINITI, Moscow, p 229 (in Russian)
Olsufiev NG, Dunaeva TN (1970) Natural focality, epidemiology and prevention of tularemia. Moscow: Medicine, p 270 (in Russian)
On the results of WNF supervision during the epidemic season of 2012. A Rospotrebnadzor letter dated 05.12.2012 No. 01/13890-12-32 (in Russian)
Ozherelyev VV, Potseluev AN (1987) The assessment of the tributaries importance in the Ob'-Irtysh opisthorchiasis focus functioning. In: Parasitic diseases in Western Siberia. Minzdrav RSFSR, Omsk, pp 56–63 (in Russian)
Pugachev ON (1983) Helminths of the Northeast Asian freshwater fish. In: The research on morphology and faunistics of the parasitic worms: the works of the Zoological Institute of USSR Academy of Sciences, vol 181, pp 93–123 (in Russian)
Reshetnikov YS (ed) (2002) Atlas of freshwater fish of Russia, vol 1. Moscow: Nauka, p 379 (in Russian)
Reshetnikov YS (ed) (2003) Atlas of freshwater fish of Russia, vol 2. Moscow: Nauka, p 253 (in Russian)
Rumiantsev VYu, Danilenko AK (2012) Small-scale mapping of terrestrial vertebrates distribution for complex atlases. Vestnik MSU Ser. 5 Geogr 2:49–54 (in Russian)
Semenova TA, Dovgalyov AS, Avdyukhina TI (2011) The Intermediate hosts of diphyllobothriasis pathogens and their infection rate. In: Relevant questions of infectious diseases epidemiology (Coll. of scientific works), vol 10. Moscow: Hygiene, pp 750–754 (in Russian)
Sidorov EG (1983) Natural focality of opisthorchiasis. Nauka, Almaty, p 240 (in Russian)
Soldatov MS (2011) Mapping of the spatial distribution of opisthorchiasis hosting fish in Western Siberia. In: Environmental epidemiology. The association of doctors of science in public administration, Kharkiv, pp 108–128 (in Russian)
Stepanova TF (2001) Epidemiological manifestations of actual risk of public infection in the combined foci of zoonotic parasitoses. In: Ecological bases of Western Siberian combined natural foci of endemic parasitoses. Tyumen State University, Tyumen, pp 129–138 (in Russian)
Stepanova TF (2002) Opisthorchiasis. Tyumen State University, Tyumen, p 196 (in Russian)
Tarasevich IV, Fetisova NF, Makarova VA, Fatalieva SF, Mediannikov OYu (2003) Rickettsiosis. In: Natural focality of diseases: the Gamaleya RAMN Institute research. Rusaki, Moscow, pp 64–98 (in Russian)
Tarasov VV (2002) The epidemiology of the vector-borne diseases. Moscow State University, Moscow, p 336 (in Russian)
The forecast of the epidemiological situation of West Nile Fever for 2013. http://34.rospotrebnadzor.ru/ epidemiologic_situation/93788/. Accessed 10 June 2015 (in Russian)
Westheide W, Rieger G, Hrsg.. Spezielle Zoologie. Teil 1. Einzeller und Wirbellose Tiere. 3. Auflage. Springer-Spektrum, Berlin-Heidelberg. 2013, pp 892
Zemskaya AA (1973) Parasitic Gamasina mites and their medical importance. Moscow: Medicine, p 166 (in Russian)
Zhang Z-Q (2011) Phylum Arthropoda von Siebold, 1848. In: Zhang Z-Q (ed) Animal biodiversity: an outline of higher-level classification and survey of taxonomic richness, Zootaxa 4138. Magnolia press, pp 99–103

Vadim Rumyantsev Ph.D. in Geography, is Senior Research Associate in the Department of Biogeography of the Faculty of Geography at Lomonosov Moscow State University. His interests include mammal ecological geography, biogeographic mapping and the use of geoinformatic technologies in biogeography. He is the author and co-author of 230 scientific publications and over 80 specialized maps and integrated national and regional atlases.

Mikhail Soldatov Ph.D. in Geography, is Senior Research Associate in the Department of Biogeography of the Faculty of Geography at Lomonosov Moscow State University. His interests include biogeography (especially geography of fish), biogeographic mapping and medical geography. He is the author and co-author of 80 scientific publications.

4 Major Natural Focal Disease Distribution

Abstract

The chapter presents textual descriptions and maps that represent the distribution, long-term average annual morbidity rate for 85 administrative subjects of the Russian Federation, and the long-term dynamics of the major natural focal diseases found on Russian territory. The chapter contains information on the distribution of 23 nosological entities (diseases), their epidemic function, conditions of population infections and oscillations in epizootic and epidemiological activity in natural foci that note seasonality and annual changes. Wherever possible, the level of danger and infection risk are evaluated. Map compilation was based on statistical data on population morbidity for infectious diseases from 1997 to 2015 and parasitic diseases from 2006 to 2015. These data were gathered by government sanitary and epidemiological monitoring services. Given the unique features of information collected based on administrative divisions, Russia's administrative subjects serve as the main cartographic units. Three integrated maps are provided at the end of the chapter: they present the morbidity structure of a network of natural focal infections, the major natural focal parasitic diseases, as well as nosological profiles of Russia's administrative subjects, which provide information on the particular set and intensity of diseases in each subject.

4.1 Tick-Borne Encephalitis

Tick-borne encephalitis (TBE) is a natural focal viral infectious disease that is mainly transmitted by Ixodidae ticks. Symptoms include central nervous system disorders, fever and general intoxication and malaise. This disease can take a large variety of clinical courses.

The TBE virus was discovered in 1937 in the Russian Far East during a comprehensive expedition led by Prof. L. A. Zilber. The same expedition also discovered that TBE is transmitted by Ixodidae ticks. The majority of the TBE virus' range is located in Russia. TBE is believed to be one of the most dangerous infections transmitted by ticks.

4.1.1 General Information

Tick-borne encephalitis is caused by an RNA virus in the *Flavivirus* genus. Today, it is universally acknowledged that the TBE virus is polytypic and is genetically and phenotypically diverse both within populations and depending on its geographic location (Korenberg and Kovalevsky 2000). Six virus subtypes have been identified, three of which are most prevalent (Ecker et al. 1999; Zlobin et al. 2001; Demina et al. 2010).

The disease's clinical course varies from asymptomatic to mild to severe; fatal outcomes are possible. The incubation period lasts between 7 and 14 days on average. The disease exhibits the traditional process of an infectious illness, beginning acutely with a rapid rise in body temperature, chills, headaches, muscle pain and nausea. Disorders of consciousness are also often reported. Typical symptoms include slight paralysis (decreased muscle tone) and shoulder and neck palsy (Cherkassky 1994).

The virus remains in the human host for a long period of time and can lead to a second wave of disease development or chronic forms of infection. It can withstand low temperatures (it remains viable at a temperature of −15 °C for up to 1 year). However, the TBE virus is sensitive to high temperatures and can be destroyed by boiling for 2 min (Korenberg et al. 2007).

4.1.2 Epidemiology and Natural Focus Characteristics

The vast majority of TBE cases are transmitted via vectors; however, consumption of unpasteurized goat's and sheep's milk or dairy products has led to food-borne transmission.

S. Malkhazova et al., *Mapping Russia's Natural Focal Diseases*, Global Perspectives on Health Geography, https://doi.org/10.1007/978-3-319-89605-2_4

Cases of transmission via cow's milk have been described, but they are considered to be quite rare (Korenberg et al. 2013). Infection as a result of tick squashing, when the virus subsequently enters the organism through the mucous membranes of the eyes or through a cut in the skin, is also possible.

The taiga tick (*Ixodes persulcatus*) and the castor bean tick (*I. ricinus*) serve as the primary vectors and natural reservoirs of the TBE virus (Fig. 4.1). According to current understanding, they are the primary vectors of the TBE virus, and the disease's range corresponds with the ticks' range (Korenberg et al. 2013). In some foci, ticks in the *Dermacentor* genus can also serve as vectors; 18 Ixodidae species were confirmed to have been spontaneously infected with the TBE virus.

Fig. 4.1 The castor bean tick *Ixodes ricinus*. *Credit* http://media0.wgz.cz/images/media0:5100b426b1bbd.jpg/klíště%20obecné%20%28%202.5mm%20%29.jpg

The range of *I. persulcatus* is almost entirely located on Russian territory and encompasses the middle and southern taiga zones, mixed coniferous forests, temperate deciduous forests and the forest steppe. The range of *I. ricinus,* which prefers warmer temperatures, spreads across nearly all of Western, Central and Southern Europe. This tick is also sometimes found in Northern Africa, the Middle East, Crimea and the Caucasus (Fig. 4.2). Both vectors are found simultaneously on a significant portion of Russia's European territory, from the northwest to the mid-Volga region. TBE has not been found in the castor bean tick's western range, meaning that the distribution of the virus is somewhat narrower than its vector's.

As previously noted, three of the six known subtypes of the TBE virus are most widespread. The Far East and Siberian subtypes are most common in the disease's eastern range: they are transmitted by the taiga tick (*I. persulcatus*). The European subtype is primarily found in the western range and transmitted by the castor bean tick (*I. ricinus*). However, different viral subtypes are not clearly confined to specific regions, so they are sometimes found beyond the

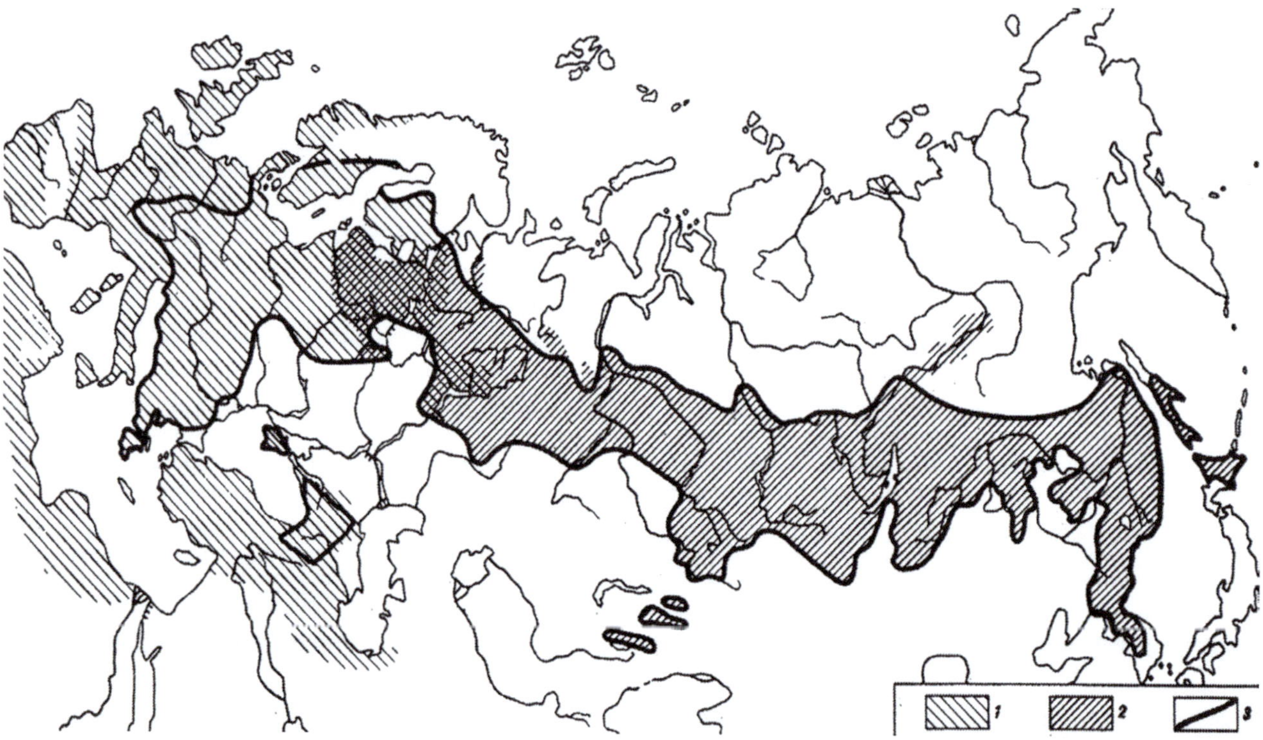

Fig. 4.2 Tick Ranges: 1—*I. ricinus*; 2—*I. persulcatus*; 3—TBE virus. *Credit* Korenberg et al. (2013)

aforementioned locations and can both circulate in the same region simultaneously (Korenberg et al. 2013). Therefore, the geographical names of these subtypes do not indicate their strict confinement to a specific region: rather, they simply reflect their relative incidence. As a result, the range of TBE is spread in an essentially unbroken band through the forest zone of Eurasia from the Atlantic and Mediterranean to the Pacific. 71 focal regions can be highlighted within this range. They are categorized into 8 groups, each of which is characterized by a unique type of natural focus and epidemiology (Korenberg and Kovalevsky 1981) (Fig. 4.3). Each focal region encompasses hundreds of individual natural foci. These regions exhibit a specific combination of biotic and abiotic factors, which in turn defines the epidemiological situation in a specific focus and the set of social factors that influence it (Korenberg et al. 2013).

During vector-borne transmission, the TBE virus undergoes several stages in a complex chain of events that are linked to the tick's life cycle; the virus can be transmitted transovarially (from a female tick to her offspring via infected eggs) and transstadially (as the egg metamorphoses into an imago). The tick life cycle is only complete if each stage is able to consume a sufficient quantity of blood, and metamorphosis always leads to a change in vectors. The virus is spread in natural foci when ticks feed on animals, many of which do not have any symptoms when infected (Korenberg et al. 2007). Mammals, especially small ones, are the main nutrient source for ticks, but birds can serve as an additional one (usually for ticks in preimaginal stages). Given that ticks in the same generation (i.e., born in the same oviposition) reach different life stages at different speeds due to a variety of factors, ticks of different stages and generations constantly exchange the TBE virus among themselves within a natural focus.

As with other natural focal infections, TBE incidence in a natural focus (i.e., the number of human infections occurring in it) depends on its epizootic potential and on the intensity of human contact with the focus (Korenberg et al. 2013).

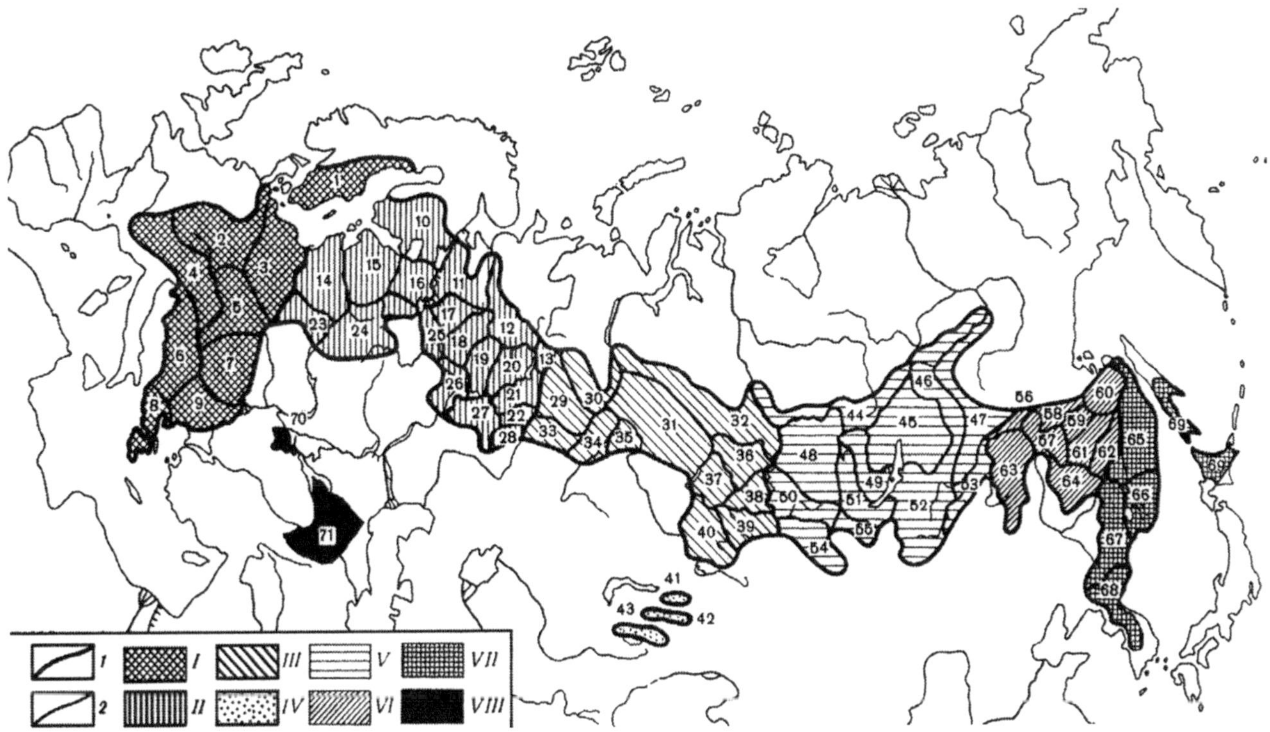

Legend
1: Viral Range Border
2: Natural Focus Region Borders
I-VIII: Natural Focus Region Groups
I – Central European and Mediterranean
II – Eastern Europe
III – West Siberian
IV- Kazakhstan and Central Asian
V – Mid-Siberian and Zabaikalsky
VI – Khingan and Amur
VII – Pacific
VIII – Crimean and Caucasian

The arabic numerals on the map indicate region numbers

Fig. 4.3 TBE range zoning. *Credit* Korenberg et al. (2013)

People who regularly come into contact with TBE natural foci rarely acquire full immunity: viral antibodies in their serum do not always prevent the appearance of clinical symptoms.

TBE morbidity is seasonal, peaking in the spring and summer, based on periods of maximum tick activity. Tick populations peak seasonally in late May and early June. Overall in Russian territory, people are typically infected in the spring and the first half of the summer; illness incidence peaks in June and the first half of July. The development of disease in the late summer and early fall is linked to prolonged activity of adult *I. persulcatus,* which is common in populations in the southern Russian Far East, as well as to the late summer and early fall appearance of *I. ricinus* nymphs and imagoes, as this species exhibits two seasonal population spikes (Korenberg et al. 2013).

The likelihood of human infection with TBE in a natural focus depends on the number of infected ticks in the area. Therefore, as long as the level of human contact with natural foci remains relatively stable, changes in morbidity are closely related to fluctuations in the overall tick population and especially the number of infected specimens. These indices, in particular, define the de facto TBE morbidity in its natural foci (Korenberg and Kovalevsky 2000).

4.1.3 Historical Background

An interest in TBE was first expressed in the early 1930s, during the exploration and development of the Russian Far East, although clinical symptoms of the disease in patients from this region were first described in the late nineteenth century. As the construction of settlements, roads, and logging camps intensified, large numbers of people began to succumb to a lethal disease. By 1937, it had become clear that this disease was an encephalitis (i.e., an inflammation of the brain) and had a viral origin, although its epidemiological characteristics were virtually unknown. Some also assumed the disease to be a toxic form of influenza. In early 1937, the People's Commissariat for Health organized an expedition led by Lev Zilber, a well-known virologist and epidemiologist (Fig. 4.4).

Thanks to his extensive and thorough research, Zilber made a brilliant proposition that was soon confirmed: that Ixodidae ticks serve as vectors for the infection. They established the primary vector—*I. persulcatus*—and isolated the virus responsible for the disease. At the time, they also proposed likely models of the virus's circulation in the wild and began to develop preventative measures against the disease (Korenberg et al. 2013; Kiselev and Levina 2004). Initially, it was assumed that encephalitis was endemic to the Russian Far East, although in the following years, before the start of the Great Patriotic War (World War II), the disease

Fig. 4.4 Lev Zilber soon after his return from his expedition to the Russian Far East, 1937. *Credit* Kiselev and Levina (2004)

was discovered in other regions of the USSR, even in its European territory. The analysis of a TBE outbreak in troops on the Volkhov front led to the identification of another tick species that transmitted TBE: *I. ricinus.* The results of this expedition served as the foundation for the main tenets of Yevgeny Pavlovsky's theory. Even today, TBE is the most significant natural focal infection in Russia. It is considered to be a classical natural focal disease and is used as a model for the study of other infections.

4.1.4 Distribution in Russia and the World

In recent decades, tick-borne encephalitis is becoming more and more of a serious issue in European countries. Diagnostic improvements have led to new data on population morbidity and the geographic distribution of natural foci. Today TBE is found in many Central European countries, in the Baltic countries and in Scandinavia. Many segments of the range exhibit a tendency towards increased morbidity and the widening of focal territories. New foci have been discovered in Denmark, Finland, Sweden, Norway, Austria, Germany and Switzerland (Suss 2011).

The majority of TBE's range is on Russian territory. Today, the epidemiological situation is characterized by somewhat of a stabilization in morbidity after its rapid rise in the 1990s. Steady growth in morbidity was first reported in the early 1970s, but the process sped up considerably beginning in 1989 (Zlobin 2006). TBE morbidity rates have increased tenfold since the late 1980s, while its fatality rate had grown from 0.3 to 2.7% (Zlobin 2006). This infection has been reported in 66 administrative subjects in the Russian Federation. Morbidity rates range from 0.01 cases per

100,000 individuals in the Moscow Oblast to over 60 cases per 100,000 individuals in the Tomsk Oblast (the maximum level of this indicator—64.2 cases per 100,000—was recorded in 1999). The highest long-term average annual morbidity rates have been found primarily in the southern part of East and West Siberia, the area surrounding the western border of the Urals and the South Urals: Tomsk Oblast, Krasnoyarsk Krai, Altai Krai, Khakassia, Tyva, Udmurtia and Perm Krai (Map 4.1).

In a variety of regions, the dynamics of TBE morbidity rates in Russia from 1997 to 2010 are characterized by nonlinear long-term oscillation (decreases, increases, oscillation with both significant and insignificant amplitudes). The greatest oscillations in morbidity are observed in regions in the South Urals, West Siberia and South Siberia. On Russia's European territory, morbidity rates are fairly stable and do not demonstrate significant changes year to year. Moreover, morbidity rates clearly show a tendency to decrease towards the end of the period in question, which corresponds with conclusions about falling morbidity after the peak in the late 1990s (Map 4.2).

Annual dynamics of TBE morbidity are represented on a dynamic chart (Map 4.3). Five levels of morbidity severity, ranging from non-existent to high, are shown for each subject of the Russian Federation for each year in the 17-year period in question (1997–2013). The regions can be divided into three groups based on their morbidity patterns: regions with sporadic and consistently low morbidity (this is mostly the northern part of the Central Federal District and the southern Primorsky Krai) regions, where low to average morbidity rates are reported practically every year (the entire Northwestern Federal District), and regions where consistently high or very high morbidity rates are reported yearly (the entire Ural and Siberian Federal District, Primorsky and Khabarovsk Krais and Amur Oblast).

Researchers have mentioned numerous times that the distribution of TBE morbidity resembles a kaleidoscope: the situation changes annually within the nosogenic territory and is combined with periodic unsynchronized spikes in a variety of regions. Meanwhile, within the disease's nosogenic territory there are always a few regions that have a high number of infected individuals for a long period of time (Korenberg et al. 2013). The most stable foci in Russia's European territory are found in Perm Krai, Udmurtia and Kirov Oblast, while in Siberia Krasnoyarsk Krai has the most stable foci (to be exact, these foci are in its southern portion, which is within the range of *I. persulcatus* and the nosogenic territory of TBE itself), Tomsk Oblast, Altai Republic, Tyva Republic and Buryatia Republic. In other parts of the nosogenic territory, morbidity rates vary from year to year.

TBE distribution can be most clearly observed on a map showing morbidity rates only for administrative districts, where natural TBE foci have been found (Map 4.4). Rospotrebnadzor (Federal Service for Surveillance on Consumer Rights Protection and Human Wellbeing) publishes a list of such territories annually (The list of territories… 2014). This image allows for a more precise understanding of morbidity in subjects that do not entirely fall within the vectors' range. For example, TBE morbidity in Arkhangelsk Oblast, when looked at for the entire region, is not very high as it is concentrated in the southern part of the Oblast (see Map 4.1). However, when computed for the populations of districts where cases have been recorded, morbidity rates become much higher. In a similar manner, the fairly high TBE morbidity rate in Krasnoyarsk Krai becomes comparable with that of Tomsk Oblast, Altai Republic and Khakassia, when only districts with reported cases are taken into account. This fact places this territory among the most infected regions in Russia.

The most vivid characteristic of TBE epidemiology in Russia in the recent years is the change in morbidity structure. While previously TBE had been thought of as a primarily occupational disease that affected people whose livelihood was tied to forests or the oil industry, as well as geologists and construction workers, beginning in the 1990s, TBE began to primarily affect people living in cities. Now they make-up 70% of all cases. Meanwhile, up to 30% of all TBE cases are found in children. This change is primarily caused by city dwellers' rapid development of plots of land in rural areas, many of which fall on natural foci. Infection occurs during hiking trips, visits to the woods, mushroom and wild plant gathering and gardening outside the city. In some regions, up to 40% of individuals were affected by tick bites that occurred within city limits (Zlobin 2006). All social strata and occupations are affected, although people between the ages of 20 and 60 are most commonly infected. Significant human contact with natural foci is listed as the main reason for the change in the epidemiologic situation, as well as insufficient efficiency in epidemic prevention (Korenberg et al. 2007).

4.1.5 Prevention

Avoiding tick bites is the most effective way to prevent encephalitis while in infected regions. It is important to bear in mind that ticks typically congregate on forest trails and pathways with grass growing alongside them. There are often far more ticks in these areas than in the surrounding forest. Ticks wait for their prey on blades of grass and grab onto passers-by. The first adult ticks appear fairly early, soon after the snow cover begins to melt, and they are most active when the average temperature is between 7 and 15 °C and the humidity is over 70%.

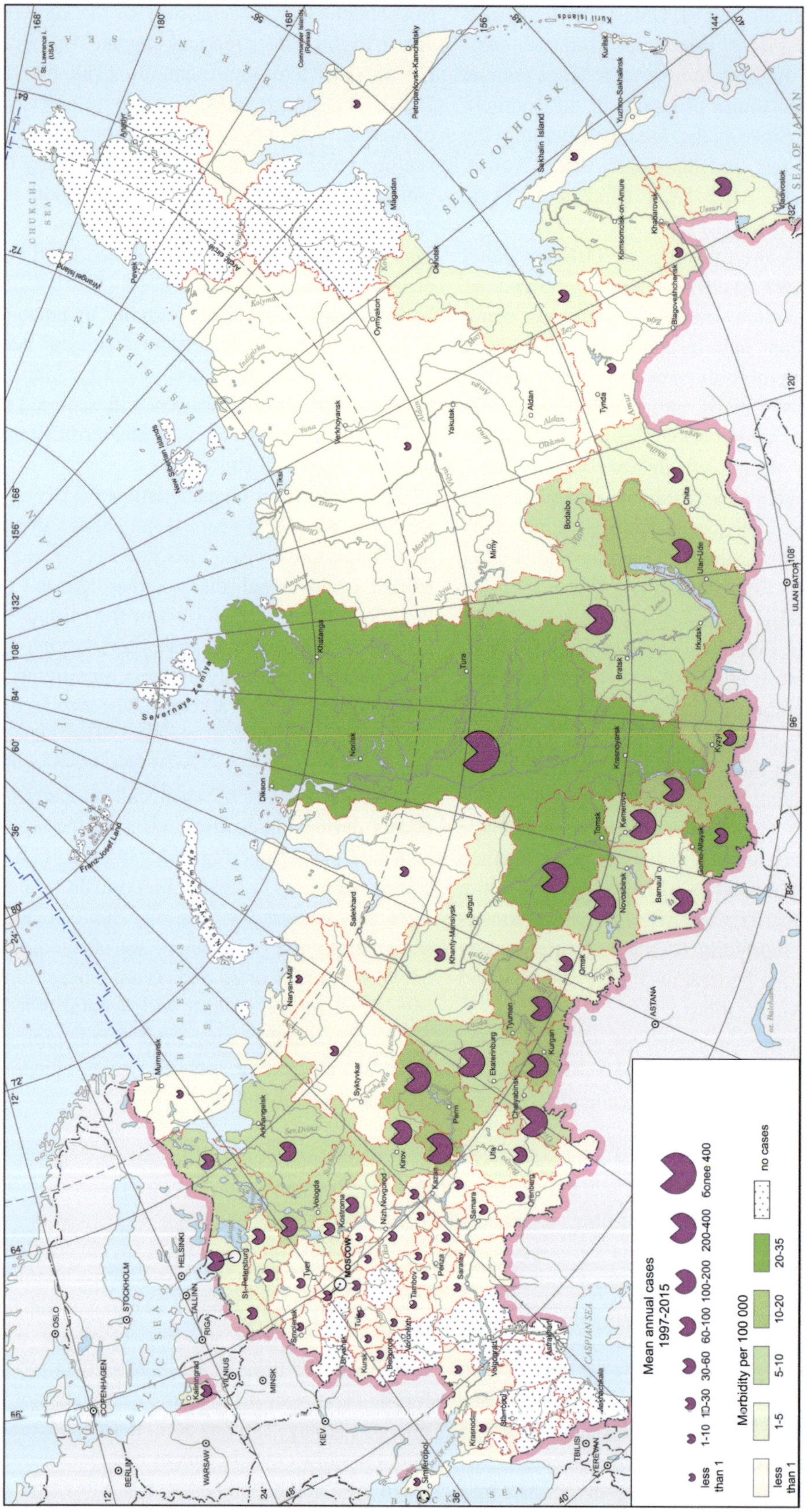

Map 4.1 TBE morbidity, 1997–2015

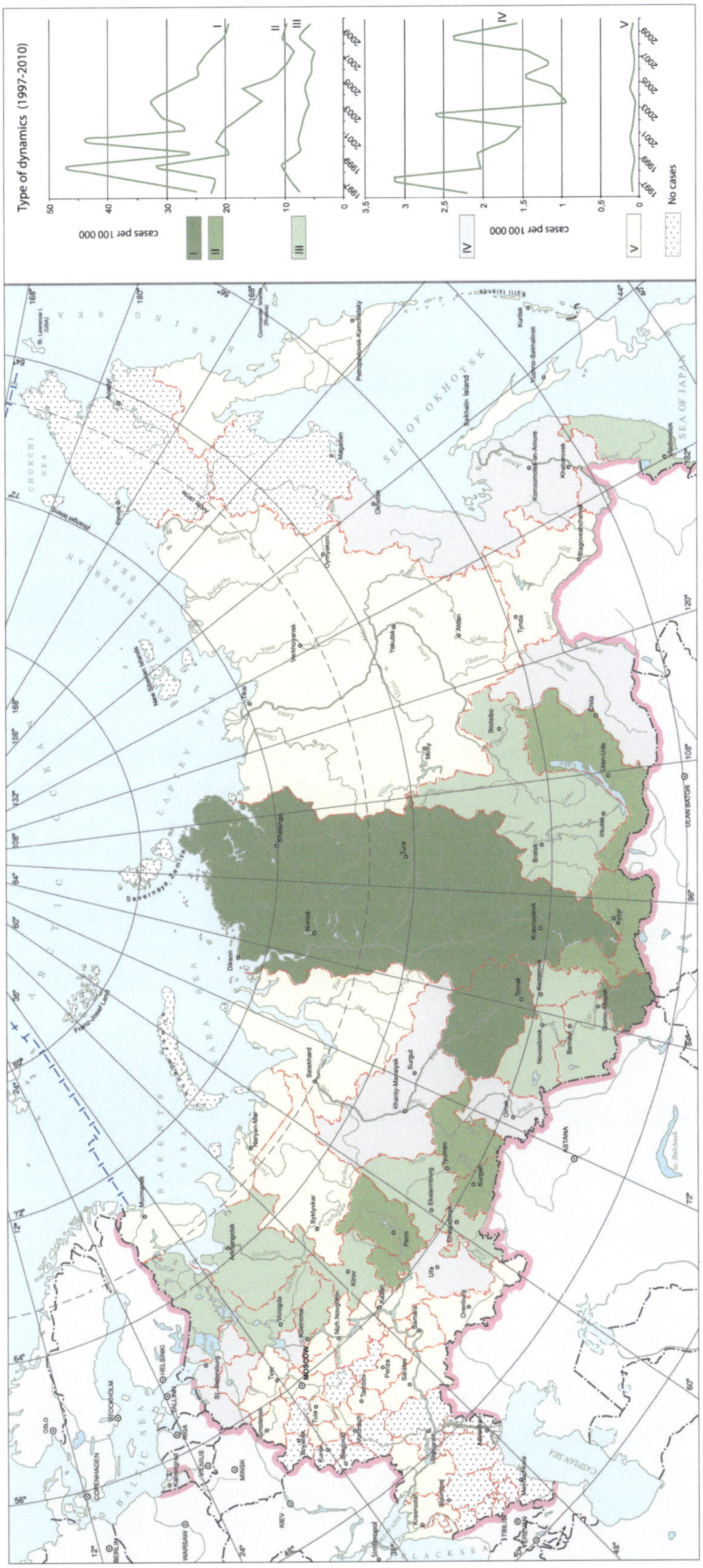

Map 4.2 Types of TBE morbidity dynamics, 1997–2010

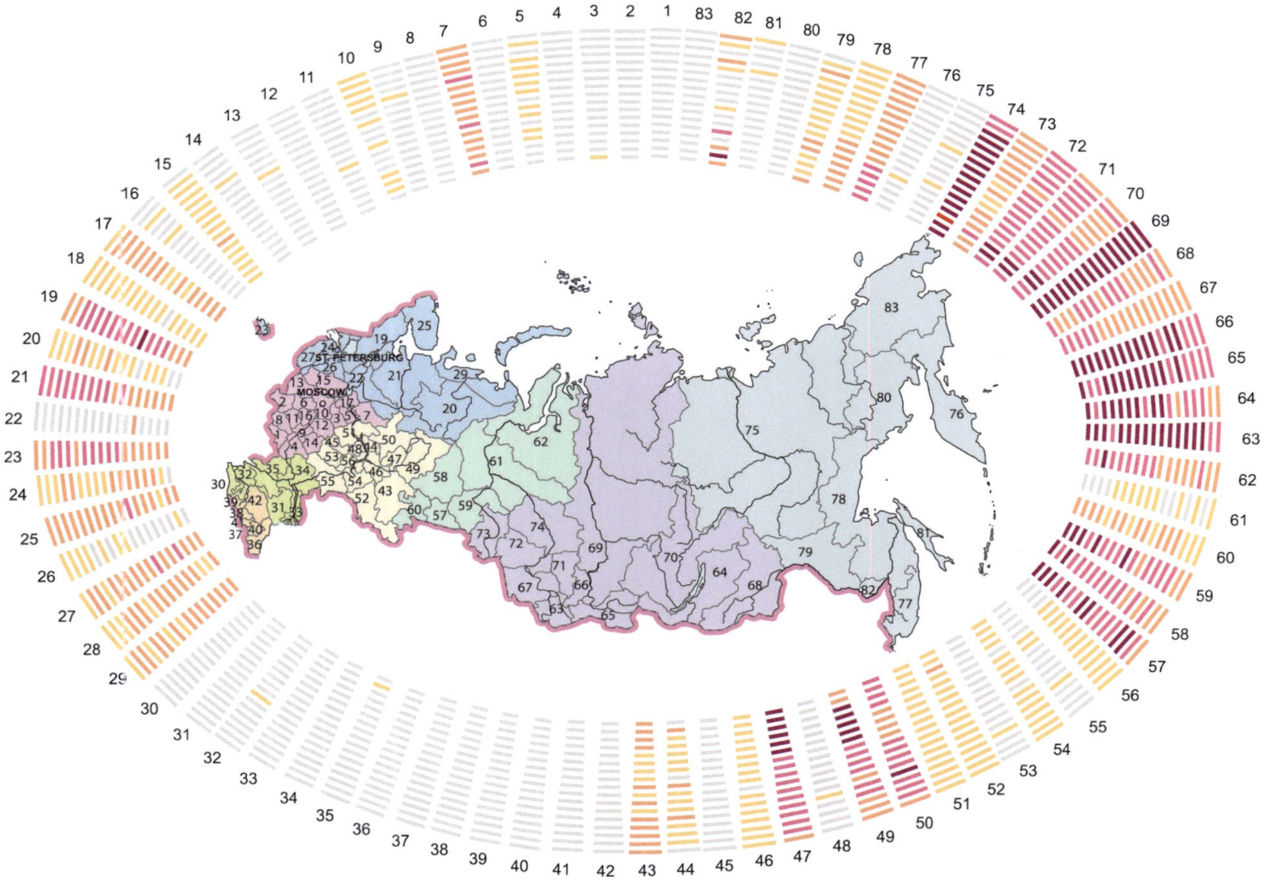

Morbidity per 100 000
- less than 1,0
- 1,0 - 5,0
- 5,0 - 15,0
- more than 15,0 (max 64,2; 1999)
- no cases

2012 2013
2010 2011
2008 2009
2006 2007
2004 2005
2002 2003
2000 2001
1998 1999
1997

1.....83 region number in legend and map

Central Federal District
1 Belgorod Oblast
2 Bryansk Oblast
3 Vladimir Oblast
4 Voronezh Oblast
5 Ivanovo Oblast
6 Kaluga Oblast
7 Kostroma Oblast
8 Kursk Oblast
9 Lipetsk Oblast
10 Moscow Oblast
11 Oryol Oblast
12 Ryazan Oblast
13 Smolensk Oblast
14 Tambov Oblast
15 Tver Oblast
16 Tula Oblast
17 Yaroslavl Oblast
18 Moscow city

Northwestern Federal District
19 Republic of Karelia
20 Komi Republic
21 Arkhangelsk Oblast
22 Vologda Oblast
23 Kaliningrad Oblast
24 Leningrad Oblast
25 Murmansk Oblast
26 Novgorod Oblast
27 Pskov Oblast
28 Saint Petersburg city
29 Nenets Autonomous Okrug

Southern Federal District
30 Republic of Adygea
31 Republic of Kalmykia
32 Krasnodar Krai
33 Astrakhan Oblast
34 Volgograd Oblast
35 Rostov Oblast

North Caucasian Federal District
36 Republic of Dagestan
37 Republic of Ingushetia
38 Kabardino-Balkar Republic
39 Karachay-Cherkess Republic
40 Republic of North Ossetia-Alania
41 Chechen Republic
42 Stavropol Krai

Volga Federal District
43 Republic of Bashkortostan
44 Mari El Republic
45 Republic of Mordovia
46 Republic of Tatarstan
47 Udmurt Republic
48 Chuvash Republic
49 Perm Krai
50 Kirov Oblast
51 Nizhny Novgorod Oblast
52 Orenburg Oblast
53 Penza Oblast
54 Samara Oblast
55 Saratov Oblast
56 Ulyanovsk Oblast

Ural Federal District
57 Kurgan Oblast
58 Sverdlovsk Oblast
59 Tyumen Oblast
60 Chelyabinsk Oblast
61 Khanty-Mansi Autonomous Okrug (Yugra)
62 Yamalo-Nenets Autonomous Okrug

Siberian Federal District
63 Altai Republic
64 Republic of Buryatia
65 Tyva Republic
66 Republic of Khakassia
67 Altai Krai
68 Zabaykalsky Krai
69 Krasnoyarsk Krai
70 Irkutsk Oblast
71 Kemerovo Oblast
72 Novosibirsk Oblast
73 Omsk Oblast
74 Tomsk Oblast

Far Eastern Federal District
75 Sakha Republic (Yakutia)
76 Kamchatka Krai
77 Primorsky Krai
78 Khabarovsk Krai
79 Amur Oblast
80 Magadan Oblast
81 Sakhalin Oblast
82 Jewish Autonomous Oblast
83 Chukotka Autonomous Okrug

Map 4.3 Long-term TBE morbidity dynamics, 1997–2013

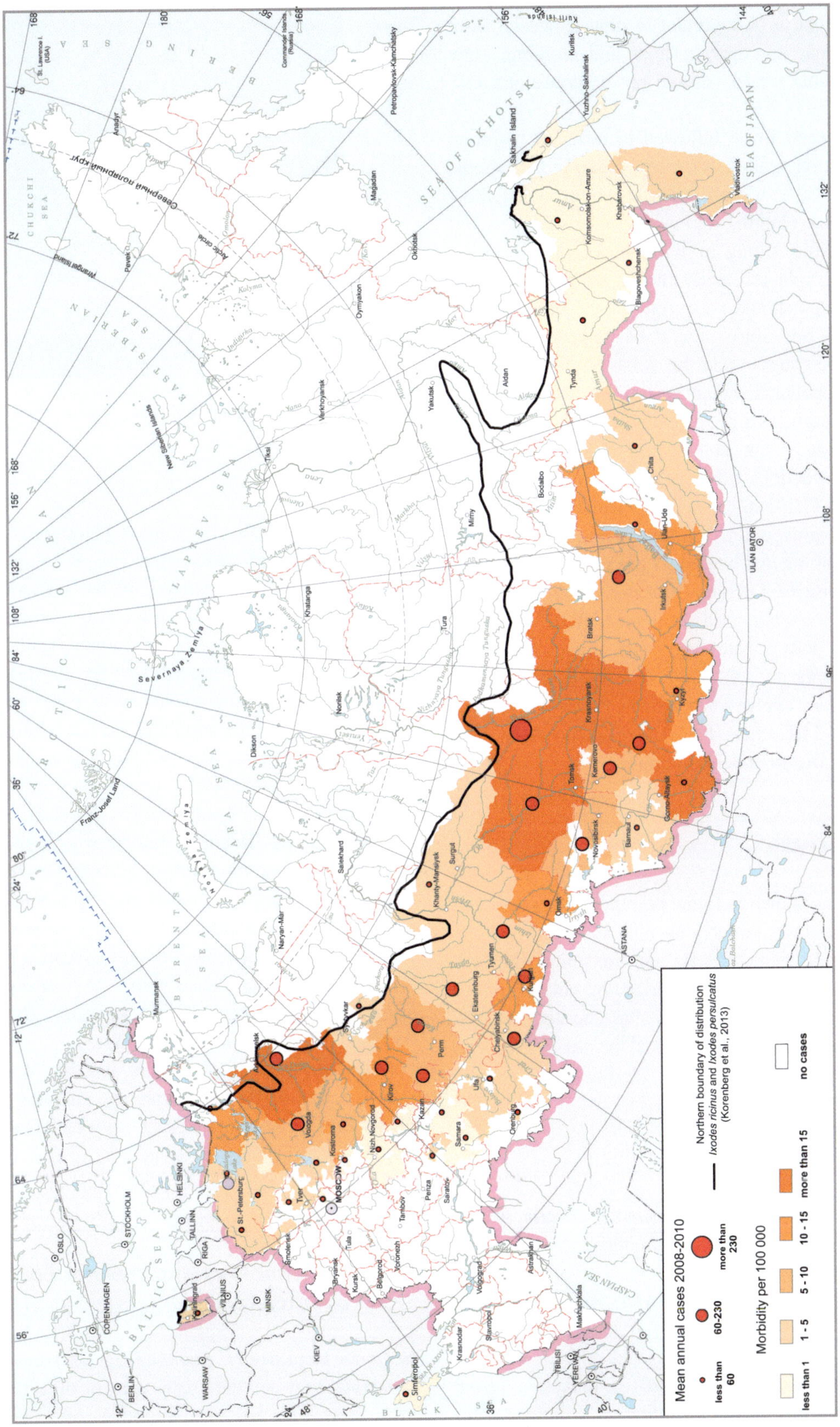

Map 4.4 TBE distribution

Given that the majority of infections are related to entering natural foci, it is important to protect oneself from tick bites and wear appropriate shoes and clothing. Raw milk, especially goat's milk, should not be consumed in a natural focus.

Certain rules need to be followed in order to maintain individual protection against tick bites. While in the woods, one must periodically inspect themselves for ticks, thoroughly check their clothing and other objects ticks could land one, and avoid sitting or lying down on the grass. Monochrome and light-coloured clothing should be worn in the forest; pant legs should be snugly tucked into boots or tight-fitting socks and shirts should be tucked into pants; cuffs must also be snug around the wrists. Acaricidal products and insect repellents should be used. In the event of a tick bite, it is important to get medical help.

Immunoglobulin—a concentrated and purified solution of immunoglobulin fractions separated from donor plasma or serum-containing TBE antibodies using alcohol—is used for emergency TBE prevention. Immunoglobulin is administered on the day of the tick bite under physician supervision. This lowers the risk of TBE development.

Vaccination is widely used to prevent TBE. Licensed state, municipal and private healthcare organizations all provide vaccinations free of charge. The vaccines used for TBE prevention are approved in Russia. Preventative vaccination against TBE must be performed no later than 21 days before a trip to endemic regions.

4.2 Lyme Disease

Lyme disease (also known as Lyme borreliosis) is a group of vector-borne natural focal infections, both acute and chronic. It is caused by spirochete bacteria and transmitted by Ixodidae ticks. Lyme disease is characterized by polymorphous clinical symptoms and can affect the skin, the nervous system, the circulatory system, the liver and the musculoskeletal system.

4.2.1 General Information

Lyme disease is caused by spirochetes in the *Borrelia* genus.

On Russian territory, two Palearctic *Borrelia* species—*B. garinii* and *B. afzelii*—are the main causal agents of Lyme disease, while the causal agent of what was traditionally defined as Lyme disease, *B. burgdorferi* sensu stricto, has a Holarctic distribution and is rarely found only in Russia's European territory (Korenberg et al. 2013). In Russia, Lyme borreliosis refers to this traditional definition, while the other forms are called Ixodidae tick borreliosis. A few other *Borrelia* species are also human pathogens, but they are more sparsely distributed.

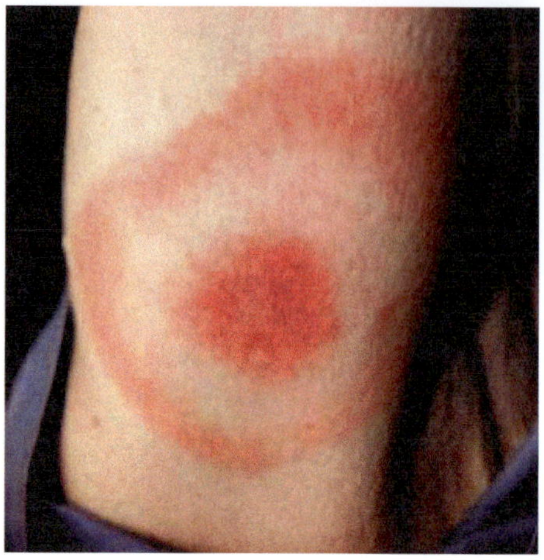

Fig. 4.5 Erythema migrans (cdc.gov). *Credit* www.cdc.gov Image 9874

Borreliosis is characterized by the bacteria's ability to remain in the host's organism for long periods of time and attack a variety of systems and organs, leading to chronic disease. The disease process occurs in three successive stages: (1) local, during which most cases exhibit disease around the tick bite location and the characteristic erythema migrans rash appears (Fig. 4.5); (2) disseminating, during which the bacterium spreads across other organs and tissues and a more systemic syndrome; (3) persistent, which occurs after the *Borrelia* bacterium has been present in the host's organism for a long time and is a chronic form of the disease. This general model is not found in every single case of Lyme disease, and the disease can take a variety of courses, making accurate diagnosis difficult. Because of this, Lyme disease has been 'associated with "'protean' manifestations' and called 'the great clinical imitator' (Nadelman and Wormser 1998; Korenberg et al. 2013).

It has been established that the majority of tissue damage in Lyme disease is polymorphous, affecting a variety of organs. Erythema migrans is the only conclusive marker of Lyme disease. However, it is not always present—in a number of cases the disease can be non-erythematous.

4.2.2 Epidemiology and Natural Focus Characteristics

The castor bean tick (*I. ricinus*) (Fig. 4.6) and the taiga tick (*I. persulcatus*) are the main vectors of Lyme disease in Eurasia. *Borrelia* has been found in other tick species as well, but their epidemiological role is extremely small.

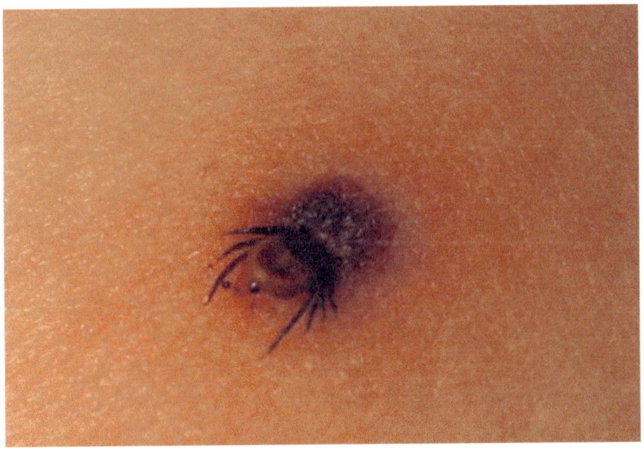

Fig. 4.6 An *I. ricinus* tick that has bitten a human. *Credit* V.A. Mironova

Therefore, the two aforementioned species are responsible for the spread of Lyme disease.

Ixodidae ticks have a fairly high *Borrelia* infection rate. In Western Europe, the infection rate fluctuates between 5 and 20%, sometimes reaching 30–40%. In areas inhabited by both tick species, *I. persulcatus* ticks are usually more infected than *I. ricinus,* making the former a more dangerous vector.

The primary reservoir hosts of *Borrelia* are small mammals, such as northern red-backed voles (*Myodes rutilus*), the tundra vole (*Microtus oeconomus*), Eurasian field mice (*Apodemus*) and other species. *Borrelia* has also been found in other species, such as dogs, wild and domesticated hoofed mammals, birds, etc.

People become infected when bitten by a tick that contains *Borrelia* in its salivary glands, or when they smash a tick on their skin. People are highly susceptible to *Borrelia*, and the disease has been found in members of every age group. People who spend time in forests are at higher risk of contracting the disease. The highest risk of infection with Lyme disease is in mixed coniferous forests and the southern taiga.

Given that Lyme disease is ecologically linked to Ixodidae ticks, this disease's distribution, like that of tick-borne encephalitis, is completely dependent on its hosts' ranges.

I. persulcatus is found in different types of taiga and mixed coniferous forests, as well as the mountainous taiga. In northern regions, thanks to the cooler climate and high humidity, these ticks are only found in especially well-drained and well-heated areas in flat interfluves or large river valleys. South of the middle taiga subzone, the taiga tick is ubiquitous in interfluvial forest communities, excluding overly humid sections. *I. ricinus* is common in the moderately hygrophilous and mesophilous biotopes of European lowland and mountain forests, especially in their northern sections. It favours especially warm areas in well-lit forests, as well as clearings and forest edges. South of the border of the temperate deciduous forest zone, the tick is found in intrazonal conditions, in river valleys and lowlands covered in tree and bush growth (Filippova 1977).

4.2.3 Historical Background

Various clinical manifestations of Lyme disease have been known to humans since the late nineteenth century, when the physician Alfred Buchwald first mentioned a unique chronic skin condition, later termed as acrodermatitis chronica atrophicans. In the early twentieth century, Arvid Afzelius, a Swedish dermatologist, described chronic erythema migrans, now known to be a characteristic symptom of Lyme disease. For most of the twentieth century, a number of conditions linked to tick bites were described by researchers in multiple countries. However, only in 1974 did Klaus Weber, a German scientist, propose that all these illnesses may be caused by *Borrelia* bacteria (Weber 1974). In 1977, an American rheumatologist, Allen Steere, along with his colleagues, described unique cases of arthritis among the residents of the town of Lyme, Connecticut (Fig. 4.7), and linked them to Ixodidae ticks (Steere et al. 1977). The illness was named after the town. A few years later, in 1982, Willy Burgdorfer and his colleagues identified a previously unknown *Borrelia* species as the cause of the disease (Burgdorfer et al. 1982). It was later named *B. burgdorferi.* Later, similar diseases were described in Europe and Russia.

As a result, by the mid-80s the biological cause of a disease whose various clinical symptoms had been known for a century under different names finally became clear. These discoveries gave rise to a host of large-scale studies in the USA, Europe and Russia, which specified the species of Lyme disease-causing bacteria in Europe, identified vectors

Fig. 4.7 Near the town of Lyme, Connecticut, USA (Google Earth)

and warm-blooded hosts, and described the distribution and features of natural foci.

For a fairly long period of time, it was thought that *B. burgdorferi* was the sole pathogen in Lyme disease. As a result, a disease found in North America and Eurasia had one name: Lyme disease. Soon after, however, it became clear that Lyme disease is made up of a group of etiologically independent diseases. Today in Russia, Lyme disease is referred to as Ixodidae tick borreliosis (Korenberg et al. 1990).

4.2.4 Distribution in Russia and the World

Lyme disease is the most common tick-borne disease in Russia, Europe and the United States. According to data from the Center for Disease Control (CDC) in the USA, in 2009 it was the fifth most commonly reported disease with nearly 30,000 cases (9.8 per 100,000 individuals) (www.cdc.gov). In North America, as in Eurasia, the disease's nosogenic territory matches the distribution of its main vectors and numbers three regions: the Northeast (which has the highest morbidity rates), some areas in the Midwest and the West Coast (Fig. 4.8). Lyme disease (the specific kind) has also been reported in Canada and isolated cases have been described in Mexico. In Europe, Lyme disease is reported from southern Scandinavia to the northern Mediterranean, and it is most widespread in Central and Eastern Europe. Lyme disease is found in Japan, Korea and China.

Lyme disease morbidity rates have been reported in Russia since 1992, although it is clear that it was widespread among the population before this. According to current understanding, Lyme disease natural foci are tied to forest zones from the Baltic countries to Sakhalin. Given that Lyme disease vectors also carry tick-borne encephalitis and other infections, cases of co-infections with two or three pathogens are common (Korenberg et al. 2013).

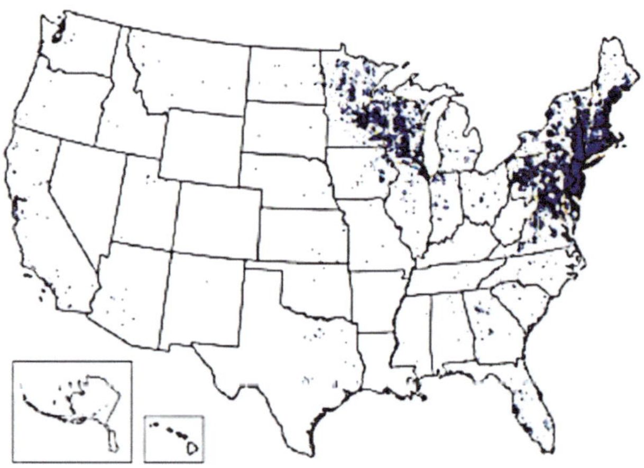

Fig. 4.8 Lyme disease distribution in the USA, 2011. *Credit* www.cdc.gov

A significant portion of the global range of Lyme disease is located within Russia. The disease has been reported in 77 administrative subjects of the Russian Federation, with morbidity rates ranging from 0.02 (Rostov Oblast) to 71.7 (Tomsk Oblast) per 100,000 individuals. The long-term average annual morbidity rate varies from 25 to 40 cases per 100,000 individuals (Map 4.5). Morbidity is most severe in the zone located between 52 and 61°N, which includes the Yaroslavl, Vologda, Kostroma and Kirov Oblasts in European Russia and Tomsk Oblast in West Siberia. The greatest number of cases are found in 15 administrative subjects, located near the Kama River, the Ural Mountains and the southern part of West Siberia. Here, the risk of infection remains high year to year.

Only in eight administrative subjects have never reported a case: North Caucasian Republics, Astrakhan Oblast and Chukotka Autonomous Okrug.

Lyme disease is most prevalent in the spring and summer: this is related to tick activity. In areas where *I. persulcatus* is the main vector, morbidity peaks in the spring and the first half of the summer. In foci where *I. ricinus* dominates, two seasonal peaks have been identified, one from late April to mid-June and the other from late August to mid-September. This is related to this species' second period of heightened activity. Variations in morbidity from year to year are related to climate conditions, tick population sizes, as well as the population sizes of ticks' main nutrient sources. It has been established that the epizootic process of Lyme disease is activated, leading to higher morbidity risks, in years with abundant snowfalls in the winter months, followed by a dry spring and a warm summer (Kashuba et al. 2014).

Lyme disease's morbidity rates in Russia from 1997 to 2010 are characterized by nonlinear long-term oscillation (decreases, increases, oscillation with both significant and insignificant amplitudes) in a variety of regions (Map 4.6). The greatest oscillations in morbidity rates occur in the subzone of the southern taiga in European Russia and the southern part of West Siberia. The south part of the forest zone in European Russia, the south of Western Siberia, Zabaikalsky Krai and Khabarovsk Krai all exhibit a tendency towards higher morbidity rates. In the rest of the country, morbidity rates are quite stable and do not exhibit significant changes from one year to another.

Annual Lyme disease morbidity levels (ranked based on five levels of severity, from non-existent to high) for each administrative subject over a period of 17 years (1997–2015) are shown on a dynamic cartogram (Map 4.7). Several region types can be identified based on these levels of severity. Regions with sporadic morbidity generally span most of Yakutia and the Russian Far East, excluding its southern portion. Sporadic cases are also found in southern European Russia in administrative subjects within the Southern Federal District. Krasnodar Krai is an exception:

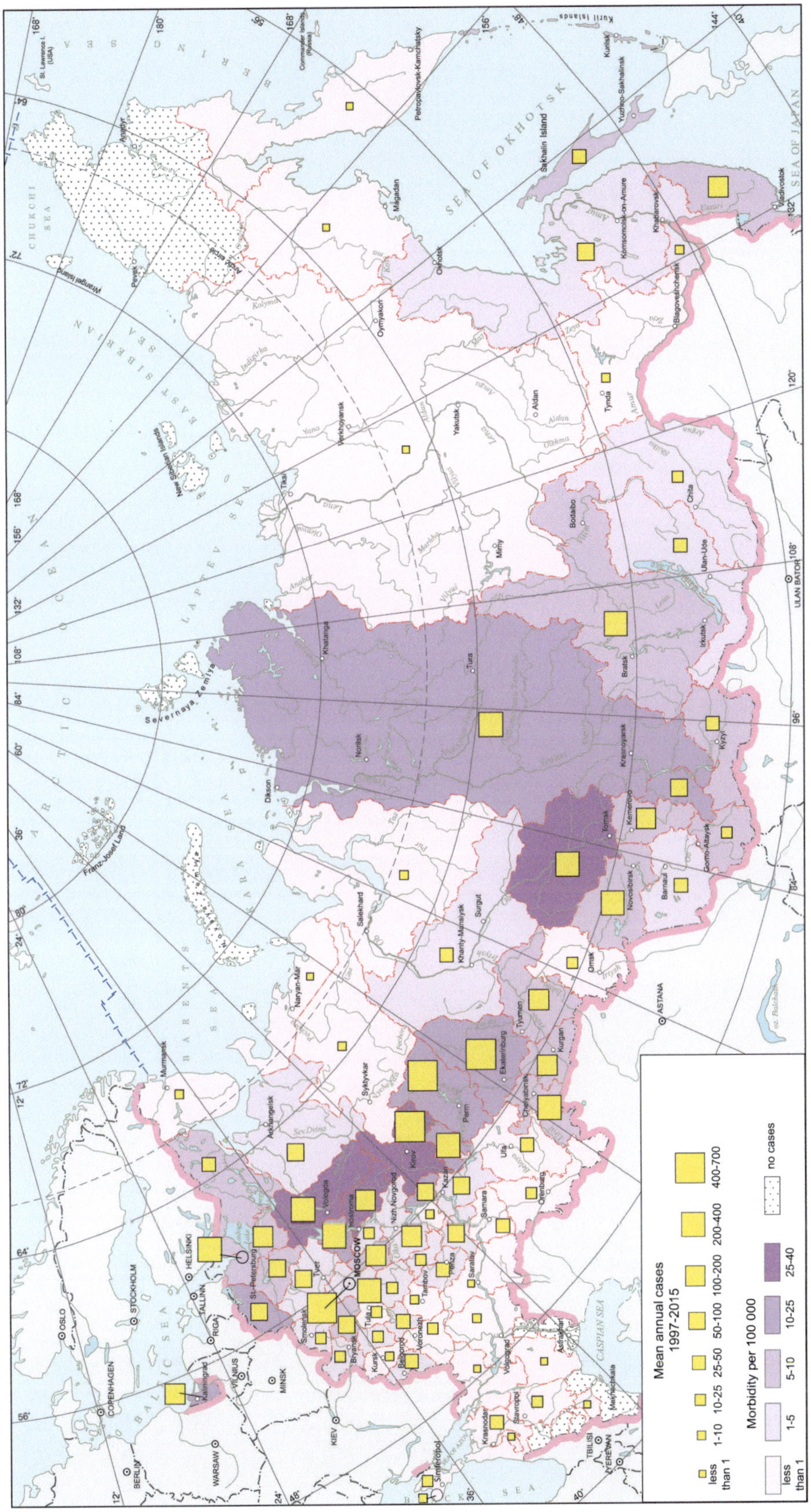

Map 4.5 Lyme disease morbidity, 1997–2015

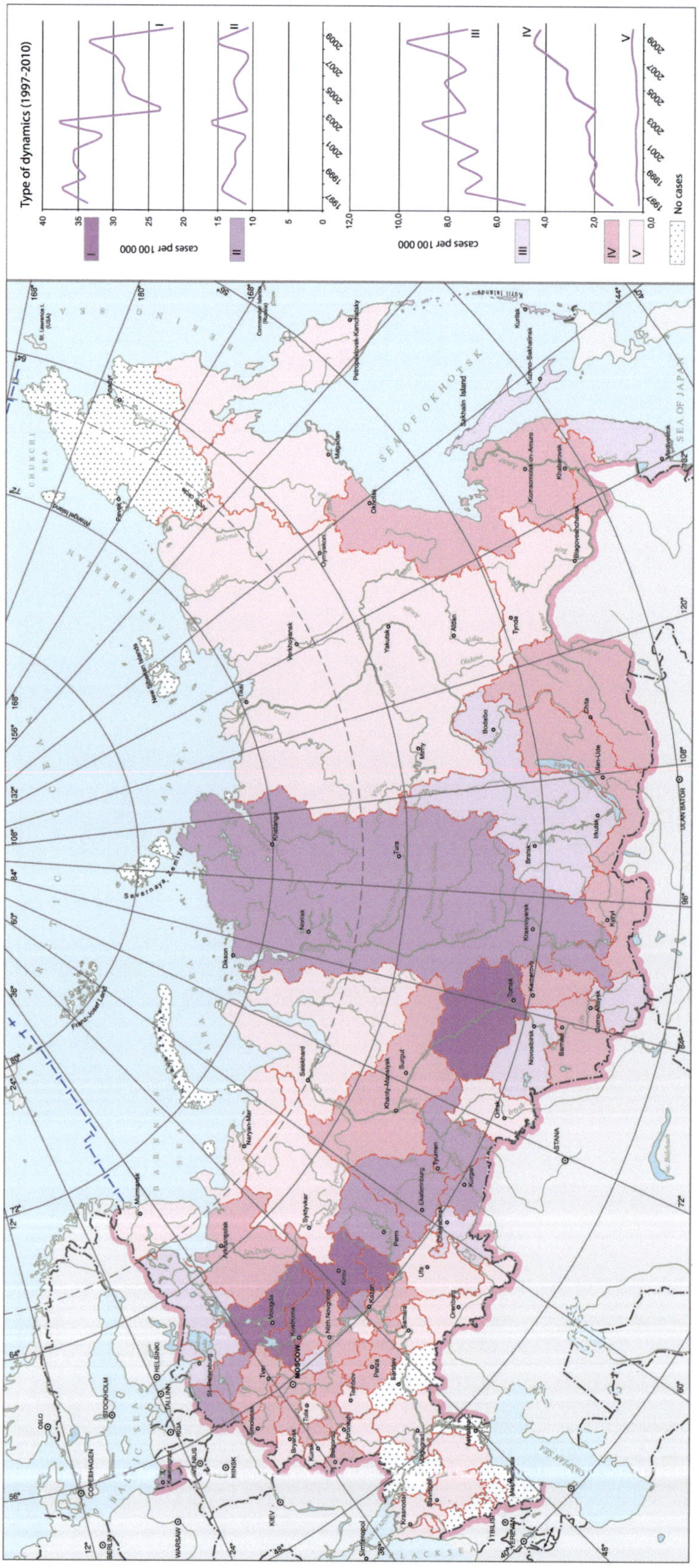

Map 4.6 Types of Lyme disease morbidity dynamics, 1997–2010

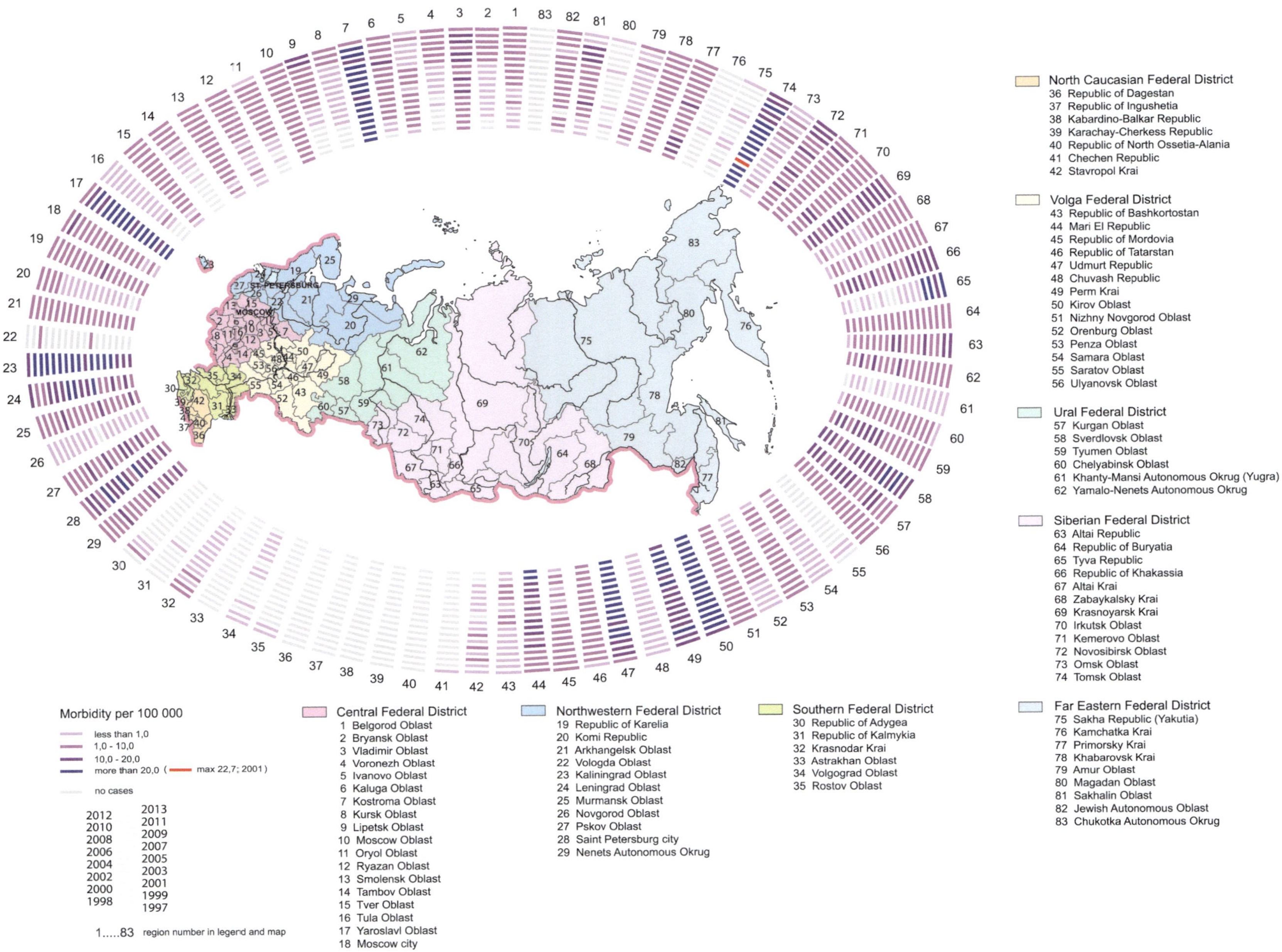

Map 4.7 Long-term Lyme disease morbidity dynamics, 1997–2013

low and moderate morbidity rates have been reported there every year. Regions with low yet consistent morbidity rates are located in the south of the Central Federal District. The north of the Central Federal District, as well as the entire Northwestern Federal District are characterized by consistently high and moderate annual morbidity. Morbidity is steadily high in the Volga region, the Ural Mountains, south Siberia and the Russian Far East.

Therefore, Lyme disease is quite evenly distributed across Russian territory, while its natural foci are stable, as revealed by annual morbidity reports. This is likely related to the high infection rate of ticks with *Borrelia* bacteria and the high level of contact between natural foci and the human population.

4.2.5 Prevention

Non-specific preventative methods aimed at protecting oneself from ticks are described in detail in the *Tick-Borne Encephalitis* section (see Sect. 4.1). Given that pathogenic *Borrelia* and the TBE virus have the same main hosts in Eurasia, it is important to bear in mind the high probability of co-infection as a result of one tick bite (Korenberg 2001). Moreover, Ixodidae ticks carry a number of other infections and have been proven to periodically contain multiple pathogens simultaneously. Given that combined vaccines are not currently available, and other methods of specific prevention are significantly limited thanks to a lack of research on co-infection, specialists recommend means of individual protection as the most trustworthy prevention methods. This approach prevents tick bites and thus any transmittable infections.

Lyme disease vaccines are in development, but they have yet to be made widely available, so emergency antibiotic therapy is the only available method of specific prevention after a tick bite in areas where Lyme disease is common (Bacterial Diseases 2014). If it is confirmed that the tick carried *Borrelia,* preventative antibiotic therapy is administered. This form of prevention in the early incubation period is seen as a possibility to reduce drug dosages to lower amounts than would be required at the peak of illness. Antibiotics are given only by prescription and self-medication is unacceptable.

4.3 North Asian Tick Typhus

North Asian tick typhus (also known as North Asian tick fever or Siberian tick fever) is a natural focal obligate vector-borne disease that belongs to the spotted fever group of diseases. North Asian tick typhus is the most widespread rickettsial disease transmitted by ticks in Russia (Tarasevich 2005). Its natural foci are located primarily in the Asian part of Russia and its surrounding nations (Rudakov et al. 2011). The growth in morbidity recorded in the past decade places this rickettsiosis among re-emerging infections.

4.3.1 General Information

Rickettsia refers to a vast group of gram-negative microorganisms in the Rickettsiales order, which is ecologically tied to arthropods. It includes two families. The *Rickettsia* genus has a unique role, as it is assumed to be evolutionarily related to eukaryotic mitochondria (Rudakov 2009). Forty-five species of parasitic microorganisms carried by arthropods have been identified in the *Rickettsia* genus (Rudakov et al. 2011). According to current levels of understanding, rickettsia is one of the most ancient groups of bacteria and is in the alphaproteobacteria group.

Rickettsia sibirica, which causes North Asian tick typhus, is an obligate intracellular parasite that enters the cytoplasm and nucleus of infected cells. It is transmitted to humans via tick bites. The disease has an incubation period of approximately 3–7 days. The beginning of the illness is usually acute and characterized by a high body temperature, chills and increasing intoxication and malaise. Over the course of 2 days, body temperature reaches 39–40 °C. The most characteristic symptom is an abundant rash, which usually appears between the second and fourth days of illness. A small non-tender lump, called a primary effect or complex, forms over the tick bite. Potential complications include aseptic meningitis, pneumonia and myocarditis. The prognosis is good, although outlying fatal outcomes have been described.

4.3.2 Epidemiology and Natural Focus Characteristics

Infected Ixodidae ticks are the main pathogen's reservoirs and vectors. Overall, rickettsia ecology is based on its ability to parasitize on a wide variety of phylogenetically distant hosts: blood-sucking arthropods and warm-blooded animals (birds and mammals). Rickettsia's extensive distribution among and high level of adaptation to arthropods suggests a lengthy phase of coevolution between the two groups. Today arthropods are believed to be the primary hosts for rickettsia.

Fourteen species of Ixodidae tick have been found to be infected with the causal agent of North Asian tick typhus, *Rickettsia sibirica.* Ticks in the *Dermacentor* genus, such as *D. nuttalli, D. silvarum, D. marginatus* (Fig. 4.9), as well as some species in the *Haemaphysalis* (usually *H. concinna*), *Hyalomma* and *Rhipicephalus* genera are most epidemiologically significant. Ticks can host rickettsia for a long time and transovarially and transstadially transmit them to their

Fig. 4.9 *Dermacentor marginatus* (ornate sheep tick). *Credit* http://de.wikipedia.org/wiki/Schafzecke#mediaviewer/File:Dermacentor_marginatus,_male.JPG Фото Lucarelli Creative Commons Attribution-Share Alike 3.0 Unported license

Fig. 4.10 A typical *D. marginatus* Habitat (Altai Krai). *Credit* N.G. Kadetov (permission obtained)

offspring, determining the pathogen's circulation patterns in the natural focus (Rudakov et al. 2011). Larvae and nymphs of these tick species tend to feed on small mammals, mostly rodents, whereas adult ticks feed on large mammals, including livestock and humans. The nymphs of certain tick species can also bite humans.

Ticks in the *Dermacentor* genus primarily inhabit steppe and forest steppe landscapes, except for *D. silvarum,* which, like *H. concinna,* another one of North Asian tick typhus's hosts, is found in inundable forests and brush growth in riverbeds. Therefore, the risk of human infection is characterized by substantially non-uniform spatial distribution. The epidemiological characteristics and regional distribution of the infection are determined by the local host fauna, the nature of distribution in the territory, the availability of suitable habitats, rickettsia infection rates and the frequency of population–vector contact.

Rickettsia circulation in a natural focus is organized in a chain: Ixodidae ticks—small wild mammals—Ixodidae ticks. The infection caused by *R. sibirica* is asymptomatic in mammals. No epizootic cases have been described. Ixodidae ticks' biological characteristics make the disease seasonal (Tarasevich et al. 1990). Morbidity rises in the spring and peaks in May–June, followed by a less pronounced peak in late summer and early autumn. This situation is found in forest steppe foci, where ticks in the *Dermacentor* genus are the main vectors (Fig. 4.10) in foci where *H. concinna* is the main hosts, morbidity peaks in mid-summer (Rudakov et al. 2006).

Natural focus classification is landscape based and takes into account the main vector species. Four types of natural focus have been identified on Russian territory: *Dermacentor* steppe and forest steppe, *Haemaphysalis* forest steppe, *Hyalomma* semi-desert and desert and *Dermacentor, Haemaphysalis.* Natural focus types are named after the specific vectors, underscoring the crucial role of ticks in shaping the epidemiologic characteristics of a focus (Rudakov et al. 2011). *Dermacentor* steppe and forest steppe foci are the most widespread and epidemiologically significant. They are linked with ticks in the *Dermacentor* genus. These foci are characterized by oscillations in epidemiological activity from 1 year to another.

Human infection occurs after an infected tick attaches itself to the individual in a natural focus. There are certain age-related and occupational risk groups, and morbidity is related to the intensity of risk group contact with different types of natural focus (Rudakov et al. 2006).

People become infected in the steppe zone, anywhere animal husbandry is common, close to human settlements, in windbreaks and in recreational areas. The age and occupational distribution of those infected depend on the type of agricultural activity and the distance between human settlements and natural foci of the specific population. In places where natural foci are located close to homes, most victims are children, whereas in other territories rural adult populations are most affected.

4.3.3 Historical Background

North Asian tick typhus was first described in 1936 in Primorsky Krai by E. M. Mill', after cases of a previously unknown infectious disease that occurred after tick bites and caused a high fever, a rash and neurological symptoms were noticed in the region in 1934 and 1935. In other Russian territories, various researchers (N. I. Antonov, A. A. Naishtat, M. D. Shmatikov and M. A. Velik), described the disease as 'Primorye tick fever', 'Far East tick rash fever' and 'tick

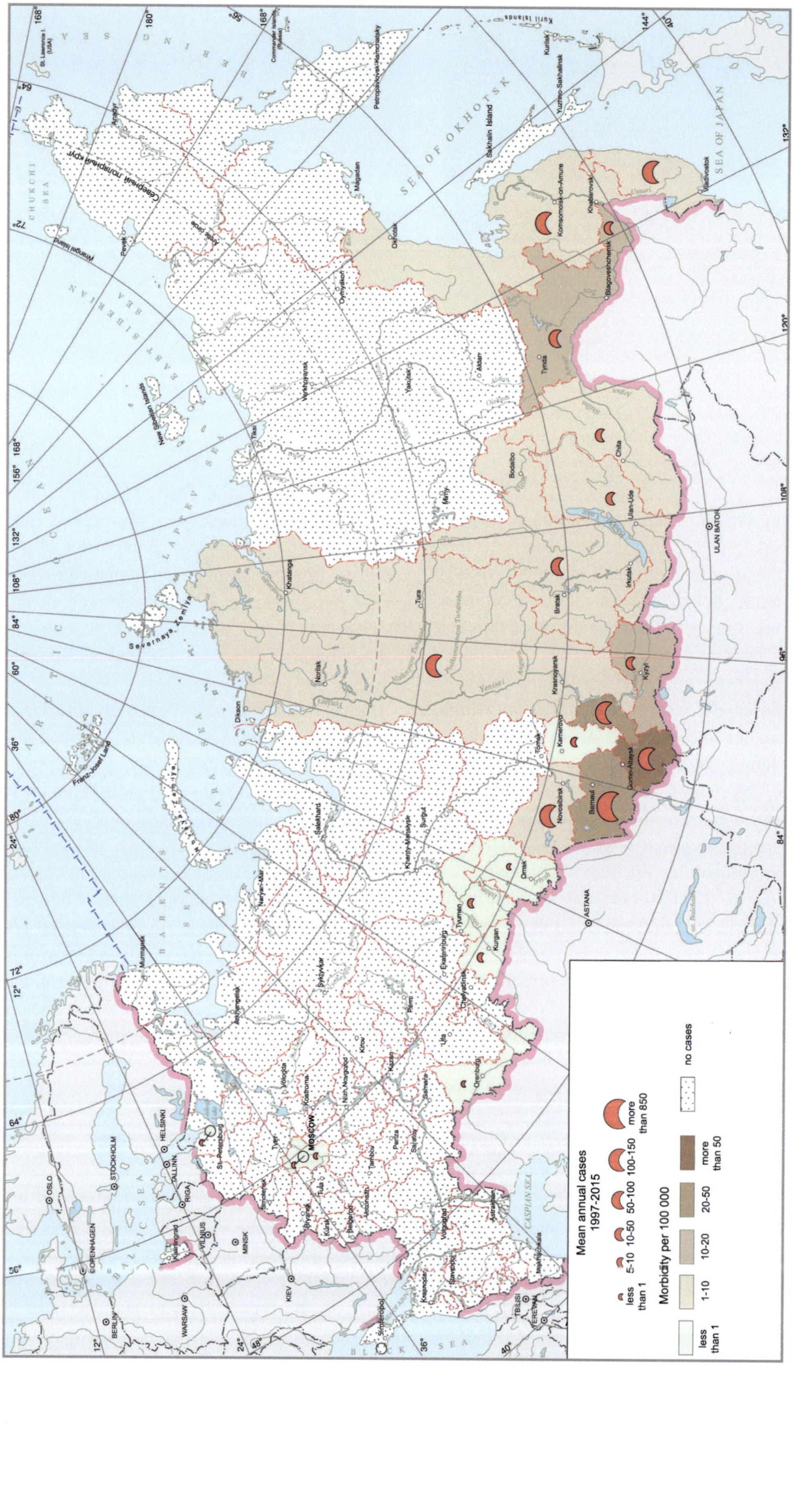

Map 4.8 North Asian tick typhus morbidity, 1997–2015

rash fever' (Rudakov et al. 2011). The All-Union Institute of Experimental Medicine began detailed research in Krasnoyarsk Krai in 1937, leading to the identification of the disease's causal agent—*R. sibirica*—from both humans and ticks in 1938 (Tarasevich 2005). The disease was defined as a separate nosologic form. Research conducted by the Central Institute of Epidemiology and Microbiology in Khabarovsk Krai confirmed the role of ticks in the *Dermacentor* and *Haemaphysalis* genera in this disease's distribution. In 1948, Elena Golinevich isolated a *Rickettsia* strain from *D. nuttalli* ticks in Krasnoyarsk Krai.

In the late 1940s, Yevgeny Pavlovsky came to the conclusion that tick typhus fevers are natural focal diseases and proposed a model of the circulatory pathways of the infectious agent in the biocoenosis of the natural focus (Tarasevich et al. 1990).

By the 1980s, natural foci had been found in Krasnoyarsk Krai, Altai Krai, Primorsky Krai, Khabarovsk Krai, Kemerovo Oblast and Tyumen Oblast.

4.3.4 Distribution in Russia and the World

Outside of Russia, North Asian tick typhus is found in North and East Kazakhstan, Central Asia, Mongolia and China, as well as the Trans-Caucasus, where natural foci were found quite recently. In the 1970s, North Asian tick typhus morbidity significantly increased, along with the morbidity rates of other spotted fevers.

North Asian tick typhus morbidity rates rose substantially in Russia and its neighbouring countries in the 1980s and 1990s and remained high for some time, leading it to be defined as a re-emerging infection (Rudakov et al. 2003). In the 1990s, morbidity rates increased in Russia, peaking in 2001 (2.4 cases per 100,000 individuals). Later the morbidity rate fell to 1.1 and has remained stable since.

The Russian nosogenic territory of this disease stretches from Kursk Oblast to Primorsky and Khabarovsk Krais on the eastern side and includes South Siberia, the Amur region and Primorsky Krai. Today new strains and natural foci are constantly being identified.

The infection has been reported in 21 regions of the Russian Federation. Cases in regions west of Kurgan Oblast are evidently brought in from elsewhere. Long-term average annual data demonstrate that the highest morbidity rates are found in Altai Republic, Khakassia, Altai Krai, Tyva, Khabarovsk Krai and the Jewish Autonomous Oblast (Map 4.8). When morbidity for three separate consecutive years (2011, 2012, 2013) is examined, the infection is clearly extremely stable both in its geographic distribution and its morbidity.

4.3.5 Prevention

As with other tick-borne infections, North Asian tick typhus prevention consists of preventive measures against tick–human contact: wearing protective clothing in natural foci, checking one's body for timely tick discovery and insect repellent use. Visits to natural foci are to be avoided during peaks in tick activity.

4.4 Tularemia

Tularemia is a highly dangerous natural focal zoonotic infection. Tularemia's causal agent is extremely pathogenic in humans: the appearance of 10–50 bacteria in the body leads to the development of disease. As a result, this is a Category A priority pathogen, listed among potential bioweapons and agents of bioterrorism (Mescheryakova 2010).

4.4.1 General Information

Tularemia is caused by the *Francisella tularensis* bacterium. According to today's classification, this bacterium is the only member of the Francisellaceae family. This pathogen is highly persistent, can withstand low temperatures and high humidity for long periods of time (it survives at −30 °C and can remain in ice for up to 10 months). It is less tolerant to desiccation (it remains in the pelts of rodents who died of tularemia for up to 1.5 months at room temperature (20–25 °C) and up to 1 week at 30 °C). It remains viable at a temperature of 10 °C for up to 9 months in freshwater and 2.5–4 months in soil. It is less tolerant of high temperatures (it dies within 5–10 min at 60 °C and 1–2 min at 100 °C) (Olsufiev and Rudnev 1960; Mead and Petersen 2009).

The tularemia pathogen varies depending on what part of its range it is located, allowing it to be divided into individual subspecies differing in virulence and certain biochemical properties. Today, the following four subspecies of the tularemia bacterium have been identified: the non-arctic *Fr. tularensis tularensis* and Holarctic *Fr. t. holarctica*, which are most common, as well as the North American *Fr. t. novicida* and Central Asian *Fr. t. mediaasiatica* (Svensson et al. 2005; Sjostedt 2007).

The non-arctic subspecies is widespread in the USA and Canada. In humans, it is lethal (without medical intervention) in 5–6% of cases transmitted by direct contact. The Holarctic subtype is common in Europe and Asia, as well as North America, and it is lethal in less than 0.1% of cases transmitted by direct contact, i.e., it is less lethal than the non-arctic subspecies.

(a)

(b)

Fig. 4.11 Tularemia's main hosts: the common vole (**a**) and the European water vole (**b**). *Credit* A. фото Dieter TD, https://commons.wikimedia.org/wiki/File:Feldmaus_Microtus_arvalis.jpg. This file is licensed under the Creative Commons Attribution-Share Alike 3.0 Unported license. B. фото Nick Ford www.flickr.com

4.4.2 Epidemiology and Natural Focus Characteristics

Tularemia is characterized by humans who have nearly 100% susceptibility to the disease, as well as a multiplicity of transmission modes:

- Vector-borne: via the bites of infected blood-sucking arthropods (mosquitoes, horseflies, ticks)
- Direct contact: via damaged skin and mucous membranes while touching infected humans or dead rodents, rabbits, etc.
- Oral: via ingestion of infected food and water that did not undergo preventive heat treatment.
- Airborne: via inhalation of droplet nuclei produced by grain processing and relocation of hay and straw infected by diseased rodents, as well as inhalation of droplet nuclei caused by washing and chopping of animal feed contaminated by the waste products of rodents infected with tularemia.
- Water-borne: via damaged skin while swimming in contaminated bodies of water Mammals are the main link in tularemia's epizootic network. Animals are divided into three groups based on their susceptibility to tularemia (Olsufiev and Dunaeva 1970).

The first group consists of mammals that are highly susceptible and sensitive to tularemia: the European water vole, the muskrat, every species of small mouse-like rodent, except for the striped field mouse, lagomorphs and eulipotyphla, except for hedgehogs, water shrews and Russian desmans (Fig. 4.11).

The second group consists of highly susceptible but not sensitive mammals: striped field mice, grey, black and short-tailed bandicoot rats, five spermophilus species (the little ground squirrel, red-cheeked ground squirrel, russet ground squirrel, Asia Minor ground squirrel and the long-tailed ground squirrel), squirrels, chipmunks, forest dormice, water shrews, lesser white-toothed shrews and European hedgehogs.

The third group consists of minimally susceptible and practically insensitive mammals: the majority of carnivorous mammals and livestock (sheep, cattle, horses, pigs, reindeer and camels), and domesticated cats and dogs.

4.4.3 Historical Background

George McCoy and Charles Chapin first described the tularemia pathogen in 1912. The microorganism was named *Bacterium tularense* as it was derived from zoological material found near Tulare Lake in California, USA (Fig. 4.12).

Fig. 4.12 Tulare lake region (California, USA). *Credit* Фото J. Stephen Conn (https://www.flickr.com/photos/jstephenconn/7936654338/in/photostream) Attribution-NonCommercial 2.0 Generic (CC BY-NC 2.0)

Fig. 4.13 Tularemia's first researchers: George McCoy (1876–1952), Charles Chapin (1856–1941), and Edward Francis (1872–1957) https://www.aai.org/About/History/Notable_Members/Presidents/McCoy_George.html https://ihm.nlm.nih.gov/luna/servlet/detail/NLMNLM11101411857171459:Charles-V–Chapin,-M-D-?qvq=w4s:/what/Portraits/;lc:NLMNLM11&mi=1043&trs=25660# https://ihm.nlm.nih.gov/luna/servlet/detail/NLMNLM11101415192177584:-Edward-Francis-

Chapin and his assistant were infected while working in the lab, proving that the pathogen posed a danger to humans. In 1925, H. Ohara isolated the same microbe, which caused a disease which he called as *yato-byo,* 'rabbit fever'. Later on, Edward Francis concluded that these microbes were the same and reported that the infectious disease now termed as tularemia is transmitted to humans from rodents by insects. This researcher's name is now included in the Latin name of the pathogen (*Francisella*), as well as one of the synonyms of tularemia (Francis' disease) (Fig. 4.13).

4.4.4 Distribution in Russia and the World

Tularemia natural foci are located in a variety of terrestrial ecosystems and are associated with various landscapes. The persistence of natural foci depends on the following factors: the high ecological plasticity of the tularemia pathogen, its reliance on multiple hosts and vectors, its ability to remain in the outside environment for long periods of time, especially at low temperatures, its hydrophilicity (the ability to survive in water for long periods of time and potentially to inhabit aquatic invertebrates), its persistence (ability to remain functionally active) in the organisms of highly sensitive animals, its ability to remain in Ixodidae ticks throughout metamorphosis and its ability to survive in unfavourable conditions in a dormant state.

Six main landscape types of natural foci have been identified in the Russian Federation: steppe, forest, tundra, field and meadow, floodplain and swamp, brook and synanthropic (urban) (this type of focus is distinct from the others) (Olsufiev and Dunaeva 1970; Mescheryakova 2003).

These various landscapes are not equivalent in the favourability of conditions created for natural foci: the forest steppe and steppe, floodplains, river deltas, lake coasts and swamps are ideal, while most foci in the tundra, taiga and deserts are linked to the intrazonal landscapes of river valleys or lake coasts (Fig. 4.14).

Tularemia as a disease affecting wild animals and humans has been verified within the northern hemisphere in Europe, Asia and America. The pathogen has been reportedly discovered in South America (Venezuela) and Africa (Tunisia), but these cases are thought to have been acquired on other continents (Olsufiev and Dunaeva 1970; Sjostedt 2007).

Tularemia was first discovered in the USSR near Astrakhan in 1926, when pathogen cultures were isolated from infected people. However, there is a reason to believe that the disease had been seen before that but assumed to be a different disease (for instance, as a mild form of the plague (*Pestis ambulans*)). For instance, an outbreak of a disease that caused inflammation of the lymphatic glands in 1877 has been retroactively identified as tularemia. Two disease outbreaks, in Tyumen (1921) and Voronezh (1925) are also thought to have been caused by tularemia. The pathogen itself was isolated in 1926 by M. M. Voronkova, S. V. Suvorov and A. A. Volfera (Khatenever 1943; Olsufiev and Dunaeva 1970). In the USSR, in the 1940s, N. A. Gaisky and B. Y. Elbert created a live tularemia vaccine, which remains the main method of tularemia prevention to this day.

Tularemia has been found in every region of Russia (Figs. 4.15, 4.16, 4.17). Beginning in 1950 (the first year in which the disease was officially reported in the USSR), tularemia morbidity has tended to decrease, but an outbreak occurred in 1957, totalling 5488 cases. The outbreak is explained by the mass reproduction of the European water vole in USSR's European territory and insufficient vaccination rates. Every following year, morbidity decreased or remained stable at 100–200 cases per year. Rarely (in 1978,

Fig. 4.14 Typical tularemia terrain. *Credit* N.G. Kadetov (permission obtained)

1981, 1993, 1995) small tularemia outbreaks occurred (totalling fewer than 500 cases a year). 75% of all cases occurred in the central and northern parts of Russia's European territory, as well as West Siberia (Olsufiev and Dunaeva 1970; Mescheryakova 2003; Popov et al. 2012).

In the early 2000s, the number of people infected with tularemia increased, thanks to a drop in planned vaccination and re-vaccination against tularemia, as well as low levels of immunity among urban dwellers in risk groups, late diagnoses and a lack of funding for tularemia prevention. As such, there were 123 tularemia cases in 2004, while two large outbreaks occurred in 2005 and 2013 (880 and 1063 cases, respectively) (Fig. 4.18).

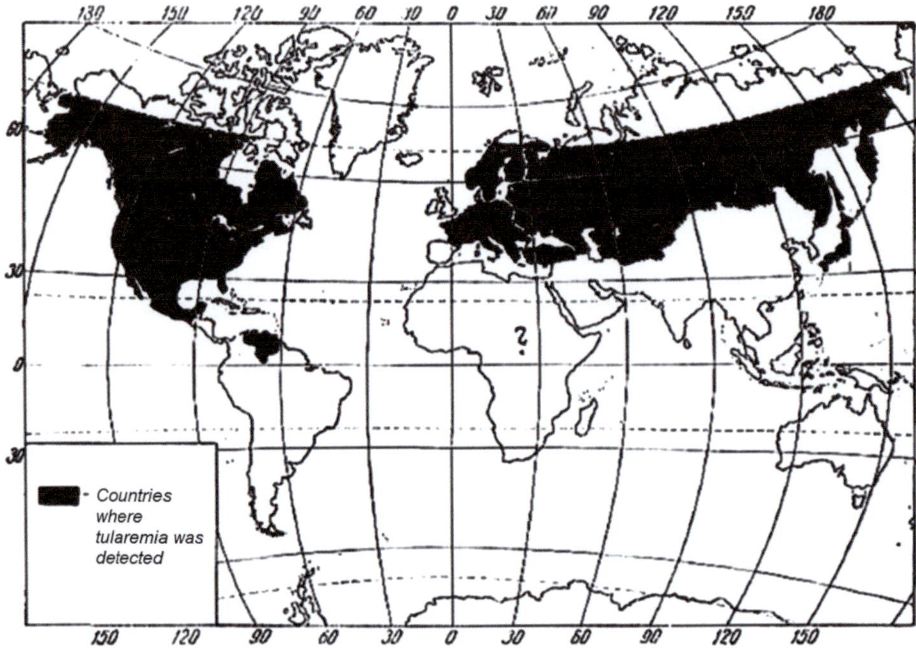

Fig. 4.15 World distribution of tularemia. *Credit* Geography of Natural focal Diseases… 1969

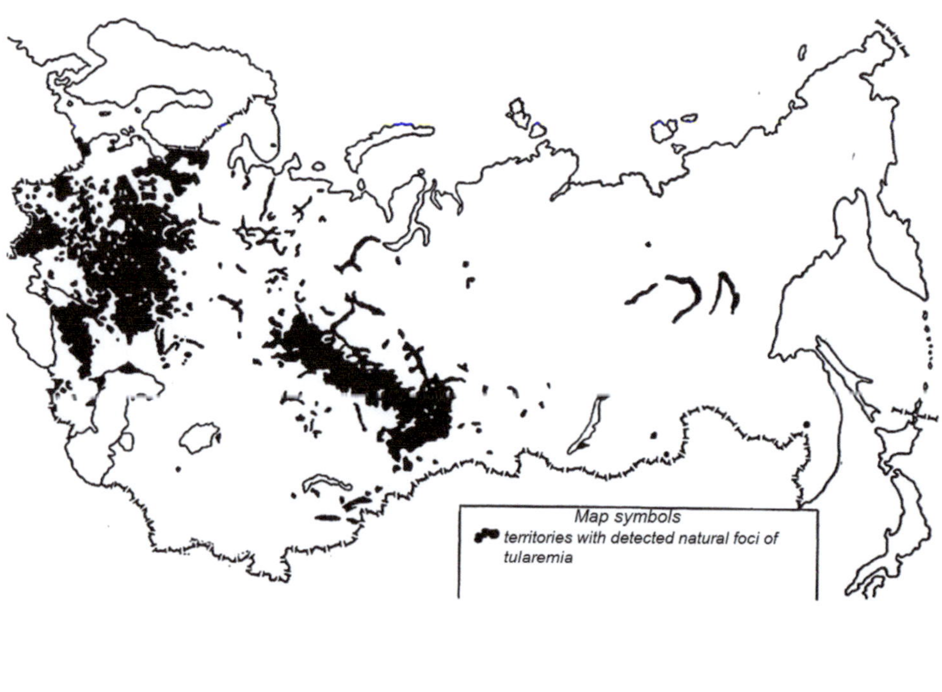

Fig. 4.16 Tularemia natural foci distribution in the USSR. *Credit* Geography of Natural focal Diseases… 1969

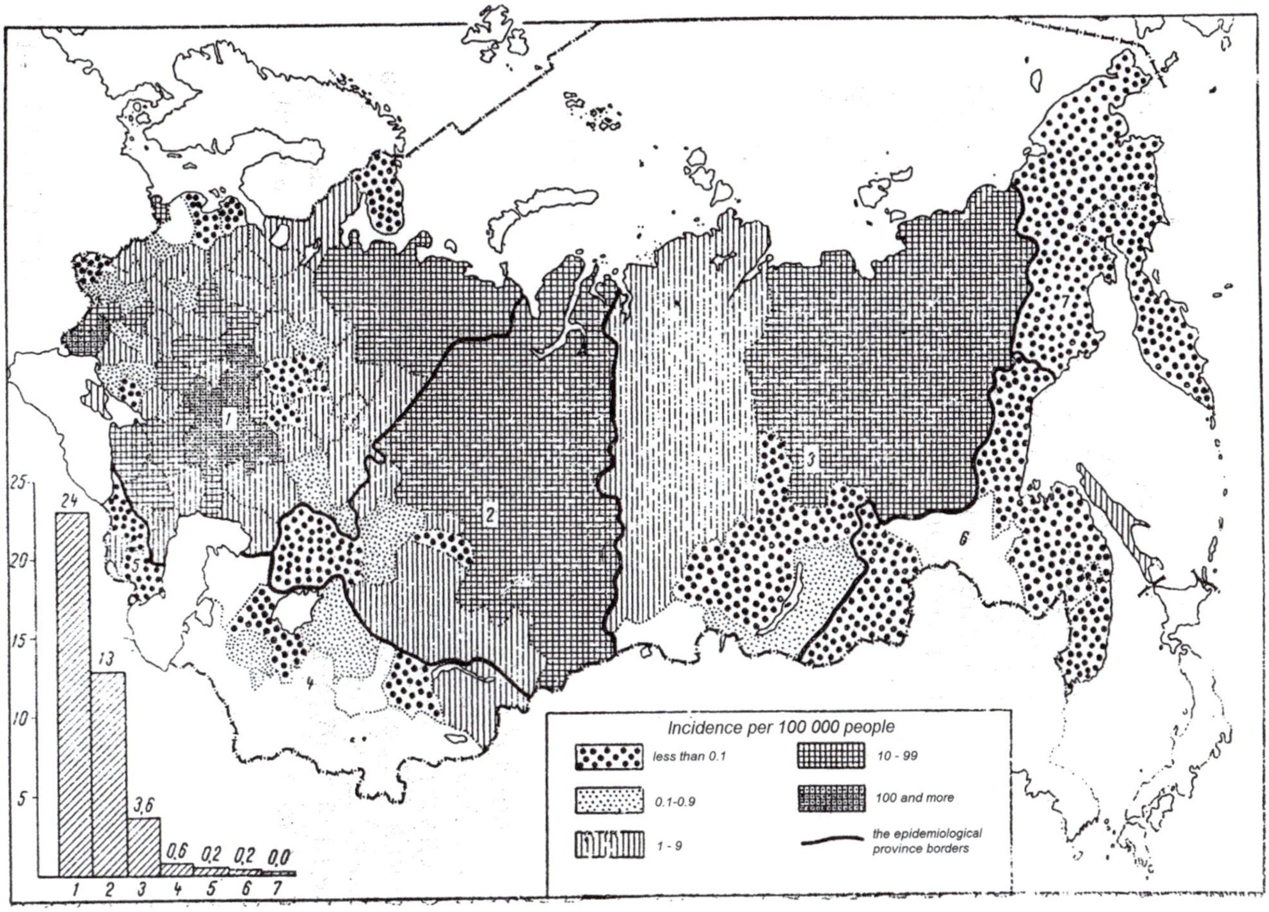

Fig. 4.17 Tularemia morbidity rates in the USSR, 1941–1966. *Credit* Olsufiev and Dunaeva 1970

Fig. 4.18 Reported cases of tularemia in Russia, 1950–2015

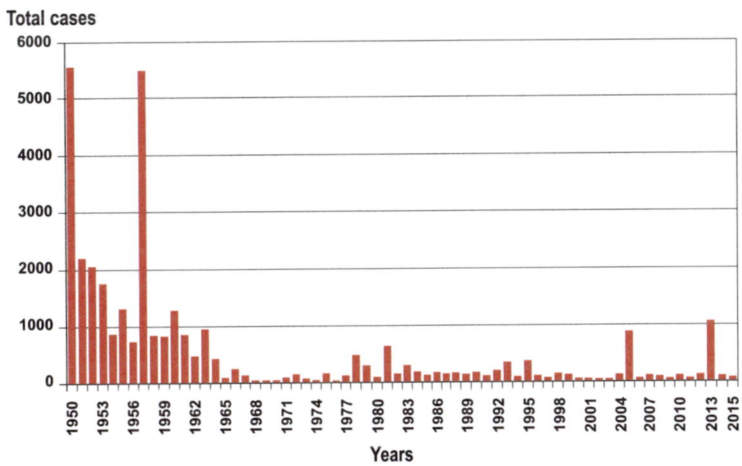

The long-term annual average (1997–2015) tularemia morbidity rate in Russia is 0.12 cases per 100,000 individuals. The highest morbidity rate (3.53 cases per 100,000 individuals) was reported in Khanty-Mansi Autonomous Okrug. This was caused by the large outbreak in 2013 (Map 4.9). Long-term annual average morbidity rates ranging from 0.20–1.00 have been reported in a large number of oblasts in the Central Federal District and in Arkhangelsk and Omsk Oblasts, as well as in Dagestan. No tularemia cases were reported in the Baikal region, in Zabaikalsky

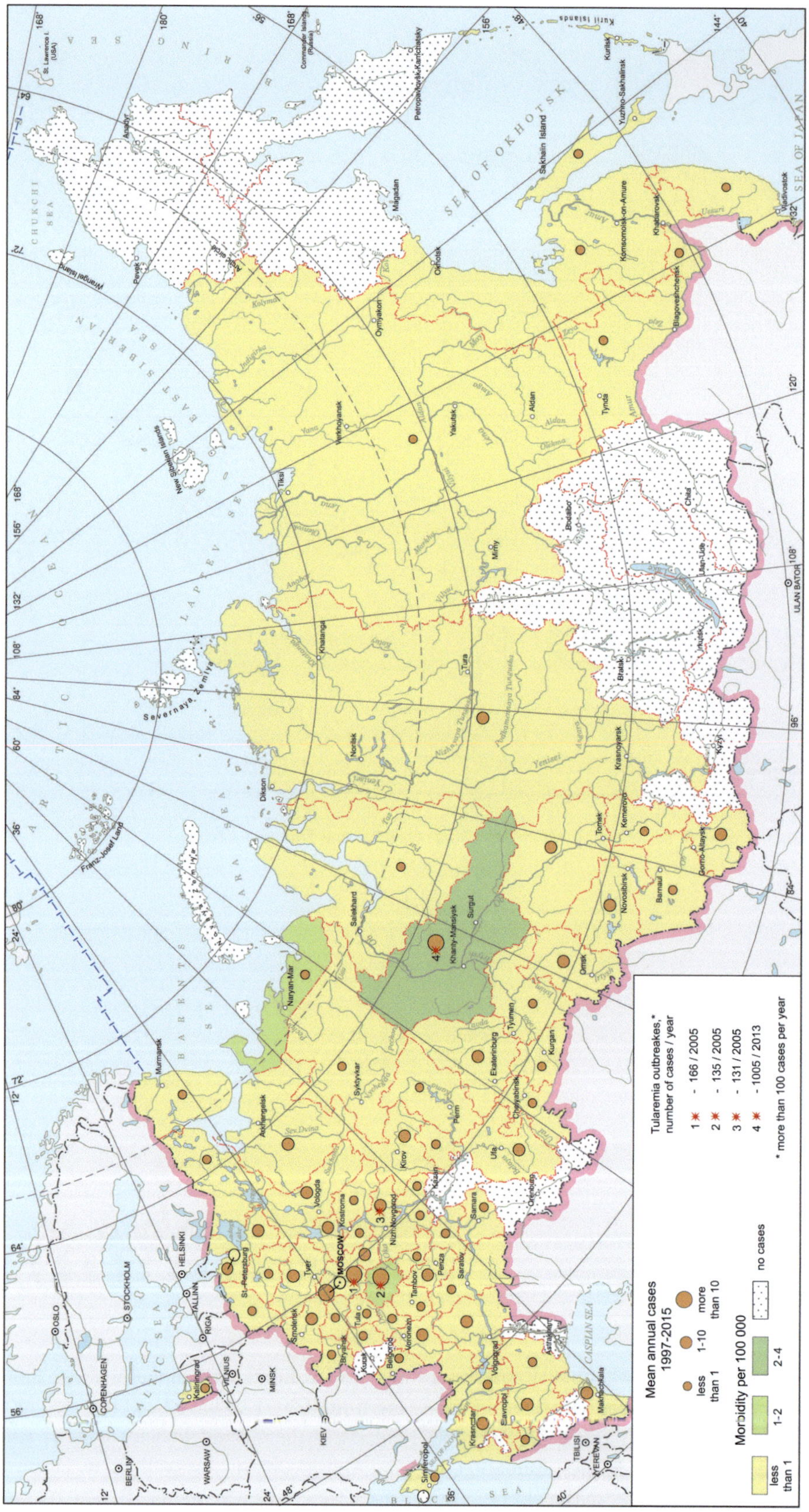

Map 4.9 Tularemia morbidity, 1997–2015

Krai, in the Russian Far East (Kamchatka and Chukotka), the Southern Urals, or in a number of North Caucasian republics between 1997 and 2015.

Tularemia morbidity rate dynamics in a variety of regions in the Russian Federation from 1997 to 2010 demonstrate long-term nonlinear oscillation (decrease, increase, oscillations with varying (both significant and insignificant) amplitudes) (Map 4.10). Explosive dynamics (characterized by sharp oscillations in the morbidity rate) are common in Arkhangelsk and Ryazan Oblasts, Nenets Autonomous Okrug and Altai Republic. Other administrative subjects with reported cases of tularemia have stable morbidity dynamics.

Tularemia morbidity dynamics are reflected on a dynamic cartogram (Map 4.11), which depicts three levels of long-term average annual morbidity for each administrative subject: low (less than 1.0 cases per 100,000 individuals), moderate (1.0–2.0 cases per 100,000 individuals) and high (over 2.0 cases per 100,000 individuals), for each year from 1997 to 2013. Based on the distribution of morbidity rates, three groups of regions can be identified: regions with sporadic morbidity, which is always low (these are primarily regions in the Russian Far East, East Siberia and some regions in West Siberia and near the Volga River), regions, where moderate morbidity rates are reported almost annually (southern West Siberia, most regions in Russia's European territory), and regions with consistently high morbidity rates reported nearly annually (a number of oblasts in the Central Federal District, as well as Arkhangelsk, Vologda and Nizhny Novgorod Oblasts and Altai Republic). Tularemia's tendency towards outbreaks is demonstrated by a juxtaposition of its morbidity rates in three consecutive years (2011, 2012, 2013). During this period, cases were reported in 39 administrative subjects of the Russian Federation. However, it is clear that they are distributed unevenly, and that morbidity rates varied significantly from 1 year to another. For instance, morbidity rates were high in southern Siberia in 2011 (Tomsk, Omsk, Novosibirsk, Kemerovo Oblasts) and in various regions in Russia's European territory, especially in the north, in 2012 (Arkhangelsk, Vologda, Kirov Oblasts). Meanwhile, the long-term average annual morbidity rate in these administrative subjects was indistinguishable from that of other regions. It must be noted that the overall number of cases during these outbreaks, despite its divergence from the long-term annual average, does not exceed 10–20 cases.

In this context, the 2013 tularemia outbreak in Khanty-Mansi Autonomous Okrug, one of the largest outbreaks since the mid- twentieth century, is quite unusual. It spanned the territory of both city of Khanty-Mansiysk and Khanty-Mansiysky District and totalled over 1000 cases (6.39 per 100,000 individuals). The outbreak was caused by a combination of factors, the main two of which were favourable conditions for the breeding of mosquitoes in the *Aedes* genus and high population numbers of tularemia's main carrier, the European water vole (Mescheryakova et al. 2014). The mosquito's ability to suck blood multiple times over the course of one season allowed for vector-borne transmission of the pathogen from infected rodents to humans. This was exacerbated by more dangerous epizootic conditions in the water vole population. Aside from this, low vaccination rates, low population immunity rates, unsatisfactory rodent extermination procedures in the natural focus and ineffective anti-mosquito interventions all contributed to the outbreak (Situation… 2013). Therefore, tularemia's multi-host and multi-vector nature, the complexity of its parasitic systems, and the dependence of natural foci on a number of natural and anthropogenic factors confirm tularemia's position among especially dangerous and unpredictable infections.

Despite anti-tularemia campaigns, the infection's natural foci continue to actively function throughout the Russian Federation and pose a constant epidemiological risk.

4.4.5 Prevention

Epizootic and epidemiological tularemia control includes constant monitoring of morbidity rates among both humans and animals in tularemia natural foci and pathogen circulation between animals and blood-sucking arthropods, as well as control of human immunity rates (Epidemiological monitoring… 2005).

Neutralization of tularemia pathogen sources, transmission factors and disease hosts, as well as vaccination, serve as the foundation of tularemia prevention. Preventive measures vary depending on the method of transmission. Vector-borne transmission can be prevented with insect repellents, protective clothing and limiting access to unfavourable territories for unvaccinated populations. Oral transmission can be prevented by avoiding swimming in freshwater and boiling drinking water. Hunters must disinfect their hands after skinning and gutting rabbits, muskrats, moles and water voles. Rodent extermination and disinfection campaigns are also quite significant (Epidemiological monitoring… 2005; WHO 2007).

Populations living in regions where tularemia is endemic should be vaccinated. As was mentioned earlier, the live tularemia vaccine that is used was developed by N. A. Gaysky and B. Y. Elbert in the 1940s. Vaccination leads to persistent long-term (5–7 year) immunity. Booster vaccines are administered every 5 years.

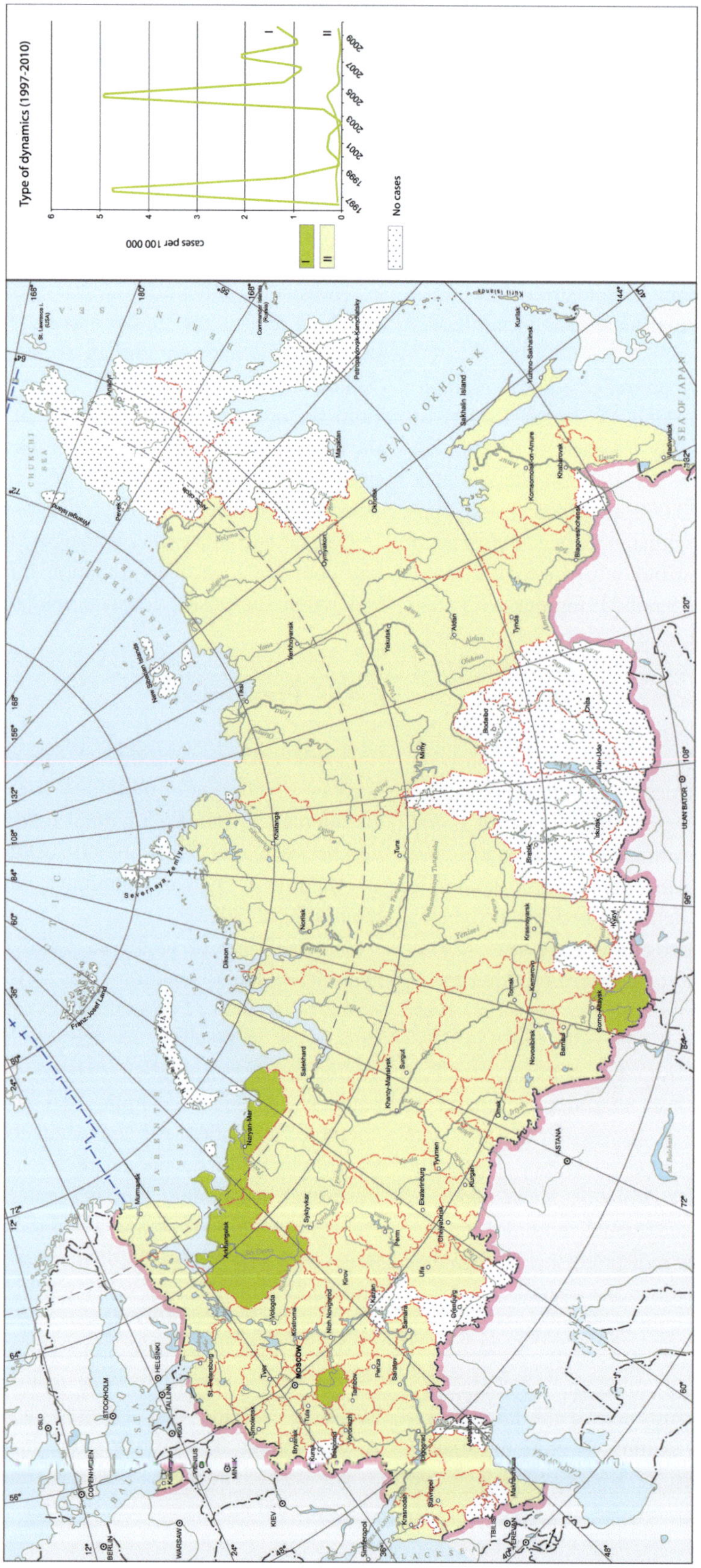

Map 4.10 Types of tularemia morbidity dynamics, 1997–2010

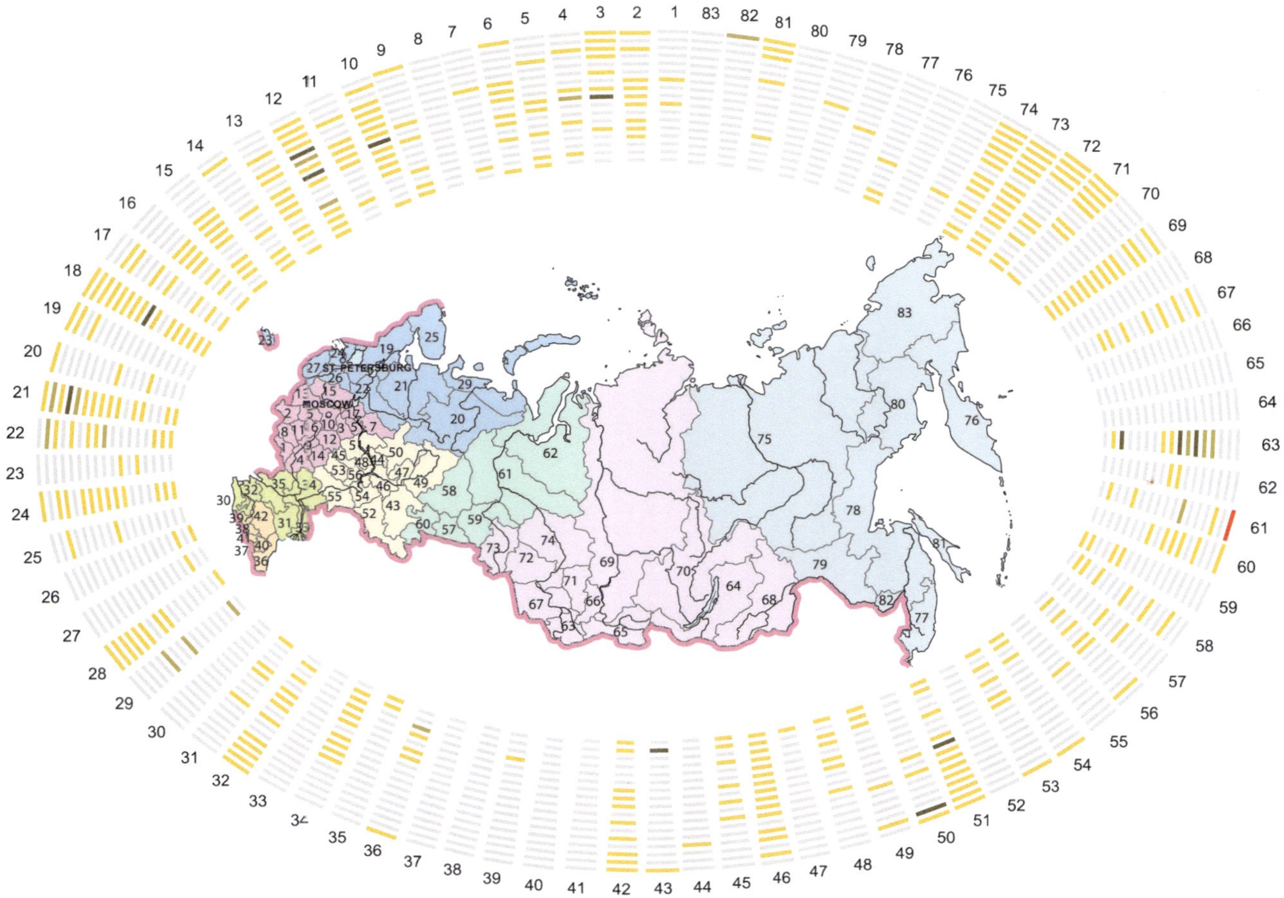

Morbidity per 100 000
- less than 1,0
- 1,0 - 2,0
- more than 2,0 (max 63,9; 2013)
- no cases

2012 2013
2010 2011
2008 2009
2006 2007
2004 2005
2002 2003
2000 2001
1998 1999
1997

1.....83 region number in legend and map

Central Federal District
1 Belgorod Oblast
2 Bryansk Oblast
3 Vladimir Oblast
4 Voronezh Oblast
5 Ivanovo Oblast
6 Kaluga Oblast
7 Kostroma Oblast
8 Kursk Oblast
9 Lipetsk Oblast
10 Moscow Oblast
11 Oryol Oblast
12 Ryazan Oblast
13 Smolensk Oblast
14 Tambov Oblast
15 Tver Oblast
16 Tula Oblast
17 Yaroslavl Oblast
18 Moscow city

Northwestern Federal District
19 Republic of Karelia
20 Komi Republic
21 Arkhangelsk Oblast
22 Vologda Oblast
23 Kaliningrad Oblast
24 Leningrad Oblast
25 Murmansk Oblast
26 Novgorod Oblast
27 Pskov Oblast
28 Saint Petersburg city
29 Nenets Autonomous Okrug

Southern Federal District
30 Republic of Adygea
31 Republic of Kalmykia
32 Krasnodar Krai
33 Astrakhan Oblast
34 Volgograd Oblast
35 Rostov Oblast

North Caucasian Federal District
36 Republic of Dagestan
37 Republic of Ingushetia
38 Kabardino-Balkar Republic
39 Karachay-Cherkess Republic
40 Republic of North Ossetia-Alania
41 Chechen Republic
42 Stavropol Krai

Volga Federal District
43 Republic of Bashkortostan
44 Mari El Republic
45 Republic of Mordovia
46 Republic of Tatarstan
47 Udmurt Republic
48 Chuvash Republic
49 Perm Krai
50 Kirov Oblast
51 Nizhny Novgorod Oblast
52 Orenburg Oblast
53 Penza Oblast
54 Samara Oblast
55 Saratov Oblast
56 Ulyanovsk Oblast

Ural Federal District
57 Kurgan Oblast
58 Sverdlovsk Oblast
59 Tyumen Oblast
60 Chelyabinsk Oblast
61 Khanty-Mansi Autonomous Okrug (Yugra)
62 Yamalo-Nenets Autonomous Okrug

Siberian Federal District
63 Altai Republic
64 Republic of Buryatia
65 Tyva Republic
66 Republic of Khakassia
67 Altai Krai
68 Zabaykalsky Krai
69 Krasnoyarsk Krai
70 Irkutsk Oblast
71 Kemerovo Oblast
72 Novosibirsk Oblast
73 Omsk Oblast
74 Tomsk Oblast

Far Eastern Federal District
75 Sakha Republic (Yakutia)
76 Kamchatka Krai
77 Primorsky Krai
78 Khabarovsk Krai
79 Amur Oblast
80 Magadan Oblast
81 Sakhalin Oblast
82 Jewish Autonomous Oblast
83 Chukotka Autonomous Okrug

Map 4.11 Long-term tularemia morbidity dynamics, 1997–2013

4.5 Leptospirosis

Leptospirosis (Weil's disease) is a group of natural focal non-vector-borne diseases caused by *Leptospira* bacteria. The diseases have similar clinical signs and symptoms and epidemiological and epizootic characteristics (Ananyina and Petrov 2006).

4.5.1 General Information

Leptospirosis is caused by spirochetes in the Leptospiracae family, which, like other microorganisms in this group, are quite ecologically diverse and can thus colonize a variety of habitats (Ananyina 2009). Leptospira can be both non-pathogenic (saprobionts, including halophiles) and pathogenic, causing diseases in humans and animals. It is thought that free-living Leptospira are some of the most widespread oligotrophic microorganisms in the world.

Leptospira are corkscrew-shaped gram-negative microorganisms that can move forward and backward linearly and rotate. Their ends are typically bent like hooks (Pokrovskiy 1983). According to the current classifications, the *Leptospira* genus includes two species: *L. interrogans* sensu lato and *L. biflexa* sensu lato. The former includes pathogenic Leptospira, while the latter are non-pathogenic saprobionts that live in bodies of water (Ananyina 2010).

Pathogenic Leptospira are ecologically, antigenically and genetically diverse: today they are divided into 17 genomic types and approximately 250 serotypes based on their affiliation with specific hosts, organotropism (affinity for specific organs), ability to survive in the outside environment and the duration of that survival, and other important clinical and epidemiological characteristics (Ananyina 2010). Antigenically close serotypes are combined into serologic groups that do not have an independent taxonomic status (Cerquiera nad Picardeau 2009). In Russia pathogens in the *Pomona, Tarassovi, Grippotyphosa* and *Sejroe* groups are most common, while *Icterohaemorrhagiae* and *Canicola* are less common. In natural foci, the most widespread serologic groups are *Grippotyphosa, Pomona, Sejroe, Javanica* and *Icterohaemorrhagiae*, while in urban foci *Icterohaemorrhagiae* and *Canicola* are the most prevalent serologic groups (Ananyina and Petrov 2006).

Leptospira can survive for over 30 days in open freshwater and for up to 9 months in moist soil. They are hydrophiles and need high humidity and pH levels within 7.0–7.4 in order to survive (Pokrovskiy 1983). They tolerate low temperatures well and remain viable if they are frozen in ice for prolonged periods of time, but die quickly from heating, desiccation and direct sunlight.

Leptospirosis is one of the most clinically severe and fatal natural focal diseases, and it has some of the worst long-term complications. Leptospirosis causes fevers, general intoxication, and liver, kidney and nervous system damage. Hemorrhagic symptoms, such as nosebleeds and hemorrhagic rashes, are also common.

4.5.2 Epidemiology and Natural Focus Characteristics

Leptospira's main hosts are numerous moisture-preferring rodent species in natural foci and livestock and dogs in anthropurgic foci. The disease carries such great epidemiological and epizootic risk because of the close relationship between humans and wild and domesticated animals in the environment. The pathogen prefers environments with ample bodies of water and a significant number of days with non-freezing temperatures annually. Leptospira follow faecal–oral transmission mechanism via water. Humans are usually infected during contact (it can be rapid) between their skin or mucous membranes with water contaminated with animal waste (Lobzin 2000). Leptospira enter the host's organism directly through mucous membranes and wounded skin, even if the wound is extremely small. Infection occurs while people are swimming, fishing, doing agricultural work or construction and gardening.

Pathogenic Leptospira continuously circulate in nature thanks to their ability to cause chronic infections in rodents. Rodents can carry these parasites throughout their entire lifetimes.

Pathogenic Leptospira are host specific, i.e. pathogens of specific serotypes circulate in populations of the same species or a limited number of species, which serve as the infection's main reservoir (the *copenhageni* serotype in the brown rat, *hardjo* in cattle, *canicola* in dogs, *tarassovi* in pigs and cattle, *grippotyphosa* in *Microtus* voles (Fig. 4.19), cattle, etc.) (Ananyina and Petrov 2006). Muridae and Cricetidae play a crucial role in preserving and spreading leptospirosis (Ananyina 2003). Various moisture-preferring rodent species—tundra voles, *Microtus* voles, striped field mice and house mice—are the main carriers of Leptospira in natural foci in Russia. However, other species participate in the epizootic process quite often and serve as additional sources of the pathogen.

As for many other natural focal diseases, in recent years leptospirosis morbidity has been characterized by an increase in the fraction of the urban population among infected individuals. As mentioned above, the *Icterohemorrhagiae* and *Canicola* serologic groups are most prevalent in urban foci. Their main sources and reservoirs are brown rats and dogs respectively. Persistent and active leptospirosis natural foci can found in urban parks and woodlands, near ponds, along gutters covered in plants, etc. In these foci, striped field mice, common voles, common shrews and other small mammals serve as the main infection reservoirs.

Fig. 4.19 The field vole (*Microtus agrestis*), which carries the most common *Leptospira* serotype in Russia, *grippotyphosa*. *Credit* Фото David Perez, https://ru.wikipedia.org/wiki/%D0%A2%D1%91%D0%BC%D0%BD%D0%B0%D1%8F_%D0%BF%D0%BE%D0%BB%D1%91%D0%B2%D0%BA%D0%B0#/media/File:Microtus_agrestis_02_by-dpc_(cropped).jpg. David Perez - собственная работа, CC BY 3.0, https://commons.wikimedia.org/w/index.php?curid=42999353

Thanks to human agricultural activity, leptospirosis aetiology is constantly changing. As a result, in Russia in the 1980s there was an increase in highly virulent pathogens of the *Icterohaemorrhagiae* serologic group, whose main reservoir is the brown rat (Fig. 4.20). The growth of the brown rat population was directly linked to an anthropogenic activity that led to an increase in brown rat populations both in rural and urban terrain (Karaseva et al. 1986; Ananyina and Petrov 2006). Most people who become infected in such foci come into contact with synanthropic rodents or environments they contaminated due to their occupations. This includes gardeners, hikers, fishermen and hunters. Beginning in the 1990s, a type of leptospirosis caused by bacteria in the *Canicola* serologic group, which is primarily linked to dogs, became more common in people. This was caused by an increase in feral and domesticated dog populations. However, cattle and pigs, aside from dogs, were also involved in the epizootic process.

Leptospirosis usually causes sporadic morbidity combined with local outbreaks, which are usually related to swimming in prohibited areas. These are usually zones where livestock drinks, grazes or is vaccinated against leptospirosis (On the status… 2012).

4.5.2.1 Historical Background

Research on leptospirosis began in the late nineteenth century after the German physician A. Weil described four cases of acute liver diseases that affected the kidneys and spleen. He categorized them as an individual group. In 1888, the Russian physician Nikolai Vasiliev described a few cases of a fever that attacked the liver, kidneys and brain simultaneously. Later, this disease was called Vasiliev-Weil Disease (or Weil Disease) (Fig. 4.21). The pathogen was isolated in 1914 by a Japanese researcher, A. Inado, and his co-authors.

In 1927, in the USSR, Viktor Bashenin described a new nosologic unit, which he termed water fever. He had studied mass outbreaks of the disease in the Dubna River floodplain. Water fever turned out to be an independent form of

Fig. 4.20 The brown rat (*Rattus norvegicus*), the primary carrier of *Leptospira* in anthropurgic foci. *Credit* http://www.geograph.org.uk/photo/1585934 © Copyright Cameraman and licensed for reuse. Attribution-ShareAlike 2.0 Generic (CC BY-SA 2.0)

Fig. 4.21 A. Weil (1848–1916) and N. P. Vasiliev (1852–1891). *Credit* http://vkjournal.ru/doc/394810https://en.wikipedia.org/wiki/Adolf_Weil_%28physician%29#mediaviewer/File:Adolf_Weil.png

leptospirosis that did not cause liver disease. Its pathogen, *L. i. grippotyphosa* was discovered in 1928 by S. P. Tarasov and German Epstein. It was later discovered that this etiologic form of leptospirosis is most common in the former USSR and Europe (Ananyina 2003). *L. i. canicola* was isolated in 1931 in Holland, while *L. i. Pomona* was isolated in 1936 in Australia. Throughout the entire twentieth century, new pathogens were discovered within the *L. interrogans* species. In the late 1990s, the bacterium's DNA was sequenced and it was attributed to one of the most ancient phylogenetic branches of Eubacteria, bifurcating at the root of evolutionary pathways. After this, the International Committee on Systematics of Prokaryotes accepted an alternative system of *Leptospira* species differentiation. It was based on a new understanding of the species in Eubacteria. As a result, 17 genomic types of *Leptospira* were identified (Ananyina 2009).

Longitudinal studies beginning in 1950, conducted by V. V. Ananyin and E. V. Karaseva in the leptospirosis natural focus on Lake Nero (Fig. 4.22) demonstrated that leptospirosis can be analysed using the basic principles of E. N. Pavlovsky's natural focus theory (leptospirosis was the first identified group of non-vector-borne natural focal zoonotic diseases). In the 1960s and 1970s, after the role of 'wild' natural foci in leptospirosis epidemiology decreased, their anthropogenic transformation in modern conditions became a subject of interest (Karaseva 2000; Ananyina 2003).

4.5.3 Distribution in Russia and the World

Leptospirosis is believed to be one of the most widespread groups of zoonotic diseases in the world. It is found everywhere, on every continent except Antarctica, but it is especially common in humid subtropical and tropical climates, where Leptospira is most diverse. Some nosological entities are severe and highly lethal. Leptospirosis management has become more complicated in recent decades, making it possible to categorize it as a re-emerging infection (Hartskeerl et al. 2011). There is a tendency towards leptospirosis globalization, which has been linked to intensifying human migration, including international tourism (Bandara et al. 2014).

Leptospirosis is found in the majority of European countries. On average, the morbidity rate is 1.5 cases per 100,000 individuals. It has been noted that since the mid-1990s morbidity has increased in industrialized countries compared to agricultural societies: the disease's clinical course has become more severe and cases of infection are more common in urban populations. Mechanization of agriculture and widespread vaccination of livestock on the one hand and increased contact with pets that are Leptospira carriers and increasing aquatic tourism rates on the other contributed to this shift (Pavlenko 2013).

In Russia, the nosogenic territory of leptospirosis spans 74 administrative subjects (Map 4.12). The disease is spread by outbreaks. The highest long-term average annual

Fig. 4.22 Lake Nero in Yaroslavl Oblast, where a leptospirosis natural focus has been studied since 1950. *Credit* V. F.Lurie (permission obtained)

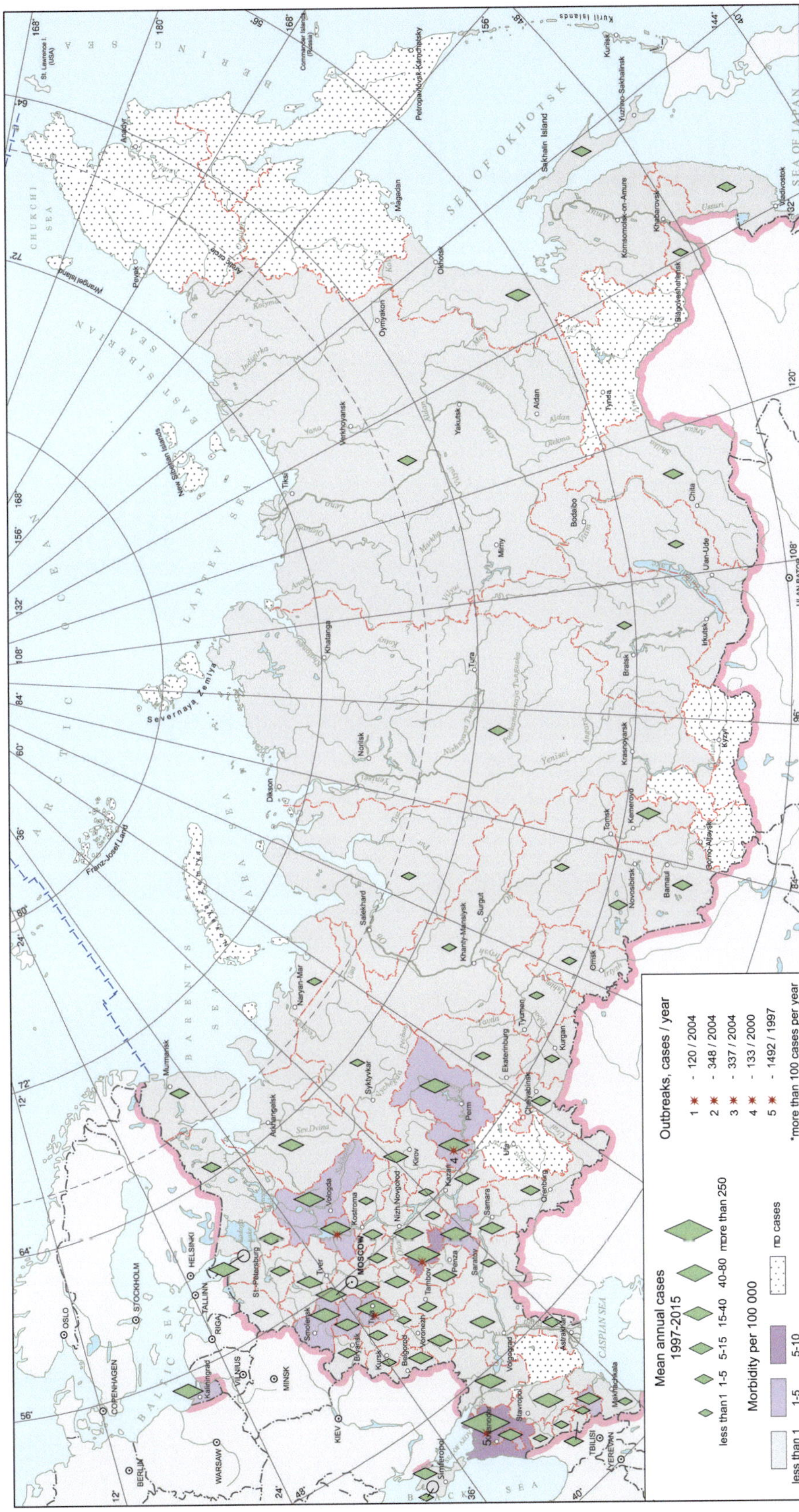

Map 4.12 Leptospirosis morbidity, 1997–2015

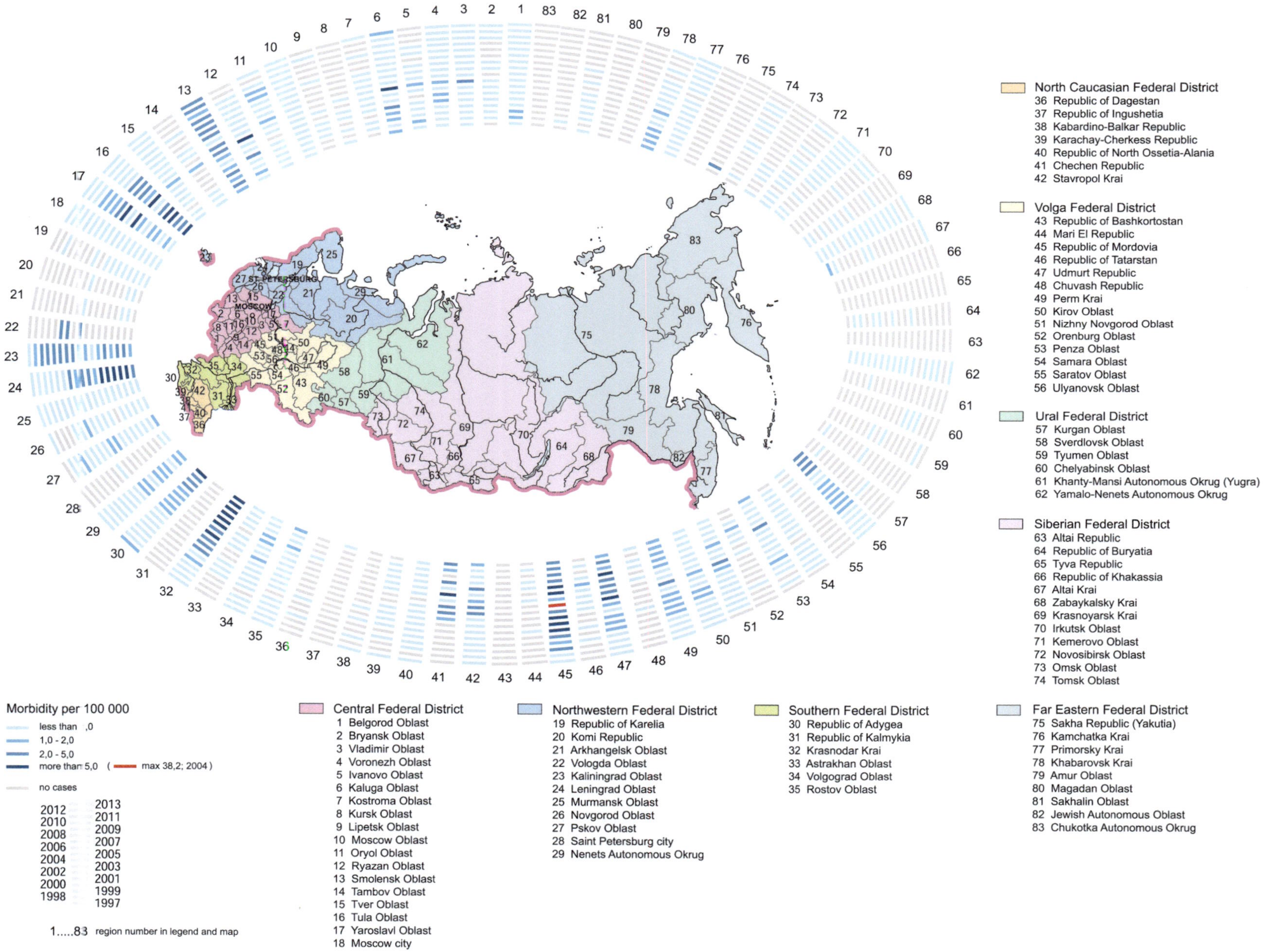

Map 4.13 Long-term leptospirosis morbidity dynamics, 1997–2013

morbidity rate (6.7–8.2 cases per 100,000 individuals) was reported in two regions: Mordovia and Krasnodar Krai. This occurred primarily because of outbreaks in 1997 (29.7 cases per 100,000 individuals in Krasnodar Krai) and 2004 (38.3 cases per 100,000 individuals in Mordovia). High morbidity rates have also been found in Russia's European territory: in Kaliningrad, Tula, Smolensk and Vologda Oblasts, as well as Udmurtia. In other regions, morbidity rates do not exceed 2 cases per 100,000 individuals.

Overall, throughout the period in question, leptospirosis morbidity in Russia has exhibited a tendency to decrease. This decrease is wave like, with insignificant spikes in 1997 (1.6 cases per 100,000 individuals), 2001–2002 (1.0 case per 100,000 individuals) and 2004 (1.7 cases per 100,000 individuals).

Annual leptospirosis morbidity rates for each administrative subject in Russia over a 17-year period (1997–2013) are reflected on a dynamic cartogram (Map 4.13). Based on the distribution of morbidity rates, regions with persistently annually reported high morbidity rates can be identified (Smolensk, Tula and Yaroslavl Oblasts in Central Russia, Kaliningrad and Leningrad Oblast in the northwest, and Krasnodar Krai in the south). Relatively high morbidity rates have also been reported in some republics in the Volga region.

Approximately 70% of leptospirosis cases occur in the urban population. The main reasons for this distribution are growth in urban populations, better diagnostics compared to rural areas and urban sprawl, which lead to contact with infected territories. Leptospirosis mortality is reported annually—from 2009 to 2011 there were approximately 40 lethal cases among adults.

Although leptospirosis morbidity rates are not high overall, local epizootics have been identified in animal populations. In the second half of 2011, leptospirosis epizootics occurred in 28 administrative subjects in Russia.

The public health service has identified the following urgent issues: cases of occupational infection with leptospirosis among agricultural workers, meatpackers and public utility workers, a decrease in rodent extermination campaigns, and low vaccination rates in high-risk groups.

4.5.4 Prevention

Leptospirosis prevention is based on a combination of medical and veterinary sanitation efforts. Rodent extermination campaigns are conducted and bodies of water are protected from animal waste in animal farms with high leptospirosis rates. To avoid infection, individuals working on these farms must wear protective clothing and use personal protective equipment. It is important to avoid swimming in unauthorized locations and small bodies of standing water. Watering holes for livestock should especially be avoided. Water from open bodies of water or stream should not be used for drinking or personal use. Personal hygiene must be carefully maintained while working with infected animals and when in a natural focus (www.rospotrebnadzor.ru).

High-risk groups (animal farm workers, vacuum truck (honeywagon) operators and exterminators) and people living in at-risk territories undergo vaccination. In human settlements primarily children are vaccinated, while in natural foci rice farmers, irrigation engineers and other at-risk individuals are vaccinated. Vaccines are administered 2 months before the start of the epidemic season. Dogs and livestock are vaccinated in at-risk territories.

4.6 Hemorrhagic Fever with Renal Syndrome

Hemorrhagic fever with renal syndrome (HFRS) (also known as Korean hemorrhagic fever, epidemic hemorrhagic fever and nephropathis epidemica) is a viral non-vector-borne zoonotic disease that is widespread around the world. In Russia, HFRS has one of the leading morbidity rates of all natural focal diseases.

4.6.1 General Information

HFRS is caused by a group of viruses called Hantaviruses, in the *Hantavirus* genus of the Bunyaviridae family. This genus includes over 40 genetically and serotypically distinct viruses, and their diversity is constantly increasing. Hantaviruses can be pathogenic or harmless to humans. The epidemiologic role of certain hantaviruses is yet to be identified. They have strong evolutionary ties with some small mammal species, primarily rodents (the Rodentia order) and insectivores (the Insectivora order). Some hantaviruses have immunologically and genetically distinct intra-species genotypes with different natural reservoirs (Tkachenko et al. 2013).

Today eight hantaviruses can be found on Russian territory: Hantaan, Amur, Dobrava-Belgrade (2 viral variants, DOBV-A and DOBV-B), Seoul, Puumala (PUUV), Khabarovsk, Tula and Topografov. Only the first five viruses have been identified as pathogenic to humans. Approximately, 98% of all HFRS cases in Russia are caused by PUUV (Bernstein et al. 2010).

Hantavirus transmission is typically airborne: warm-blooded hosts that excrete the virus into the environment serve as the source of infection for humans. HFRS has a variety of clinical presentations, which depend on the specific hantavirus the individual is infected with (Lee 2003). This can be partially explained by their varied virulence (Lvov et al. 1989). The main clinical symptoms of HFRS are fever, general intoxication and malaise (chills, aches, etc.), acute kidney failure, acute shock, low blood pressure and vascular leakage.

4.6.2 Epidemiology and Natural Focus Characteristics

Rodents are the main natural reservoir of hantaviruses, especially members of the Muridae and Cricetidae families.

The Hantaan, Amur and Seoul viruses primarily circulate in the Russian Far East; DOBV and PUUV are found in Russia's European territory (Mikhailova et al. 2001). Hantaviruses' distinct characteristic is that each individual hantavirus has its own host. In foci in Asia and the Russian Far East, the Asian subspecies of the striped field mouse (*Apodemus agrarius mantchuricus*) plays the main role in spreading and maintaining Hantaan virus. Amur virus is hosted by the Korean field mouse (*A. peninsulae*). In foci in Russia's European territory, Puumala virus and Dobrava-Belgrade virus are hosted by the bank vole (*Myodes* (formerly *Clethrionomys*) *glareolus*) (Fig. 4.23), the western subspecies of the striped field mouse (*A. agrarius agrarius*), and the Black Sea field mouse (*A. ponticus*) (Mikhailova et al. 2001, Tkachenko et al. 2013).

Today, new hantaviruses have been discovered that are linked to other groups of mammals, namely insectivores and bats. Nonetheless, current public understanding connects hantaviruses to rodents specifically, especially given how well studied this connection is. Incontrovertible evidence of host-to-host hantavirus transmission only exists for rodents. As of today, the role of other animals in preserving and spreading the virus has not been proven (Bernstein et al. 2001, 2010).

Fig. 4.23 The bank vole (*Myodes glareolus*), the natural reservoir of Puumala virus (PUUV). *Credit* A. A.Kadetova (permission obtained)

Lab and field research has shown that HFRS is a chronic asymptomatic infection in rodents that have no noticeable effect on their lives (growth, reproduction, activity, lifespan) (Tkachenko et al. 2012). Cases of rodents carrying the virus for their entire lives (up to 15 months) have been identified (Bernstein et al. 2010). The pathogen enters the environment through rodents' saliva, urine and faeces. Rodent–rodent transmission is usually airborne (during contact between animals or contact with animal nests).

The diversity of ecological ties between HFRS pathogens, its carriers and humans is shown on a diagram (Fig. 4.24). The HFRS parasitic focal system includes the pathogen, rodents and humans. Links between different components of the system are marked with arrows. Environmental conditions (terrain specifics, climate conditions, fauna, feed availability and amount) define rodent ecology and their population's status (density and age structure), which in turn influences the rodent population and its behaviour (including levels of rodent–rodent and rodent–human contact) in a given year. Evidently, the hosts' genetic characteristics are responsible for their susceptibility to hantaviruses. The virus' presence and amount in a given focus depend on environmental factors linking to microclimatic and geochemical parameters that influence the infection rate among rodents. Finally, social factors are responsible for the level and length of human population contact with natural foci. Working together, these factors define humans' risk of infection in a specific natural focus (Reusken and Neyman 2013).

Human contact with the infection occurs either in natural conditions or when infected rodents migrate into human settlements. Human infection usually occurs via an airborne pathway (while people are taking care of pets, cleaning rooms, visiting the woods or, sometimes, when they are working with infected rodents or materials—hay, straw, etc.). Faecal–oral transmission is also possible (when unwashed food that has had contact with infected rodents is eaten), but its existence has yet to be proven. An infected human is not contagious to other humans.

HFRS morbidity in PUUV foci in Russia is pretty clearly seasonal, thanks to drops in rodent populations in the winter and decreases in rodent–human contact. Most human infections occur in the late summer and fall, peaking in September and October. In other hantavirus, foci morbidity peaks in the winter.

Given that HFRS is closely linked to its main carriers, its spread does not exceed their ranges and depends on the

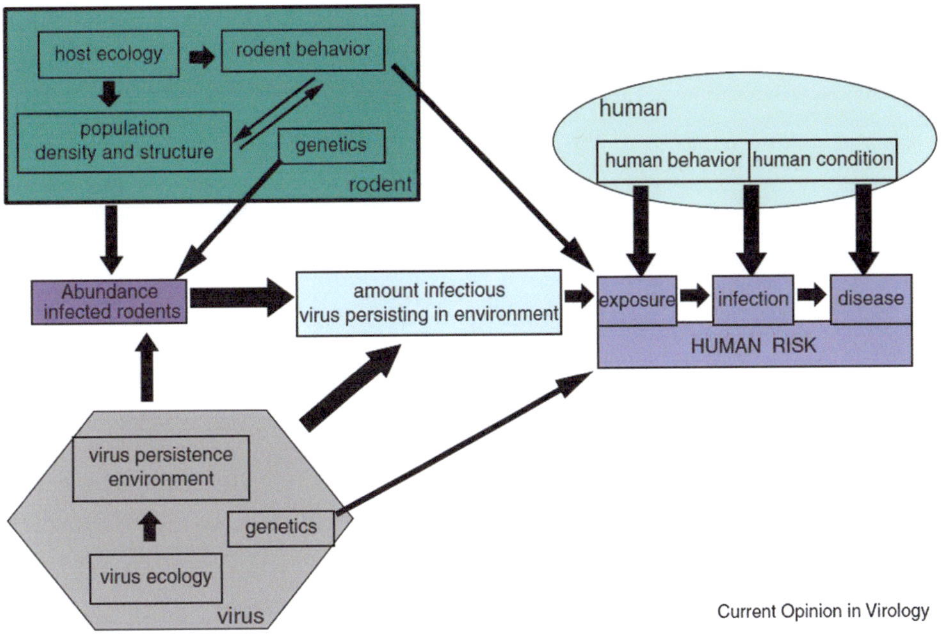

Fig. 4.24 Hantavirus circulation in natural foci and risk to humans. *Credit* Reusken and Neyman (2013)

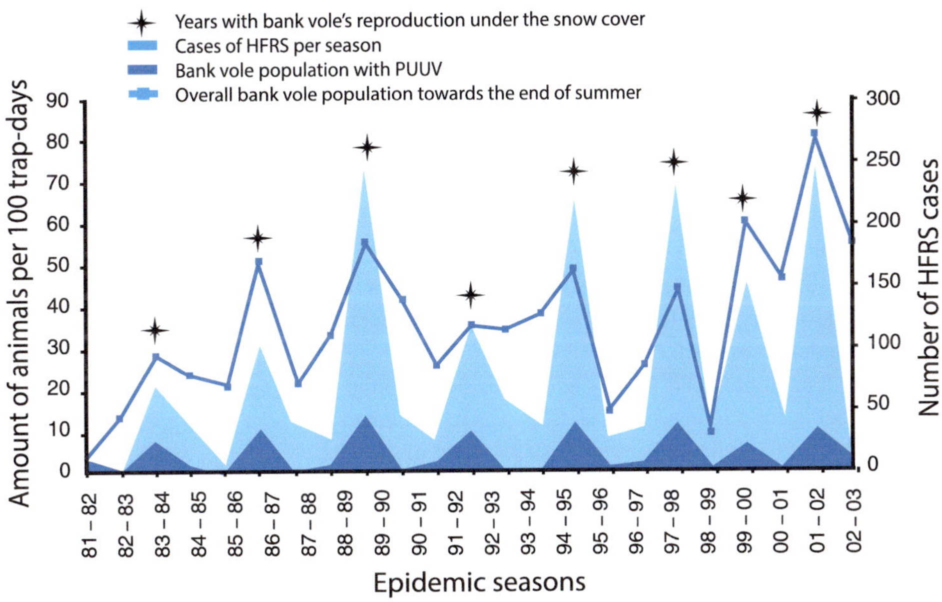

Fig. 4.25 Dynamics of bank vole populations (in their optimum range) and the Puumala natural focus activity in Udmurtia. *Credit* Bernstein et al. (2010)

condition of the host population in any given year. As such, PUUV is distributed within the bank vole's range, and most cases of infection occur in its optimum habitat—deciduous and mixed coniferous forests with high linden populations (Bashkortostan and Udmurtia).

HFRS outbreaks occur when the main hosts' population is high. For PUUV, a correlation has been found between increases in HFRS cases and bank vole population peaks in years with intensive vole reproduction under the snow cover. These peaks repeat themselves every 3–4 years (Fig. 4.25). Reproduction under the snow cover is usually linked with feed availability and favourable winter conditions (Bernstein et al. 2010).

4.6.3 Historical Background

HFRS as a distinct human disease was discovered in the early twentieth century. Nonetheless, there are earlier accounts of cases among soldiers in different parts of the world with similar symptoms. These cases were given various names.

Fig. 4.26 Anatoly Smorodintsev (1901–1986), HFRS pioneer in the USSR. *Credit* https://upload.wikimedia.org/wikipedia/commons/0/00/A.A.Smorodintsev.jpg CC BY-SA 3.0

Large outbreaks occurred among soldiers in the American Civil War, World War I, World War II, the Korean War and the Bosnian War, when the number of cases exceeds multiple thousands (Lee 2003). Diseases similar to HFRS occurred in the Finnish and German armies in Lapland, which were located 50 kilometres north of the Arctic Circle. In general, soldiers living in foxholes and trenches surrounded by lots of lemmings became infected (Somov and Besednova 1981).

The disease's modern name—hemorrhagic fever with renal syndrome—was approved by the WHO in 1978. Before this, in the USSR, the disease was called 'hemorrhagic nephrosonephritis'—this was the title given to it when it was first described by Anatoly Smorodintsev (Fig. 4.26) in the Russian Far East in 1930. He also noted that diseases with similar symptoms had been found in this region in 1913 and 1914 (Slonova 2003). Smorodintsev identified the disease's viral cause in 1940–1944, which was confirmed by M. L. Chudakov in 1956.

For a long time, HFRS was thought to be endemic to natural foci in the Russian Far East. This was re-evaluated in the 1950s, when clinically similar diseases were discovered in Russia's European territory. The virus causing HFRS was first isolated only in the late 1970s (Zee et al. 1978). After this, intensive molecular and laboratory-based research on hantaviruses began.

4.6.4 Distribution in Russia and the World

Hantaviruses are extremely widespread. Various clinical forms of HFRS are mostly found in temperate latitudes in Eurasia. Approximately, 150,000 people are hospitalized with a diagnosis of HFRS annually, mostly in China. HFRS has a fatality of 3–15%. In Europe, HFRS is a leading natural focal disease: morbidity has been rising constantly for the past 20 years, and the average annual number of cases exceeds 10,000 (Reusken and Neyman 2013). In North America hantaviruses cause a different, but also severe disease—hantavirus pulmonary syndrome.

In Russia, HFRS has been reported in 57 regions. The nosogenic range of the disease is torn (Map 4.14). Morbidity has increased nearly threefold since the disease was first reported (Bernstein et al. 2010). Morbidity rates are significantly higher in administrative subjects in Russia's European territory than in regions in the Russian Far East. Bashkortostan and Udmurtia are the most affected territories (Fig. 4.27). As such, during the 1997 HFRS outbreak, the morbidity rate in Udmurtia reached 229.3 cases per 100,000 individuals; the morbidity rate remained high in following years. After the 1997 peak (14.3 per 100,000 individuals), the long-term average annual morbidity rate in Russia overall has oscillated without exceeding 7.5 cases per 100,000 individuals.

The epidemiological situation has remained complex in Mari El, Tatarstan, Mordovia and Orenburg, Ulyanovsk, Penza and Samara Oblasts, where the long-term average annual morbidity rate is greater than 15.0 cases per 100,000 individuals. As was mentioned before, population morbidity in the Russian Far East is substantially lower: between 0.1 and 6.0 cases per 100,000 individuals annually in the Jewish Autonomous Oblast, Primorsky Krai and Khabarovsk Krai. Disease foci are linked to specific terrain: mixed coniferous forests and forest steppes. The majority of cases occur after infection during recreation in rural areas (mushroom picking, visits to summer cottages, etc.).

In a variety of regions, HFRS morbidity rate dynamics from 1997 to 2010 are characterized by nonlinear long-term oscillation (decreases, increases, oscillation with both significant and insignificant amplitudes). In the most severely affected areas (Udmurtia, Bashkortostan), morbidity rates vary significantly from one year to another. Meanwhile, the long-term average annual rate remains quite high. In less severely affected regions, such as those in the Volga Region and the southern Urals—Penza, Ulyanovsk, Samara and Orenburg Oblasts and Tatarstan—year-to-year morbidity rate variations are less pronounced, yet the long-term average annual rate remains comparatively low. However, these regions still experience outbreak years mixed with years with relatively low morbidity rates. Some regions in European Russia, where the morbidity rate is low, demonstrate a tendency towards an increasing morbidity rate and greater variability from one year to another. This indicates that outbreaks lead to more significant increases in morbidity than they did before (Map 4.15).

Annual HFRS morbidity rate levels (five total, ranging from low to high) for each administrative subject in the Russian Federation over the course of 17 years (1997–2013) are shown on a dynamic cartogram (Map 4.16). Regions in the Volga Region and the southern Urals have had stably high morbidity rates. The majority of subjects in the Central Federal District, the Northwest and the southern Russian Far

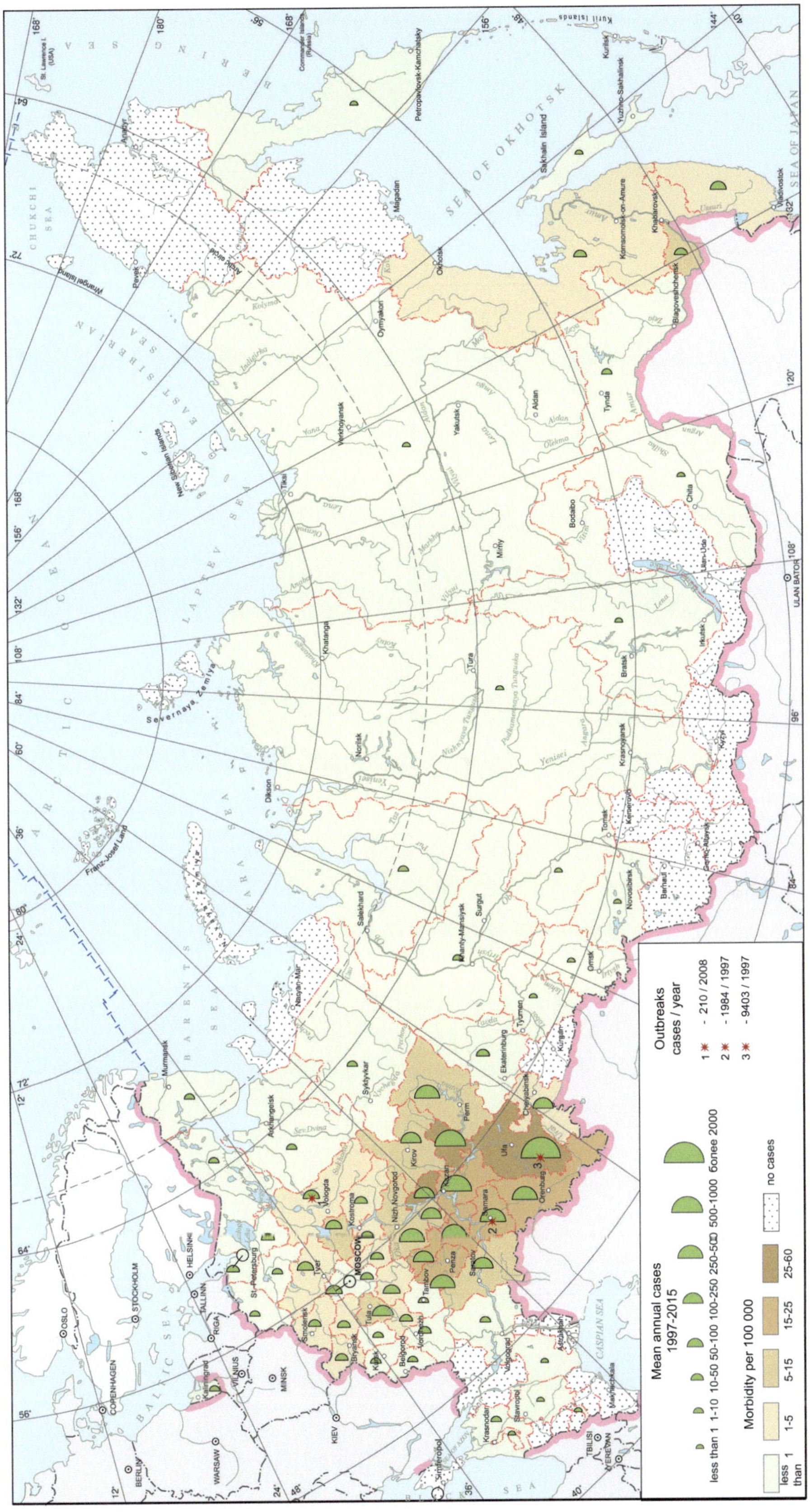

Map 4.14 HFRS morbidity, 1997–2015

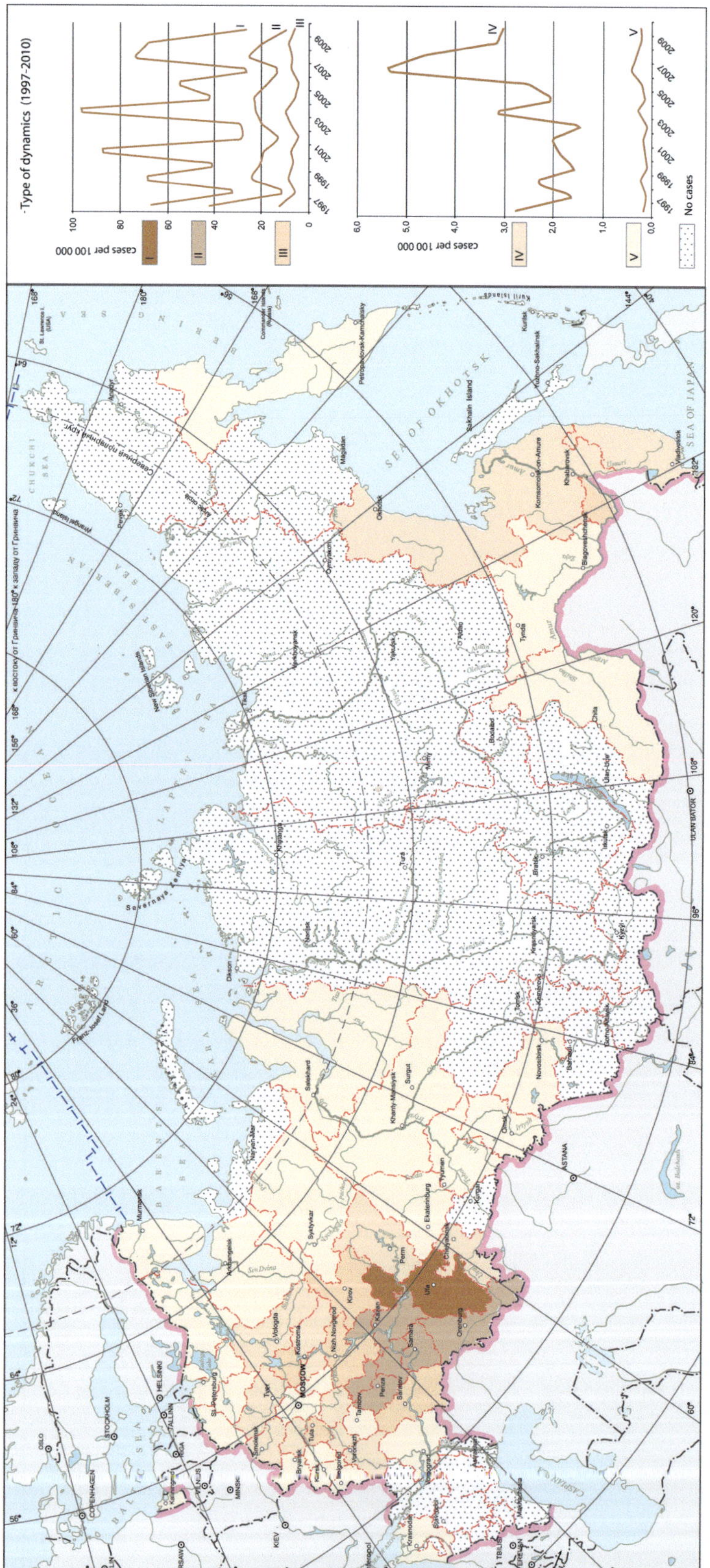

Map 4.15 Types of HFRS morbidity dynamics, 1997–2010

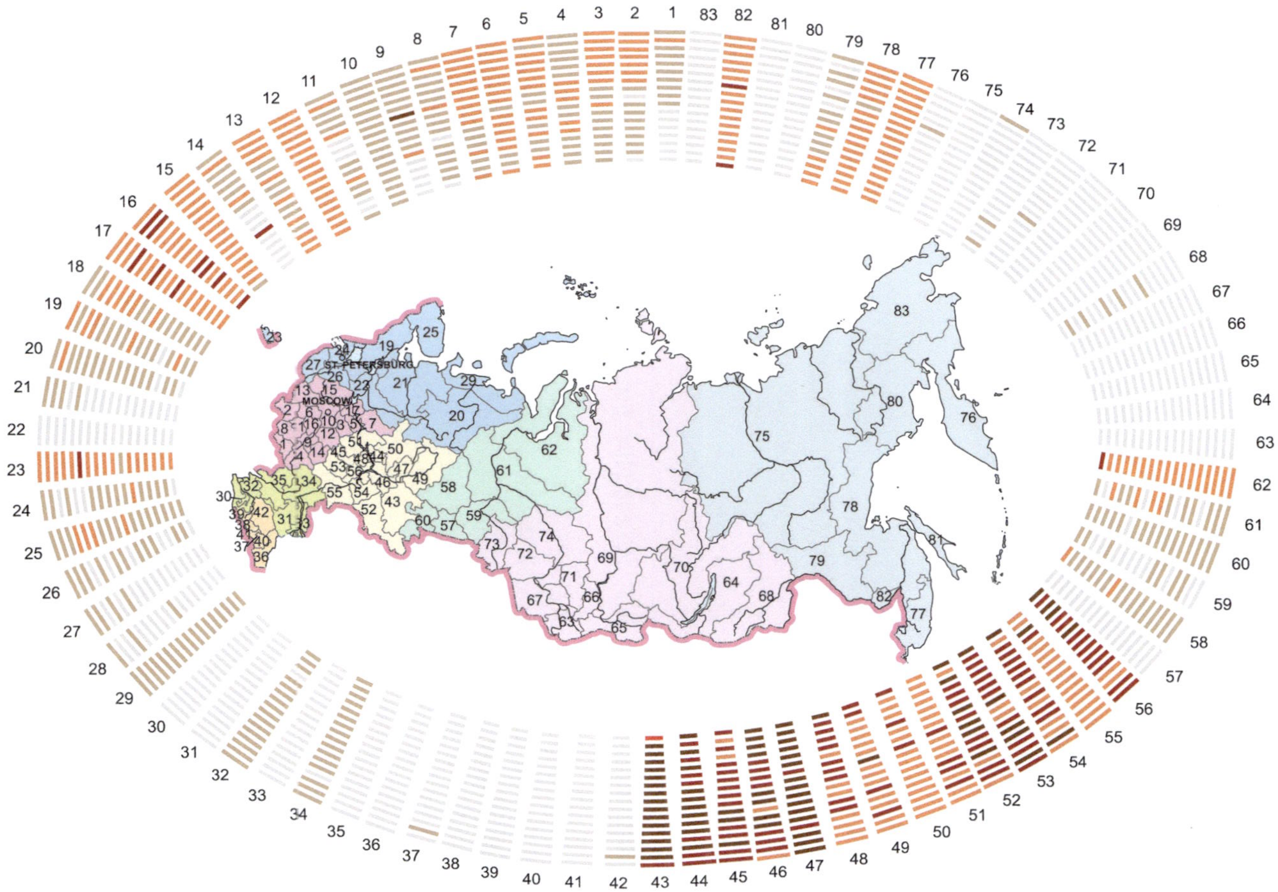

Morbidity per 100 000
- less than 1,0
- 1,0 - 10,0
- 10,0 - 20,0
- more than 20,0 (max 229,3; 1997)
- no cases

2012 2013
2010 2011
2008 2009
2006 2007
2004 2005
2002 2003
2000 2001
1998 1999
1997

1.....83 region number in legend and map

Central Federal District
1 Belgorod Oblast
2 Bryansk Oblast
3 Vladimir Oblast
4 Voronezh Oblast
5 Ivanovo Oblast
6 Kaluga Oblast
7 Kostroma Oblast
8 Kursk Oblast
9 Lipetsk Oblast
10 Moscow Oblast
11 Oryol Oblast
12 Ryazan Oblast
13 Smolensk Oblast
14 Tambov Oblast
15 Tver Oblast
16 Tula Oblast
17 Yaroslavl Oblast
18 Moscow city

Northwestern Federal District
19 Republic of Karelia
20 Komi Republic
21 Arkhangelsk Oblast
22 Vologda Oblast
23 Kaliningrad Oblast
24 Leningrad Oblast
25 Murmansk Oblast
26 Novgorod Oblast
27 Pskov Oblast
28 Saint Petersburg city
29 Nenets Autonomous Okrug

Southern Federal District
30 Republic of Adygea
31 Republic of Kalmykia
32 Krasnodar Krai
33 Astrakhan Oblast
34 Volgograd Oblast
35 Rostov Oblast

North Caucasian Federal District
36 Republic of Dagestan
37 Republic of Ingushetia
38 Kabardino-Balkar Republic
39 Karachay-Cherkess Republic
40 Republic of North Ossetia-Alania
41 Chechen Republic
42 Stavropol Krai

Volga Federal District
43 Republic of Bashkortostan
44 Mari El Republic
45 Republic of Mordovia
46 Republic of Tatarstan
47 Udmurt Republic
48 Chuvash Republic
49 Perm Krai
50 Kirov Oblast
51 Nizhny Novgorod Oblast
52 Orenburg Oblast
53 Penza Oblast
54 Samara Oblast
55 Saratov Oblast
56 Ulyanovsk Oblast

Ural Federal District
57 Kurgan Oblast
58 Sverdlovsk Oblast
59 Tyumen Oblast
60 Chelyabinsk Oblast
61 Khanty-Mansi Autonomous Okrug (Yugra)
62 Yamalo-Nenets Autonomous Okrug

Siberian Federal District
63 Altai Republic
64 Republic of Buryatia
65 Tyva Republic
66 Republic of Khakassia
67 Altai Krai
68 Zabaykalsky Krai
69 Krasnoyarsk Krai
70 Irkutsk Oblast
71 Kemerovo Oblast
72 Novosibirsk Oblast
73 Omsk Oblast
74 Tomsk Oblast

Far Eastern Federal District
75 Sakha Republic (Yakutia)
76 Kamchatka Krai
77 Primorsky Krai
78 Khabarovsk Krai
79 Amur Oblast
80 Magadan Oblast
81 Sakhalin Oblast
82 Jewish Autonomous Oblast
83 Chukotka Autonomous Okrug

Map 4.16 Long-term HFRS morbidity dynamics, 1997–2013

Fig. 4.27 Mixed coniferous forests near the Ural Mountains, where HFRS natural foci are most widespread. *Credit* N. G.Kadetov (permission obtained)

East have moderately stable morbidity rates. There are also regions with sporadic morbidity (some of these cases might have been brought in from elsewhere) and regions without any HFRS natural foci. The morbidity rate for each individual year, with the exception of 2013, does not differ substantially for the long-term average annual rate. However, in some years, such as 2013, morbidity rates drop. This may be linked with variations in rodent population size, unfavourable weather conditions or insufficient feed.

The defining characteristic of HFRS in Russia is the wave-like behaviour of long-term morbidity. Outbreak years in the Russian Far East may not coincide temporally with epidemics in Russia's European territory and vice versa.

4.6.5 Prevention

No specific measures of HFRS prevention (such as a vaccine or a specific hantavirus immunoglobulin) exist. Therefore, HFRS prevention consists of basic sanitary rodent control activities.

While in a natural focus, it is important to pay attention to accumulations of dust in potential rodent habitats. When doing dusty work (loading hay, straw or grass, unpacking boards, logs or bunches of firewood, cleaning inside buildings or building demolition), a respirator or face mask should be used. Rodent access to food must be prevented and food must be hermetically sealed. Dead rodents should not be touched without proper protective equipment. Hands must be washed carefully after contact with animals and unwashed fruits or vegetables should not be ingested. If rodent populations are too large, extermination should be conducted. Forests can also be made to resemble parks more closely.

4.7 West Nile Fever

West Nile Fever (WNF) (also known as West Nile encephalitis or meningitis or West Nile poliomyelitis) is a severe viral disease which is transmitted by mosquitoes and causes fever, encephalitis, meningitis, vomiting, diarrhoea and swollen lymph glands. The pathogen that causes WNF, West Nile virus (WNV) is an arbovirus, meaning that it is transmitted by arthropods. Birds, primarily aquatic birds and members of the crow family, are the pathogen's natural hosts, while various mosquito species are its vectors.

WNF is an emerging and re-emerging infectious disease: the nosogenic territory of WNF has been steadily expanding since the late 1990s (Petersen and Roehrig 2001).

4.7.1 General Information

West Nile virus (WNV) is an RNA virus in the Flaviviridae family and the *Flavivirus* genus. It is a member of the Japanese encephalitis antigenic serocomplex of viruses and causes symptoms typical of encephalitis: fever, headaches, neck numbness, myalgia and light sensitivity. Sometimes, it can lead to severe neurological symptoms, such as a coma.

WNF is clinically varied, and ranges from asymptomatic cases to severe outcomes. The disease often causes flu-like or hemorrhagic symptoms. People of an older age or immunocompromised individuals are at the greatest risk. The fatality among hospitalized patients is between 4 and 14% (Pauli et al. 2013).

4.7.2 Epidemiology and Natural Focus Characteristics

In natural foci, the virus circulates based on the following model: bird–mosquito–bird (Fig. 4.28). Other vertebrates become inadvertently involved and do not play a major role in maintaining natural foci.

WNV has been isolated from a large number of land and aquatic birds. The pathogen can survive in bird excrement and in various organs for 4–100 days, depending on the species of bird (Baturin et al. 2012). As a rule, wild birds do not show any clinical signs of WNF. This is thought to be the result of prolonged inter-population contact within the bird–mosquito–bird system. Many species, especially migratory birds, can be chronically infected (Reisen et al. 2006). All this allows WNV to be widely distributed across vast distances by migratory birds.

Anthropurgic WNF foci are typically secondary. Bird adaptation to the virus in such foci is usually non-existent. Experimental infection has shown that synanthropic birds exhibit clear clinical symptoms. These can cause individual lethal outcomes or mass mortality (Baturin et al. 2012; Adischeva et al. 2016).

Aside from birds, WNV has been found in a wide range of mammals and even some reptiles. Transmission between warm-blooded hosts occurs via mosquitoes of at least 40 species (Granwehr et al. 2004; Gubler 2007). Mosquitoes in the *Culex* genus play a leading role (Fig. 4.29). They are able to breed in various bodies of water, including extremely overgrown ones. They often breed in slightly flooded basements of residential buildings. They can live there year-round even in temperate climates. *Culex* mosquitoes have a large population and are opportunistic feeders and fairly aggressive bloodsuckers. Evidently, this is key in their role in the epidemiology and epizootology of WNF.

Aside from mosquitoes of the *Culex* genus, WNV has been found in mosquitoes of the *Aedes, Anopheles* and *Culiseta* genera (and more) and tick, such as *Rhiphicephalus rossicus, R. bursa, Dermacentor marginatus, Boophilus calcaratus, Hyalomma marginatum, Ixodes ricinus,* etc. (Putintseva et al. 2014). The virus easily adapts to Argasidae ticks, creating stable natural foci in the northern parts of WNV's range. This occurs thanks to the unique biological characteristics of Argasidae (long-term fasting, up to 9 years and longer, 20–25-year life cycles). For instance, islands on the coast of the Caspian Sea, which have neither freshwater nor mosquitoes, have a WNV natural focus consisting of the *Larus argenatus* seagull and the Argasidae tick *Ornithodors capensis* (Lvov et al. 1989). In the Volga Delta, WNV has been found in Ixodidae ticks (*Hyalomma marginatum*). This species contributed to viral maintenance in natural biocoenoses, especially between epidemics and in the early spring. Ticks often feed on birds in the crow family (Lvov

Fig. 4.28 Diagram of West Nile virus circulation in a natural focus

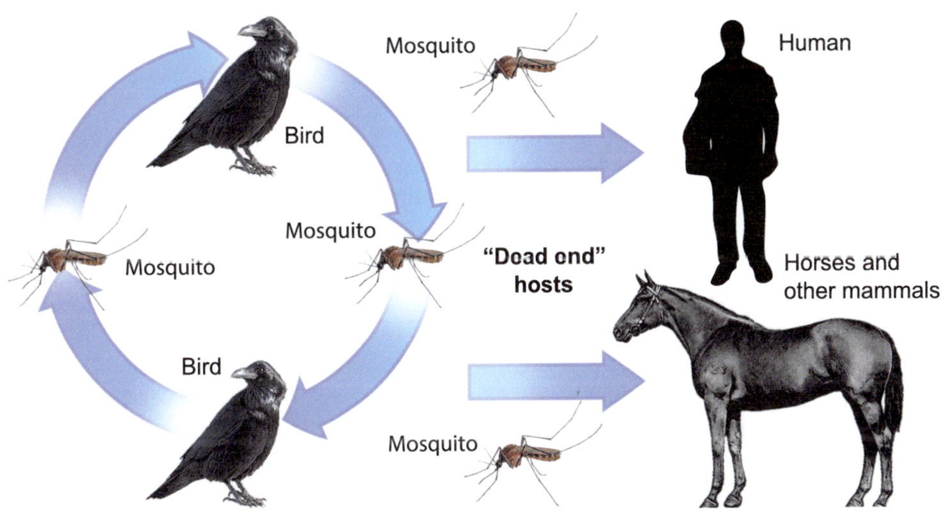

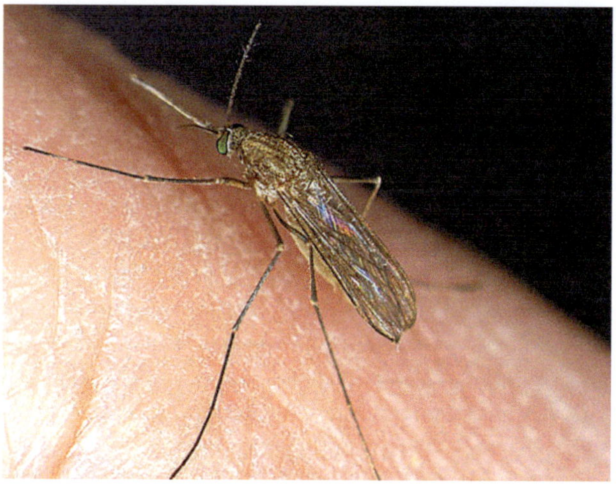

Fig. 4.29 The *Culex pipiens* mosquito. *Credit* фото Fabrizio Montarzi http://www.ncbi.nlm.nih.gov/pmc/articles/PMC3564894/ CC BY 3.0

et al. 2004, 2007). In Rostov Oblast, *H. marginatum* is joined by *Rhiphicephalus rossicus* (Moskvitina et al. 2011), allowing for viral circulation between natural and anthropurgic foci.

The two most significant natural factors in WNV's epidemiology are biotic (availability of suitable vectors and hosts) and climatic. The northern border of WNF's nosogenic territory coincides with the 16 °C July isotherm, meaning that the majority of Russia's territory is at risk (Lvov et al. 1989). On the other hand, cold winter months lower the likelihood of an outbreak during the following summer (Platonov et al. 2014). The availability of suitable bodies of water for migrating aquatic and near-aquatic birds plays an important role. The presence of a vector, the *Culex* mosquito, which quickly becomes involved in transmission, as well as conditions that allow the vectors' active reproduction create conditions for the formation of a persistent focus.

In small cities and rural settlements, private homes and farms pose a serious issue as they provide a location for mosquito reproduction in small reservoirs—rainwater barrels, pools, decorative ponds, abandoned vessels, etc.

Contemporary WNF epidemiology is characterized by a significant number of cases in urban locations, whereas it used to be considered a predominantly rural illness. This can be partially explained by the substantial number of synanthropic mosquitoes now involved in viral transmission, as well as the wide availability of breeding locations in urban areas. The fairly rapid introduction of synanthropic birds into viral circulation also allows for WNV transmission in large cities.

4.7.3 Historical Background

West Nile virus was first isolated in 1937 from a feverish patient in the West Nile province in northern Uganda, which gave it its name. Subsequently, the virus was found in a number of other countries, often very far from the territory where it was first discovered. Despite the fact that the history of WNF research is only a few decades long, there are reasons to believe that this disease was found in humans before this, beginning in the Classical Era. There is a proposition that Alexander the Great, who died at age 32 in Babylon after a severe 12-day illness, could have been sick with WNF. In any case, Alexander's symptoms (fever, loss of speech and other signs of meningoencephalitis) correspond to the symptoms of WNF. This theory is supported by Plutarch's account of a large number of dead crows in Babylon not long before Alexander's illness (Marr and Calisher 2003). According to Plutarch, Alexander saw the dying birds and interpreted them as a bad omen. Although this theory is purely a hypothesis and is unlikely to ever be proven, its authors' conclusions are intriguing.

The epidemiology and ecology of WNF were studied in detail in the 1950s and 1960s, during outbreaks in the Mediterranean. The first acknowledged WNF epidemic was an outbreak in Israel in 1951, when 123 people got sick. The disease's symptoms were thoroughly described during this epidemic. It soon became clear that, aside from people, some mammals could also get WNF. In horses the disease takes an especially severe course, often leading to fatal outcomes. Not long before this, WNV was discovered to be transmitted via vectors, and the main vector—mosquitoes in the *Culex* genus—were identified (Sejvar 2003).

However, before the 1990s, WNV outbreaks were limited and affected small territories and small numbers of people. The disease usually took a milder course and many asymptomatic cases were identified. Beginning in the late 1990s, outbreaks became larger and their geographic spread grew (Fig. 4.30). A significant number of cases began to be reported in cities. Although previously WNV was thought of as one of the least dangerous members of the Japanese encephalitis serocomplex, now a significant number of epidemics includes multiple severe cases with fatal outcomes. This has been linked to the appearance of a viral subtype in the early 1990s that is more virulent and able to expand geographically (Gubler 2007).

An outburst of scientific interest in WNV occurred when it appeared in North America. In the USA, the virus was first found in New York City in 1999, after which it rapidly spread across the United States' substantial territory, where it had never been previously reported (Fig. 4.31). The US

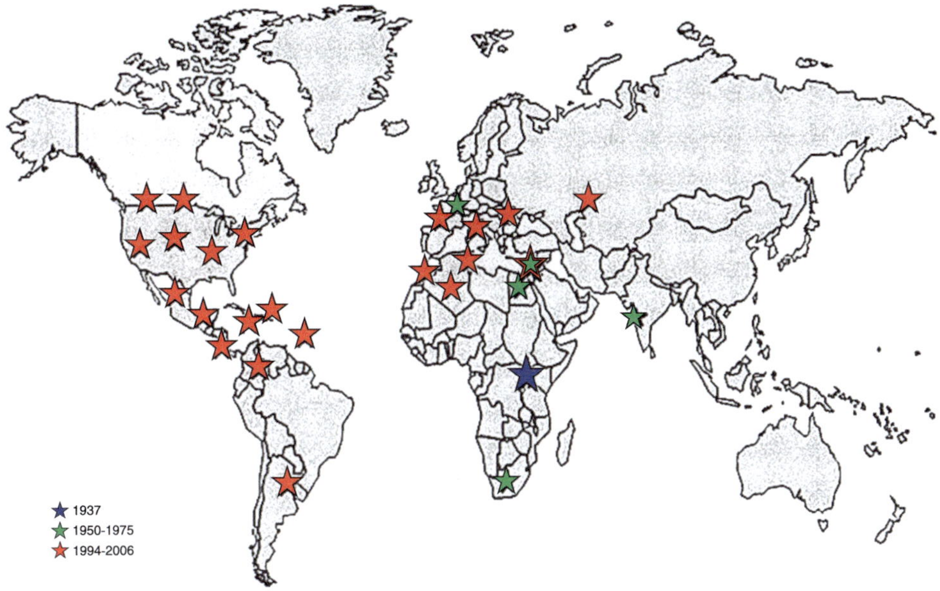

Fig. 4.30 WNF outbreaks, 1997–2006. *Credit* Gubler (2007)

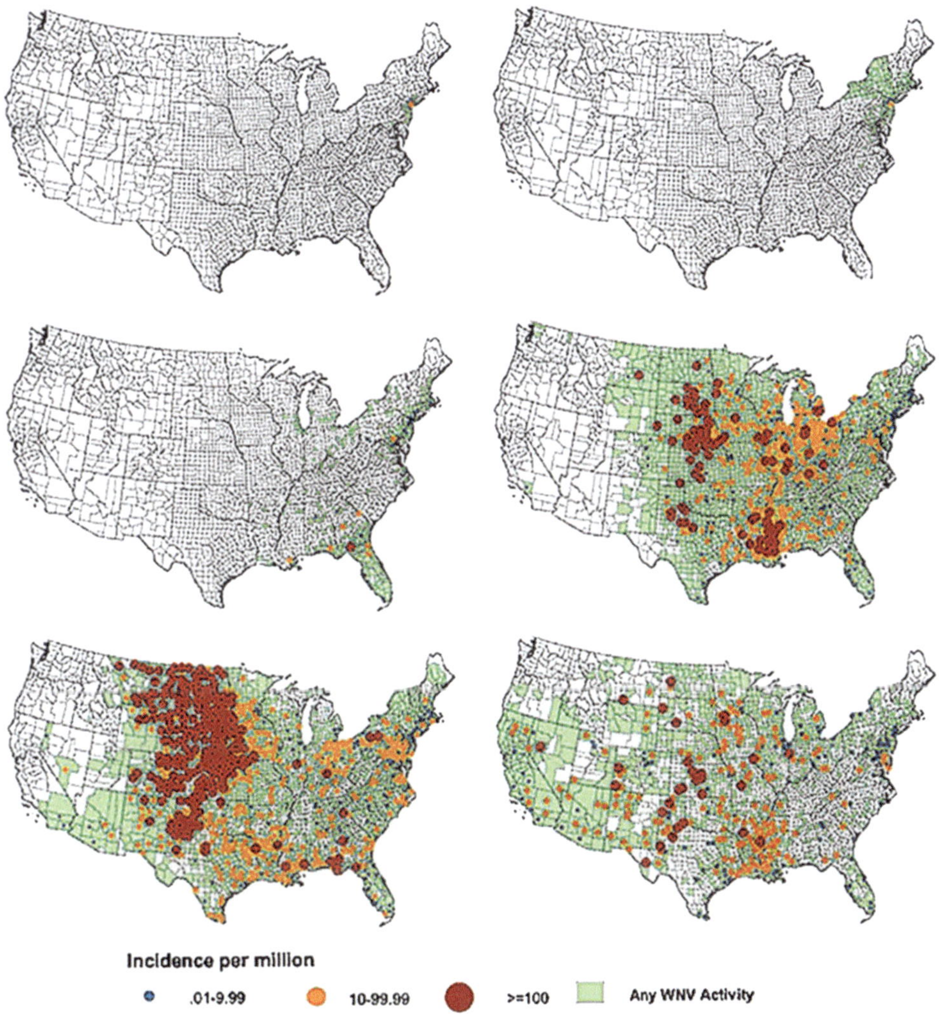

Fig. 4.31 WNF morbidity rate per 100,000 individuals, USA, 1999–2004. *Credit* Hayes et al. (2005)

epidemic was characterized by a large number of severe cases, especially among the elderly population. The virus's spread in humans was accompanied by an epizootic among horses in a number of regions.

4.7.4 Distribution in Russia and the World

WNV's range spans practically all of Africa, Southeast and South Asia, Southern Europe and some of Central Europe. It has also been found in Australia (Scherret et al. 2001). Since the 2000s, the highest number of cases in the world has been reported in the USA, and they were found across its entire territory. Originally these cases were brought in from other countries, but today nearly all of them are caused by local virus circulation (Hayes et al. 2005). WNV is rapidly spreading in South America as well, and today specialists doubt that it will be eradicated anytime soon. WNV is one of the best examples of the successful introduction of a pathogen into a new environment involving local carriers and vectors.

WNF has been reported in Russia since 1997. At first, this disease was found in 5–6 administrative subjects in the Southern Federal District—Volgograd, Rostov and Astrakhan Oblasts, Kalmykia and Krasnodar Krai—where disease foci are believed to have formed long ago (Map 4.17). The disease is characterized by outbreaks. One outbreak occurred in Astrakhan Oblast in 2005 (73 cases) and others have occurred in Volgograd and Rostov Oblasts as well. In 2010, a large outbreak occurred primarily on the territory of the Southern Federal District (Fig. 4.32). It affected not only territories that were traditionally at risk, but new ones as well (9 administrative subjects in all). According to Rospotrebnadzor, the overall number of cases was 448. Volgograd Oblast was the most affected, where 46.3% of the reported cases occurred. The infection was found in eight districts within the oblast, as well as the cities of Volgograd and Volzhsky. 90% of all cases occurred in the cities (369 individuals). In 2011 and 2012, the disease's range grew even more, and during the 2012 season, the disease was reported in 21 administrative subjects. In 2013, cases of WNF were found in 16 administrative subjects of Russia. WNF has been brought from one administrative subject to another (Putintseva et al. 2014). In 2014, the disease was reported in the Republic of Crimea (Rymarenko et al. 2014). Overall, the disease was reported in 21 Russian regions from 1997 to 2015.

The main WNF natural foci in Russia have formed in the lowlands of the Volga and Don rivers. Their deltas are rich with wetlands, which serve as a favourable habitat for migratory birds, who arrive in these river floodplains after spending the winter in places where WNF is common, such as Egypt or tropical Africa. Most WNF cases in Russia occur in the summer, beginning in mid-July. Therefore, outbreaks

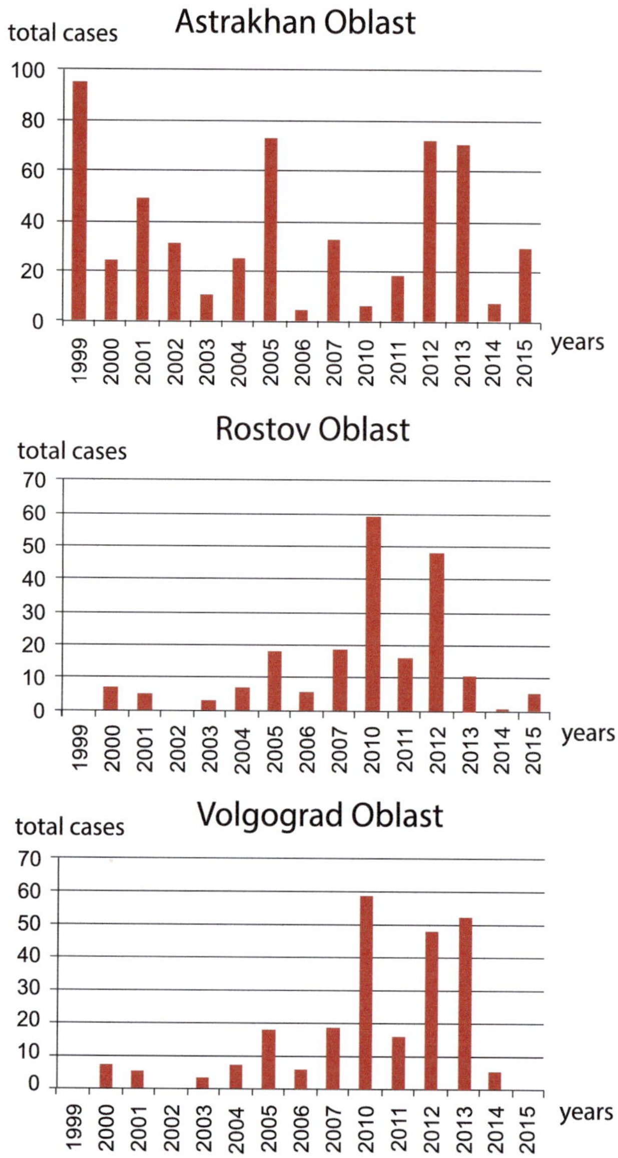

Fig. 4.32 WNF morbidity in some federal subjects of the Southern Federal District of Russia

in Russia's southern regions are first of all linked with favourable climate conditions (average July temperatures far exceed 16 °C, which allows for the rapid accumulation of virus in mosquitoes' salivary glands) and the presence of viral hosts and vectors.

4.7.5 Prevention

Anti-mosquito measures (both organizational, individual and group-based) are the main methods of WNV prevention. Organizational measures include inspecting bodies of water for mosquito breeding, landscaping, landfill eradication, weed extermination, insect extermination in pools and water tanks, desiccation, renovation of technical netting and insect

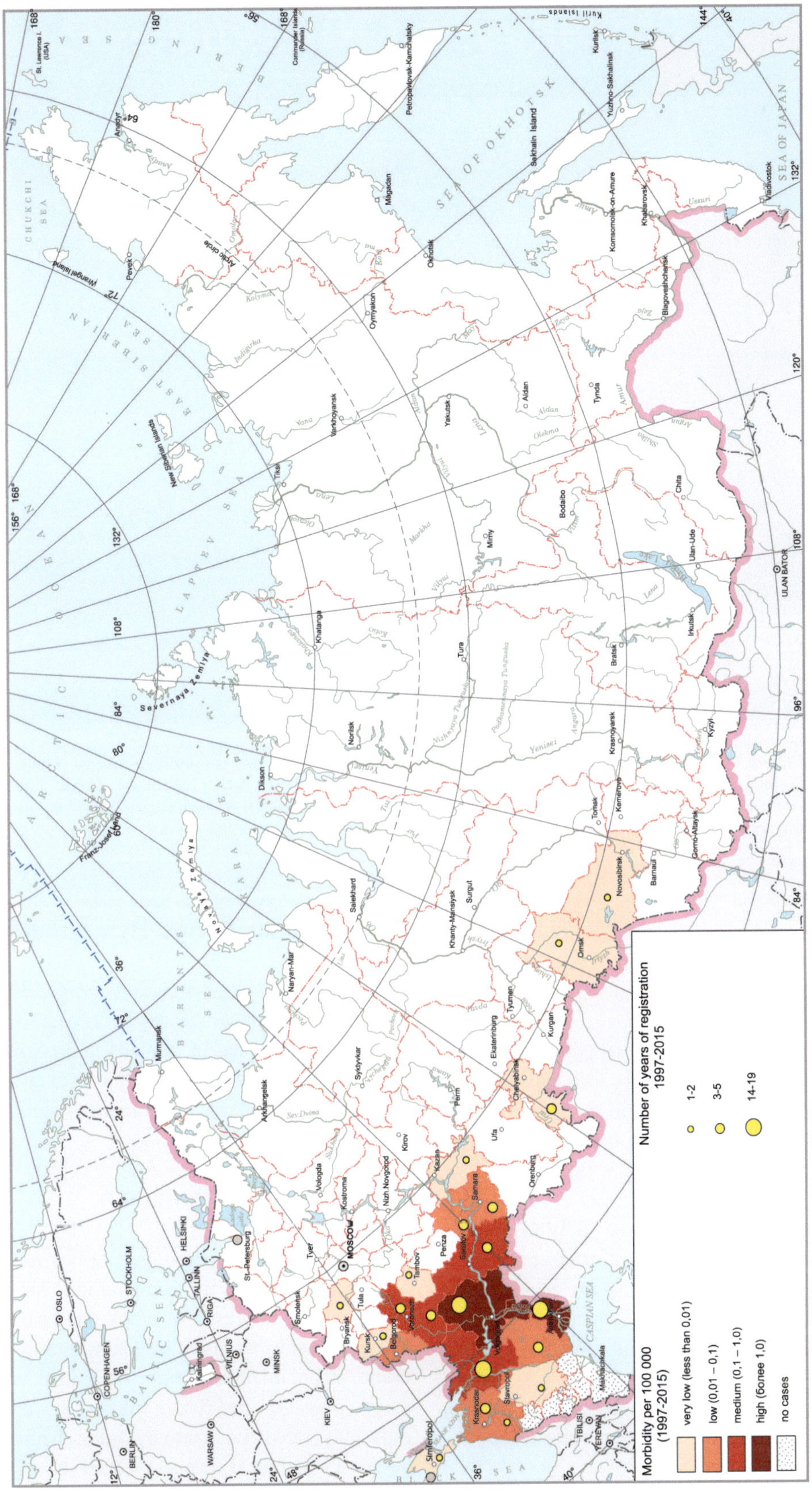

Map 4.17 WNF Morbidity, 1997–2015

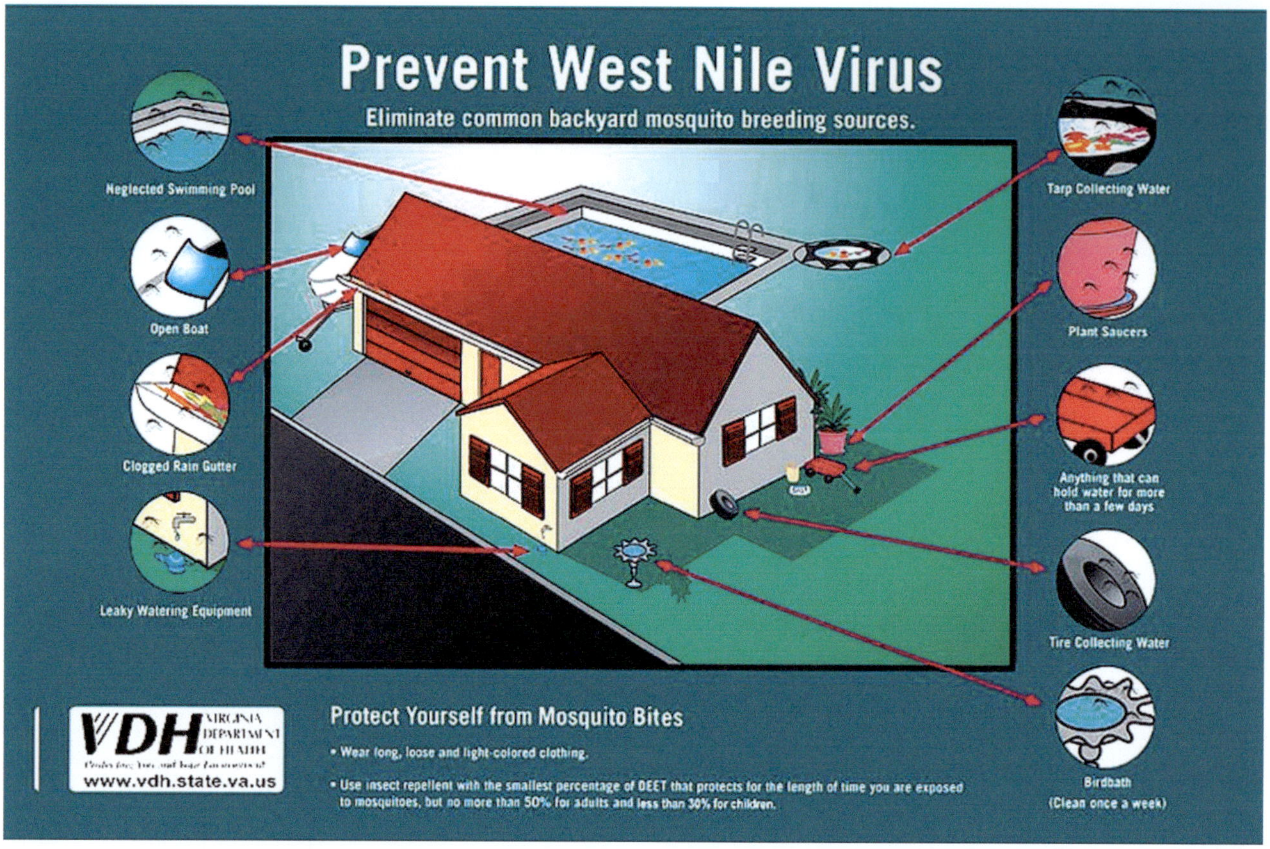

Fig. 4.33 A poster reminding residents to prevent mosquito bites due to a WNF outbreak in the USA. The poster lists places where mosquitoes tend to breed that could be found in any household: puddles, old hubcaps, abandoned vessels. The poster asks residents to watch for accidental accumulations of water, wear long-sleeved clothing, and use insect repellents. *Credit* http://www.vdh.virginia.gov/epidemiology/DEE/Vectorborne/WestNile/images/WNVhouse03L.jpg

extermination in flooded basements. Individual and group-based protection includes the placement of screens in windows and doorways, insect repellent use, etc. The following methods can be used to fight mosquitoes indoors:

- Fumigators, which slowly release insecticide
- Spirals that emit smoke that repels or kills mosquitoes
- Ultrasound devices that repel mosquitoes
- Ultraviolet light devices that exterminate mosquitoes
- Aerosols that repel mosquitoes from a room.

Population education plays a key role in WNF prevention. Although positive outcomes have been proven, in Russia, this form of prevention is insufficiently developed. In cities and suburbs in the United States, where many people own private homes, the issue of mosquito breeding is taken very seriously. Public service announcements and advertisements are directed at alerting the population to keep careful watch over bodies of water on their land (Fig. 4.33) and to take other preventive measures.

4.8 Crimean–Congo Hemorrhagic Fever

Crimean–Congo hemorrhagic fever (CCHF) (also known as Crimean hemorrhagic fever) is an acute natural focal viral infection that occurs in humans and causes fevers, general toxicosis and hemorrhagic symptoms such as a petechial rash and other hemorrhagic phenomena (Pokrovsky et al. 2003).

4.8.1 General Information

CCHF is a highly pathogenic arboviral infection in the *Nairovirus* genus of the Bunyaviridae family. CCHF infection occurs as the result of a tick bite or if a tick is smashed by one's hands. Cases of infection through contact with the blood of an infected person or animal are rarer. Infected people pose the most risk in the first few days of the disease, especially once the haemorrhaging begins (Maleev et al. 2003). Two clinical forms of CCHF have been defined: one with hemorrhagic symptoms (can take a severe, moderate or light course, depending on the extent of hemorrhagic phenomena) and one without hemorrhagic symptoms (can take a moderate or light course). The incubation period lasts between 1 and 14 days (3–6 days on average). The area around the tick bite immediately becomes warm and itchy. Three disease stages have been identified: early, peak and convalescence. The early period lasts 3–6 days. Body temperature rises rapidly. Most symptoms are similar to the general symptoms of viral infections. Fever is constant, lasting an average of 7–8 days (Guide.., 2000). The peak period is characterized by the development of hemorrhagic symptoms after a brief drop in body temperature. This leads to rapid deterioration. CCHF is fatal when the hemorrhagic symptoms are at their peak, combined with other signs of viral infection (Cherkassky 1996). Convalescence takes 1–2 months in most patients and is characterized by weakness, fatigue and neurological symptoms (Postovit 1997). The disease is severe, with a case fatality rate of 16% on average, reaching 40% in certain regions. Most commonly, death is caused by infectious and toxic shock, massive haemorrhages and kidney and liver failure (Postovit 1997; Maleev et al. 2003).

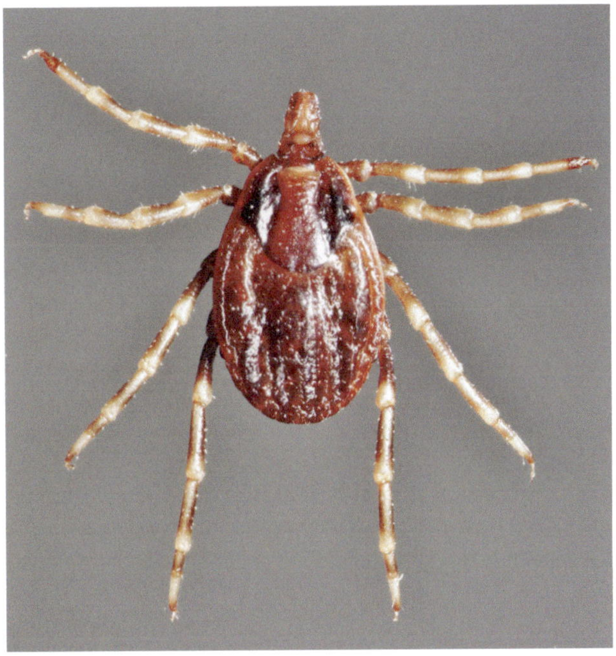

Fig. 4.34 The *Hyalomma* sp. female tick. *Credit* https://upload.wikimedia.org/wikipedia/commons/3/31/Hyalomma_tick_female_dorsal.jpg By Acarologiste (Own work) [CC BY-SA 4.0 (https://creativecommons.org/licenses/by-sa/4.0)], via Wikimedia Commons

4.8.2 Epidemiology and Natural Focus Characteristics

CCHF is typically confined to specific landscapes: semi-desert, steppe and floodplain-adjacent zones, which are primarily used for livestock grazing. Ixodidae ticks of the *Hyalomma, Dermacentor, Rhipicephalus* and *Amblyomma* genera are the principal vectors and reservoirs of the virus. They can transmit the virus to their offspring transovarially and transstadially. The *Hyalomma marginatum* tick is the main vector in Crimea and Astrakhan Oblast (Fig. 4.34). It prefers warm, arid climates. The tick's range spans southern Russia, Southern Europe, Central and Southwest Asia and North Africa (Hoogstraal 1956; Gargli et al. 2013; Maleev et al. 2003). The nymph stage prefers small animals and birds, which then become the virus' main carriers, such as the European hare (*Lepus europaeus*), the Ural field mouse (*Apodemus uralensis*), the little ground squirrel (*Spermophilus pygmaeus*) and the long-eared hedgehog (*Hemiechinus auritus*) (Guide.., 2000). In anthropurgic foci, the main carriers are cattle, sheep, goats and other livestock that serve as hosts for imago ticks. Adult ungulates acquire immunity after their first encounter with the virus and are a dead end for the infection (Maleev et al. 2003; Pokrovsky et al. 2003). Data exist pointing to the role of birds in viral transmission (Vodyanitskaya et al. 2008; Lindebor et al. 2012).

In endemic regions morbidity rates are seasonal, increasing during times of agricultural work (June–August). Morbidity also depends on climatic factors and tick activity. People who are involved in haymaking and harvest gathering, or who tend to livestock or work in slaughterhouses, or people who spend the night in the steppe are at the greatest risk of infection. Healthcare workers who are in contact with infected individuals are also a risk group, as well as laboratory workers (Pokrovsky et al. 2003). CCHF morbidity rates are sporadic and outbreaks are rare.

4.8.3 Historical Background

A pathological depiction of a disease similar to CCHF was first described in Tajikistan in 1914 by P. S. Sipovsky (Maleev et al. 2003). The disease's viral aetiology was established in 1944 and 1945 in Crimea during an expedition led by M. P. Chumakov. The expedition was organized during an outbreak of an acute hemorrhagic fever, which

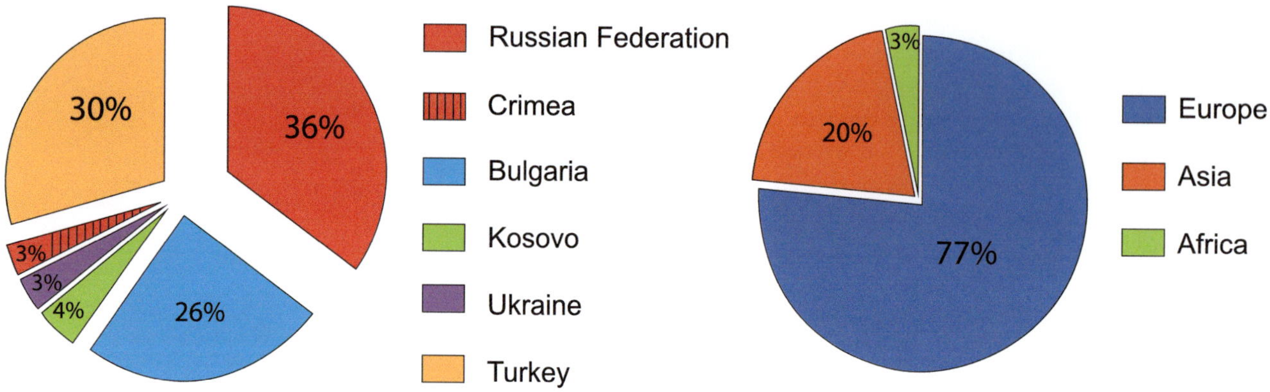

Fig. 4.35 CCHF case distribution worldwide. *Credit* Butenko and Trusova (2013)

was then named Crimean hemorrhagic fever. Between 1945 and 1969, cases with similar symptoms were recorded in Krasnodar and Stavropol Krais, Astrakhan and Rostov Oblasts and in Central Asia and Kazakhstan. In 1967, eight strains of the CCHF virus were isolated from infected individuals in these regions (Lvov et al. 1989).

Natural foci, along with a similar disease, were found in the Congo in the 1950s. In 1968, one of the USSR strains (*Drozdov*) was given to an arboviral laboratory at Yale University in the United States. This strain was found to correspond to three strains of the virus found in the Congo (Pokrovsky et al. 2003). This gave the disease the name Crimean–Congo hemorrhagic fever. The same disease has since been found in Asia, as well as Bulgaria, the former Yugoslavia, and Hungary (Maleev et al. 2003).

4.8.4 Distribution in Russia and the World

CCHF is characterized by periodic spikes in its morbidity rate of unknown causes.

Between 1944 and 2012, 6039 cases were reported in Europe, 1620 in Asia and 265 in Africa (Fig. 4.35). In the same period, 2180 cases were found in Russia, making up over 30% of all the cases found in Europe (Butenko and Trusova 2013).

The epidemiological situation is most critical in the Southern and North Caucasian Federal Districts of the Russian Federation. Between 1953 and 1970, cases were reported in Astrakhan and Rostov Oblasts, as well and Krasnodar and Stavropol Krais. A total of 623 cases was reported on Russian territory between 1948 and 1998. The majority of cases (611) occurred between 1948 and 1970. After a natural focus increased its activity in 1999, another 1775 cases, the majority of which occurred between 2005 and 2007, were reported (Fig. 4.36). In recent years, the largest number of CCHF cases has been found in Rostov Oblast and Stavropol Krai. Aside from its increased activity, the focal territory expanded north (into Volgograd and Voronezh Oblasts), south (to republics of the North Caucasus) and east (Kalmykia) (Map 4.18). The assumption is that despite acaricidal efforts, the population of CCHF's main vector, the *H. marginatum* tick, has been increasing due to favourable climate conditions. This can lead to an increase in human morbidity rates (Volynkina et al. 2015).

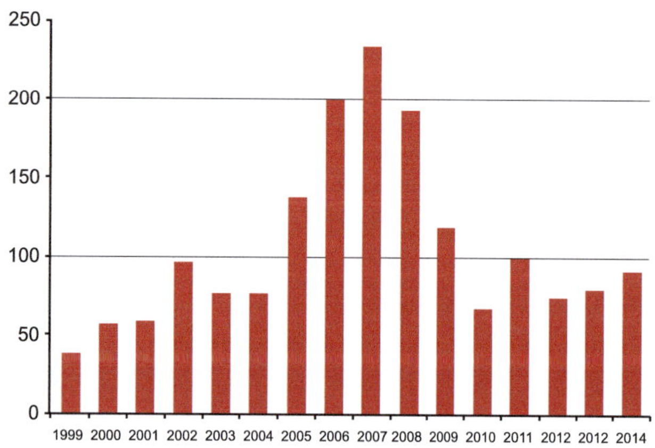

Fig. 4.36 CCHF case number dynamics in Russia, 1999–2014. *Credit* Volynkina et al. (2015)

4.8.5 Prevention

In times of heightened tick activity, it is necessary to avoid and prevent tick–human contact and reduce Ixodidae tick population sizes. For personal protection, it is important to use protective clothing when in rural areas or while performing fieldwork. Personal tick bite inspections should occur hourly, and people should inspect one another. Clothing needs to be coated with tick repellents. Repellents aimed specifically at ticks that carry CCHF are preferable. While treating animals, it is dangerous to remove or smash ticks with one's bare hands. If animals have tick bites, they need to be taken to a veterinarian and treated to prevent tick bites according to their recommendations.

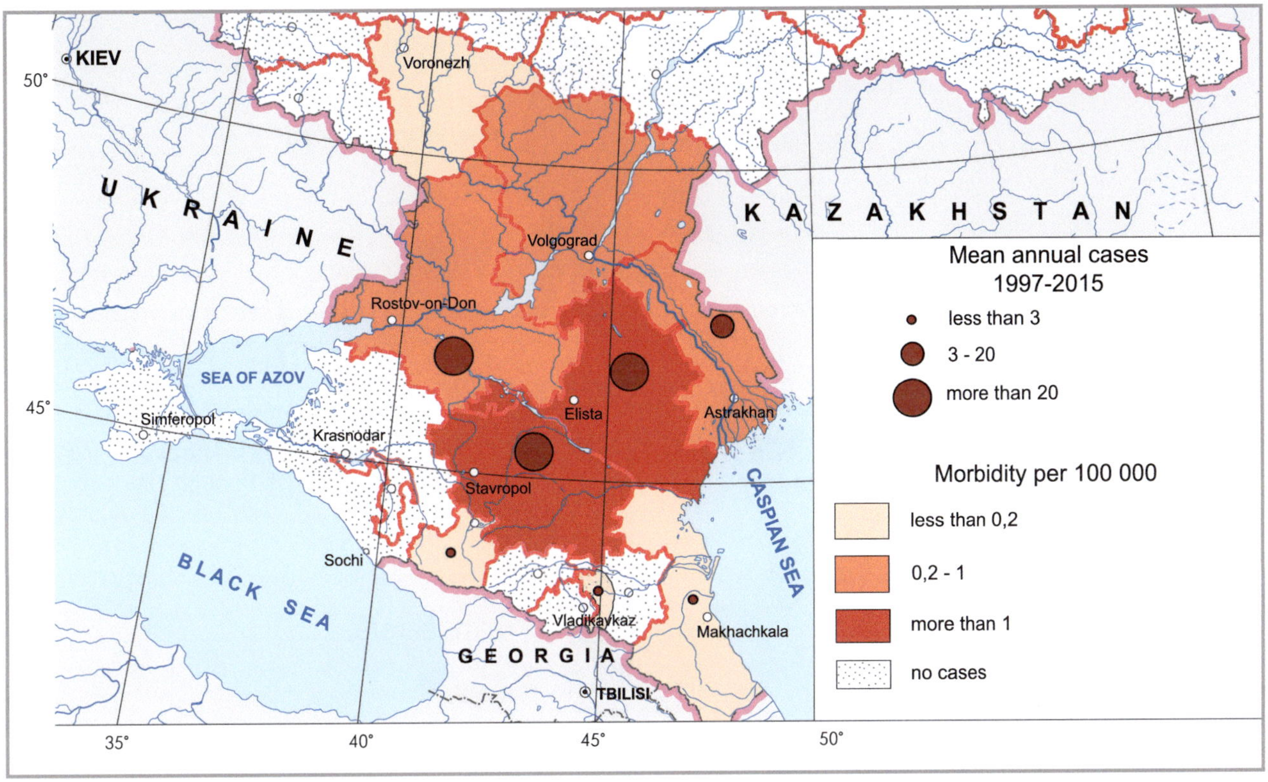

Map 4.18 CCHF morbidity, 1997–2015

Rural municipal settlements need to use acaricides seasonally on animal pastures and plan to bring livestock free of insects to pastures free of ticks. Recreational areas in suburban and rural areas should be kept sanitary: the grass should be mowed, the territory should be fenced in, and acaricides should be used.

4.9 Omsk Hemorrhagic Fever

Omsk hemorrhagic fever (OHF) is an acute vector-borne natural focal viral disease characterized by fevers and neurologic, cardiac and hemorrhagic symptoms.

OHF is a typical member of the group of emerging and re-emerging diseases, which demonstrate epizootic and epidemic activity within their previous nosogenic territory after a prolonged period of dormancy (Yastrebov and Yakimenko 2014).

4.9.1 General Information

Omsk hemorrhagic fever is caused by an arbovirus in the *Flavivirus* genus of the *Flaviviridae* family. Based on its antigenic properties, it has been classified in the same group as arboviruses of the tick-borne encephalitis complex, which includes a number of ecologically and biologically similar viruses (tick-borne encephalitis, Langat virus, Kyasanur forest disease, etc.). OHF's incubation period ranges from 2 to 10 days. The fever (reaching 40 °C) lasts 5–12 days, at times followed by a second wave of fever symptoms 10–15 days later. The disease usually ends in a complete recovery after a prolonged period of weakness. The case fatality rate is 1% on average.

4.9.2 Epidemiology and Natural Focus Characteristics

People become infected with OHF after a tick bite or after coming into contact with an infected animal (Fig. 4.37). OHF is non-transmissive in humans, i.e. people are a biological dead end for the virus.

The *Dermacentor reticulatus* tick is the most epidemiologically significant natural vector and reservoir of OHF. It was previously thought that the virus's main warm-blooded host was the narrow-headed vole (*Microtus gregalis*). Today it is known that muskrats (*Ondatra zibethicus*) are the pathogen's main reservoir. Muskrats were introduced to Western Siberia in the mid-1930s. OHFV is highly

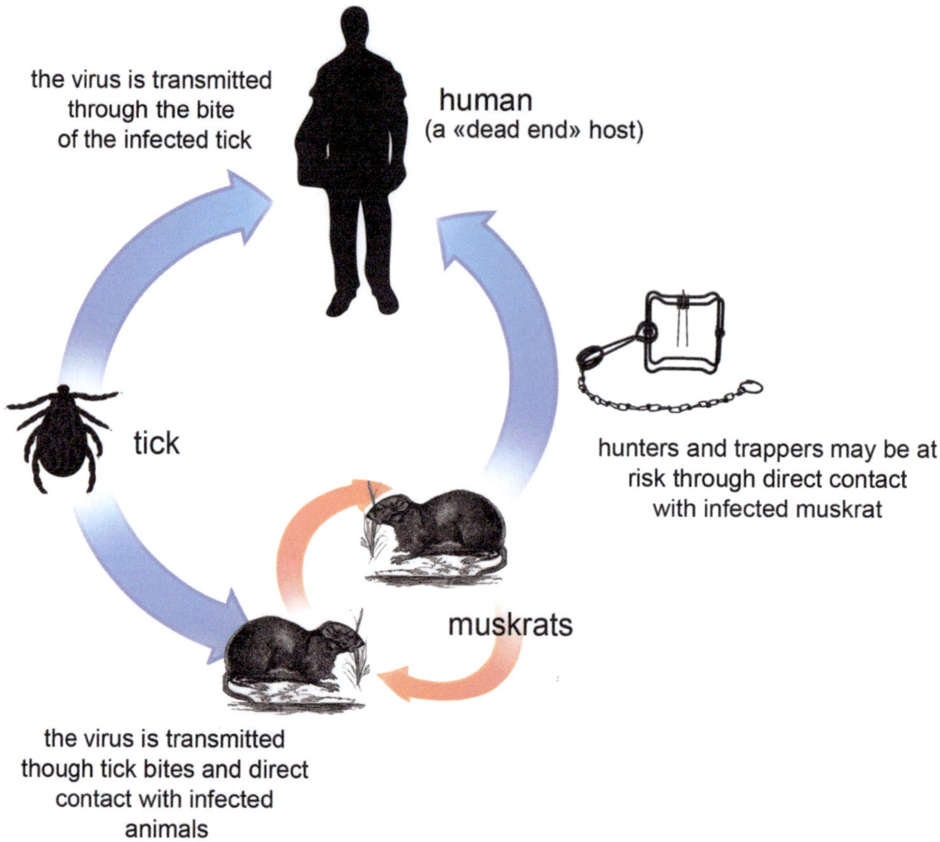

Fig. 4.37 Diagram of OHV circulation in a natural focus

pathogenic to muskrats, and many animals die during epizootics, at times annihilating populations in entire bodies of water. Based on this, locals call OHF 'muskrat disease'. In this case, people who hunt muskrats and come into contact with the pathogen while skinning them are the main OHF risk group. Experimental studies have confirmed the muskrat's leading role in disseminating the virus in the environment during epizootics and pointed at the fact that OHFV, having passed through the muskrat's organism, enters the environment with its excretions. This gives it the ability to be transmitted via air droplets and via direct contact (Yastrebov and Yakimenko 2014).

The European water vole, the tundra vole, and some other wetland mammals (especially those that come into contact with muskrats) also play an important role in the virus's circulation (Fig. 4.38). Sixteen wild vertebrate species, three Ixodidae tick species, Gamasid mites, mosquitoes and water mites *(Hydrachnidae)* have also been found to be infected with OHFV. This serves as evidence for the multiplicity of hosts and vectors in OHV natural foci (Yastrebov and Yakimenko 2014).

OHF morbidity rates are strictly seasonal and vary based on the vector-borne or non-vector-borne nature of epidemic outbreaks. In vector-borne transmission, the epidemic season lasts from April to October. The seasonal morbidity rate dynamics reflect the seasonal activity of the *D. reticulatus* tick: they exhibit two clear peaks in August and September. Aside from vector-borne cases, winter cases of OHF have been reported in hunters' families. These cases are related to contact transmission from infected muskrats (Ruzek et al. 2014).

Fig. 4.38 Muskrat *(Ondatra zibethicus)*

4.9.3 Historical Background

OHF natural foci were first discovered in 1945 in three adjacent administrative districts in Omsk Oblast (Tyukalinsky, Soldatsky and Sargatsky). All three were located in the northern forest steppe. This disease was first officially reported in 1946 (Karan et al. 2014).

OHF was first described from 1940 to 1945 by two Omsk-based scientists, G. A. Sizemova and A. A. Gavrilovskaya. Major studies of the fever were conducted from 1947 to 1949. A comprehensive scientific expedition headed by Mikhail Chumakov operated in Omsk Oblast (Fig. 4.39).

4.9.4 Distribution in Russia and the World

OHF natural foci are confined to the forest steppe regions of the West Siberian Plain. Its nosogenic territory spans the territory of Omsk, Novosibirsk, Tyumen and Kurgan Oblasts (Figs. 4.40 and 4.41). Long-term OHF natural focus dynamics reveal a variety of epidemically and epizootically distinct periods. The largest outbreaks occurred between 1945 and 1958 in Omsk Oblast. A total of 1302 cases were registered in that period. A relatively lengthy interepidemic period began after 1958.

Fig. 4.39 Mikhail Chumakov (1909–1993), one of the first researchers of OHF. *Credit* https://commons.wikimedia.org/wiki/Category:Mikhail_Petrovich_Chumakov#/media/File:Mpchumakov-portrait1.tif Public domain

Fig. 4.40 Lakes in the forest steppe regions of West Siberia serve as habitats for the muskrat, which is linked to most OHF outbreaks (a lake in Sargatsky district of Omsk Oblast is shown). *Credit* moneag, http://www.panoramio.com/photo/28883734 This file is licensed under the Creative Commons Attribution-Share Alike 3.0 Unported license

In the late twentieth century, OHF natural foci in Tyumen (1987), Omsk (1988, 1999–2007), Novosibirsk (1989–2002) and Kurgan (1992) Oblasts became epizootically active. Four administrative districts in Novosibirsk Oblast reported epidemic activity. However, the non-vector-borne way of disease transmission predominated throughout these outbreaks (Yastrebov and Yakimenko 2014).

There have been no reported cases of OHF in recent years, although the virus is constantly present in muskrat populations, even if their numbers remain relatively stable.

4.9.5 Prevention

Preventative and anti-epidemic measures include rodent extermination, tick extermination and the vaccination of populations in endemic areas.

Specific preventative measures include primarily the extermination of muskrats and the other small wetland mammals that share bodies of water with them. Systematic suppression of muskrat migration is necessary to prevent epidemiologica; contacts and break the epizootic chain that allows for viral circulation in the biocenosis. Long-lasting points of extermination and plough grooves may be used in lake basins.

Mikhail Chumakov derived a specific OHFV vaccine between 1947 and 1948. It successfully underwent clinical

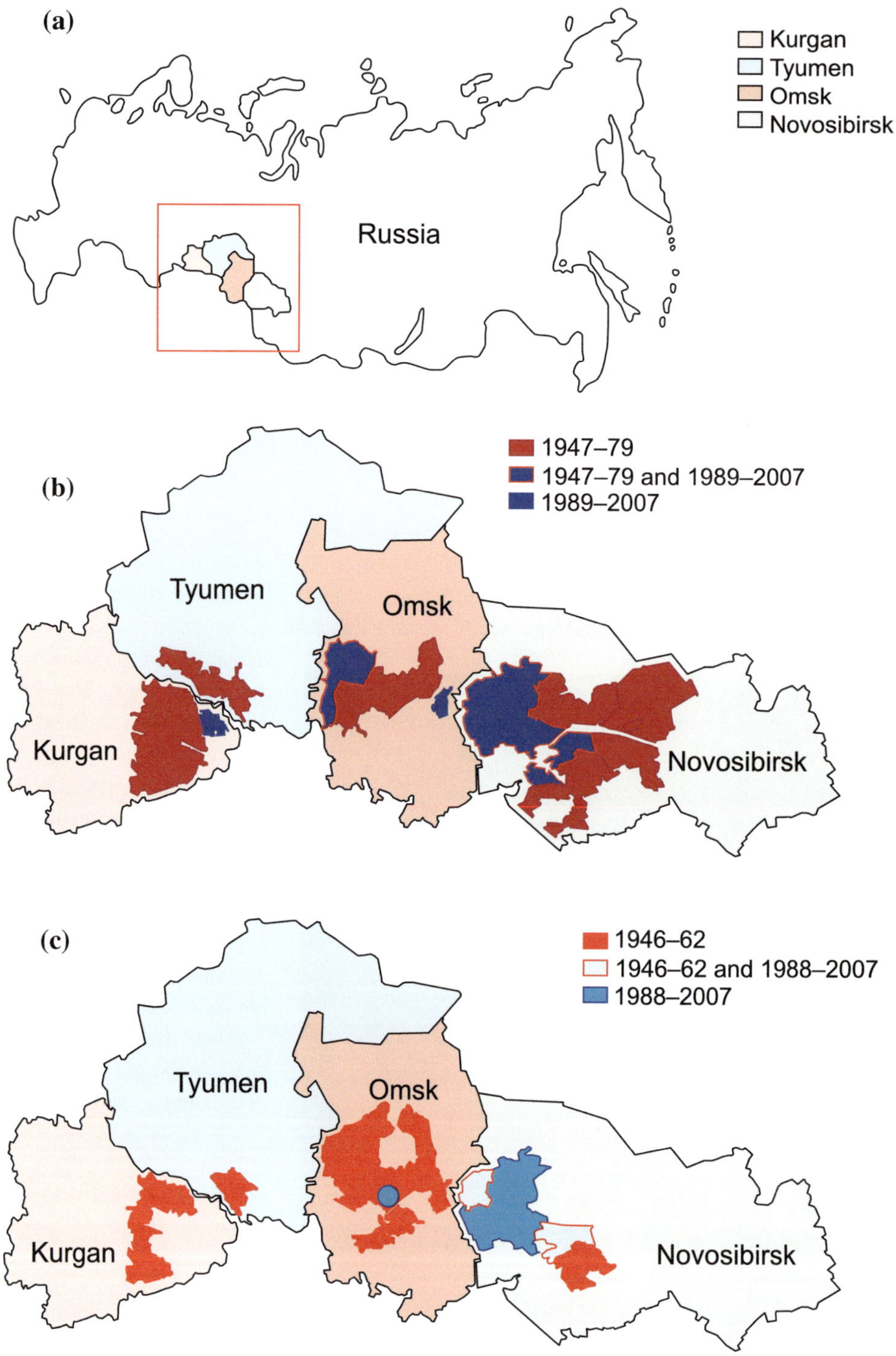

Fig. 4.41 OHF geographical distribution in Russia (A—endemic regions, B—epizootics, C—human morbidity). *Credit* Ruzek et al. (2010)

trials a year later in OHF natural foci. However, this vaccine is not regularly produced, given the rapid decrease in morbidity rates among the local population, meaning that production essentially stopped beginning in the 1950s. Today, the tick-borne encephalitis vaccine is seen as sufficient protection given the antigenic similarity of these two arboviruses. It is administered as needed.

4.10 Yersinia Pseudotuberculosis

Yersinia pseudotuberculosis infection (also known as Far East scarlet-like fever) is an acute zoonotic infectious disease characterized by fevers, general toxicosis and malaise, gastrointestinal and motor symptoms (Pokrovsky et al. 2007).

4.10.1 General Information

Yersinia pseudotuberculosis infection is caused by the *Yersinia pseudotuberculosis* bacterium. These are agile facultative anaerobes in the *Yersinia* genus of the *Enterobacteriaceae* family. They can divide in conditions ranging from 2 to 40 °C, but the optimal temperature for their growth is between 22 and 28 °C. The bacteria can divide on food in refrigerators at a temperature of 4–6 °C. They are fairly resilient to freezing and thawing and can survive for a long period of time in soil and water. These characteristics are very epidemiologically significant. In addition, *Yersinia* is sensitive to sunlight, drying, boiling and the use of general antiseptics. The pathogenic characteristics of *Yersinia* are based on their main toxins: enterotoxin, endotoxin and cytotoxins (Somov et al. 2001).

The incubation period lasts from 3 to 10 days. Disease onset is sudden. Usually, symptoms of general intoxication and malaise appear first: chills, fever, headaches, weakness, loss of appetite, insomnia, sweating and pain in the right half of the abdomen (Naktin and Beavis 1999). Pseudotuberculosis is characterized by a scarlet fever-like rash, which appears between the second and fourth days of illness, which is why it used to be called Far East scarlet-like fever (Pokrovsky et al. 2007). The rash is replaced by characteristic peeling of the skin, especially on the palms, fingers and soles of feet. Aside from the scarlet-like rash, jaundice, joint and septic rashes are seen, among others.

4.10.2 Epidemiology and Natural Focus Characteristics

More than 110 species of mammals, birds, reptiles, amphibians and arthropods are naturally susceptible to the pathogen (Gamble and Clancy 2013). Small mammals (common voles, striped field mice and rabbits) serve as the main reservoir of pseudotuberculosis, as they are widespread and reproduce quickly, meaning that their populations turn over quickly (Pokrovsky et al. 2007). Various livestock, as well as domesticated and synanthropic animals, especially pigs, large and small cattle and dogs are also susceptible to the pathogen (Souza et al. 2013). Synanthropic and other rodents are the main reservoir of the pathogen for human infections (Fig. 4.42). They are highly susceptible to Yersinia and can always infect food, water and soil, where the pathogen can not only survive for long periods of time, but also sometimes reproduce. The pathogen is passed within rodent populations through faecal–oral transmission.

The frequency of infection in small mammals depends on the conditions of infection, which are more favourable in anthropurgic foci than in nature. Seasonal migration of synanthropic rodents (rats, house mice) from buildings to

Fig. 4.42 The black rat is the main carrier of the pseudotuberculosis pathogen in anthropurgic foci (www.rg.ru). *Credit* http://www.rg.ru/2014/05/13/krisi-siteanons.html

natural habitats in the spring leads to pathogen transfer from natural foci to anthropurgic ones. This process occurs in the opposite direction in the fall. Long-term and ample microbe emission leads to the sowing of bacteria in soil, standing water, feed and food products, allowing for the infection of healthy animals (Yuschenko 2000; Tseneva 2006).

Disease transmission to humans is faecal–oral, primarily via food (Tseneva 2006). Pathogen transmission occurs during the consumption of raw or improperly cooked meat, dairy and vegetables, even those that were refrigerated. Evidently, humans are not very naturally susceptible. Severe cases generally occur in immunocompromised children. Cases of human infection directly from mammals or sick humans are unknown.

Yersinia's resilience in a variety of habitats and ability to reproduce at low temperatures allows for their accumulation in animal and vegetable products. Epidemiological studies of pseudotuberculosis outbreaks have led to the conclusion that vegetables, including root vegetables, are most significant, followed by dairy products (cottage cheese and cheese), and finally fruits (dried fruits), bread and candy. The majority of pseudotuberculosis outbreaks occur after consumption of vegetables that were kept in storehouses for prolonged periods of time. Insufficiently washed greens and salads are a common cause of pseudotuberculosis outbreaks. An epidemic that occurred in New Zealand in 2014 and led to 217 reported cases can serve as an example (Outbreak… 2014).

It is possible that insects, such as flies, cockroaches and ants can also play an epidemiological role in spreading the pseudotuberculosis pathogen as mechanical transmitters. Data exist on the role of the housefly (*Musca domestica*) in mechanical bacterial transmission. It has been shown that it is capable of harbouring viable microbes on its body 36 h after pathogen contact (Zurek et al. 2001). Houseflies inhabit human homes and settlements, feed on food waste and can come into contact with the bodily fluids of infected people, as well as waste in trash receptacles.

The specifics of pseudotuberculosis epidemiology predetermine its main distribution in cities. A number of social conditions allow for the epidemic conditions surrounding pseudotuberculosis specifically in cities. First, this epidemiological danger is caused by a large number of vegetable storehouses that do not meet sanitary norms, as well as many preschools, schools and summer camps with local kitchens, storehouses, etc. (Daiter 1985).

In the Russian Federation, clusters of pseudotuberculosis cases are most often reported in preschools, schools, boarding schools, rural children's organizations, military units, factories and educational institutions that united by a common food source.

4.10.3 Historical Background

The pseudotuberculosis pathogen was first isolated in 1883 from the guinea pig by Louis-Charles Mallasez and V. Vignal and named *Pasteurella pseudotuberculosis.* The term 'pseudotuberculosis' was based on the disease's post-mortem characteristics, which led to the formation of growths externally similar to tubercules, but different in aetiology. Later, the *Pasteurella* genus was renamed as the *Yersinia* genus, and the pathogen acquired its modern name—*Y. pseudotuberculosis.*

For a prolonged period of time, pseudotuberculosis was considered an animal disease and was rarely and usually accidentally diagnosed in humans (Timchenko and Popov 2014). The first detailed clinical description of pseudotuberculosis in humans was conducted by Masshoff and Knapp in 1954. A new period of research on the infection began in the late 1950s, when a number of servicemen in Primorsky Krai contracted a disease that caused fevers, general intoxication and rashes. It was then called Far East scarlet-like fever (Pokrovskiy 1983). Znamenskiy and Vishnyakov isolated the pseudotuberculosis bacillus in 1967. The etiologic role of the pathogen in Far East scarlet-like fever was proved by Znamensky by autoinfection (Pokrovsky et al. 2007). In 1989, Znamensky among other scholars received the USSR State Prize for his role in developing treatments for pseudotuberculosis.

4.10.4 Distribution in Russia and the World

Pseudotuberculosis is ubiquitous. Cases have been found on every continent and in every country (Pokrovskiy 1983). Today, the infection is found nearly everywhere in Russia and the Commonwealth of Independent States, although its distribution is non-uniform. Initially, pseudotuberculosis was characterized by large epidemic outbreaks in Russia's Far East, and in the 1960s and 1970s, this region essentially defined the morbidity rate for the entire former USSR. 50–85% of all morbidity could be account for by the outbreaks. Epizootics and the formation of anthropurgic foci were facilitated by the natural, socio-economic and agricultural characteristics of the eastern and northern regions of Russia (Somov et al. 2001). It was later established that pseudotuberculosis' range spans far beyond the Far East region. Cases of this infectious disease of varying intensity were reported in the northwestern, central and southern parts of Russia, in Tatarstan, Siberia, as well as Kazakhstan, Uzbekistan, Armenia, Ukraine and Belarus.

Pseudotuberculosis is found universally in Russia, although long-term average annual morbidity rates can vary quite significantly between administrative subjects (Map 4.19). The highest long-term annual average morbidity rate was reported in the south of West Siberia: in Tomsk, Kemerovo, Novosibirsk and Tyumen Oblasts, as well as Khakassia, Kamchatka Krai and St. Petersburg. High morbidity rates are also found in the northern European territory of Russia (in Arkhangelsk, Leningrad and Murmansk Oblasts), in northeastern Russia (Chukotka, Magadan Oblast), in Primorsky Krai and on Sakhalin Island.

It is noted that pseudotuberculosis morbidity rates have decreased more than sevenfold in Russia in the past 10 years. In 2013, morbidity rates decreased by 33.6% in Russia compared with 2012 (On the Status… 2014).

The outbreak nature of pseudotuberculosis spread, along with the consistency of infection in specific territories, should be mentioned. Because Moscow and St. Petersburg are independent administrative subjects and report isolated epidemiology statistics, their examples serve to explain the role of cities in the overall picture of morbidity: a substantial number of cases is reported practically every year.

4.10.5 Prevention

A pseudotuberculosis vaccine has not been developed. Measures to prevent infection among people consist of measures that fight the pathogen and sources of infection in natural and anthropurgic foci. In the interest of preventing pseudotuberculosis cases in humans, disinfection and rodent extermination events are conducted, including technical, sanitary, agricultural and forestry events. The most important measures, according to methodological guidelines (MG 3.5.1.2958–11 (2011)) are the following:

- Tight closing of doors (including covering vents with screens to prevent rodent entry).
- Cleanliness in work and storage areas, basements and enclosed territories.
- The use of sealed containers for food and waste storage.

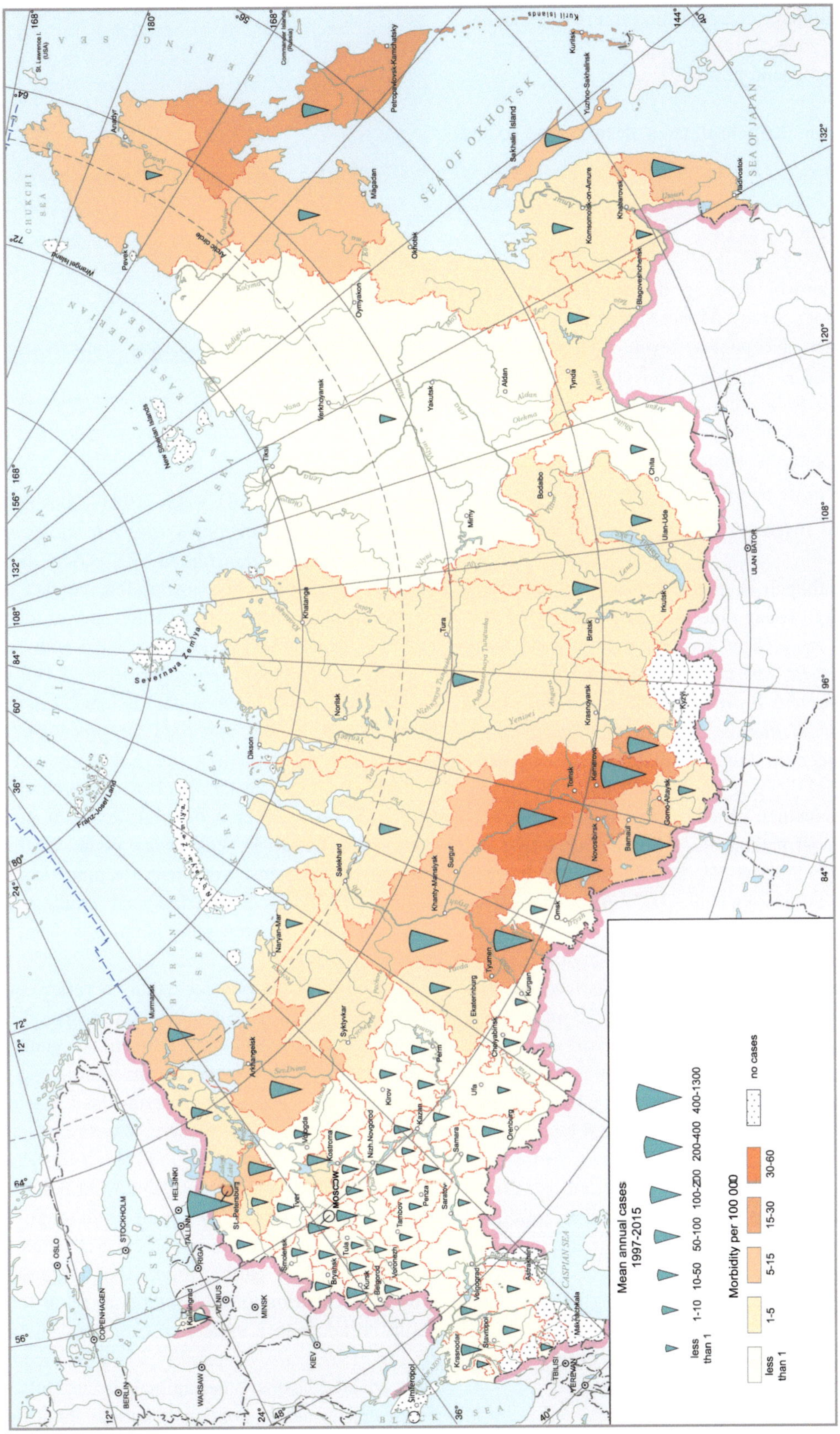

Map 4.19 Pseudotuberculosis morbidity, 1997–2015

- Sanitary cleaning of forest and park territories.
- Sanitary and maintenance tree-felling in urban parks and suburban forest zones.
- Deep ploughing of land in fields.

Extermination can help lead to a decrease in rodent populations.

4.11 Brucellosis

Brucellosis (also known as bang disease, Cyprus fever, Gibraltar fever, goat fever, Maltese fever, non-familial Mediterranean fever, Neopolitan fever, melitensis septicaemia and phthisis) is a zoonotic infectious and allergic disease, which tends to be chronic and is characterized by a long-term fever and attacks the motor, cardiovascular, nervous and urologic systems, among others (Postovit 1997).

4.11.1 General Information

The brucellosis pathogen is in the *Brucella* genus of the *Brucellaceae* family. Today nine brucellosis pathogen species are known to researchers, six of which are most common: *Br. melitensis, Br. abortus bovis, Br. abortus suis. Br. neotomae, Br. canis* and *Br. ovis.* Goats and sheep are the main carriers of *Br. melitensis,* cattle of *Br. abortus bovis,* swine of *Br. abortus suis,* desert shrub rats of *Br. neotomae,* dogs of *Br. canis* and sheep of *Br. ovis.* Brucellae are divided into biotypes based on their varied pathogenicity for humans. *Br. melitensis* is most pathogenic, while *Br. abortus bovis* and *Br. abortus suis* are less clearly pathogenic to humans (Wilks et al. 2003; Pokrovsky et al. 2007). *Br. neotomae* are non-pathogenic to humans, while the significance of *Br. ovis* is currently under study. The microbes are highly invasive and can enter the organism even via unharmed mucous membranes. Although they are categorized as intracellular parasites, they are found outside of cells and are can endure the outside environment (Turyanov et al. 1998).

The incubation period lasts 7–30 days. Clinical symptoms vary from latent brucellosis with enlarged lymph nodes and neurologic symptoms to clinically strong forms with prolonged fevers, chills, excessive sweating and attacks on various organ systems (Fig. 4.43). Post-infection immunity usually lasts between 6 and 9 months. Recurrences are seen in 2–7% of cases (Turyanov et al. 1998).

4.11.2 Epidemiology and Natural Focus Characteristics

Bacteria in the *Brucella* genus have been found in over 60 vertebrate species, but the role of wild animals in human infection is small (Pokrovskiy 1983). Sheep, goats, cattle and swine are the infection's reservoir and main source.

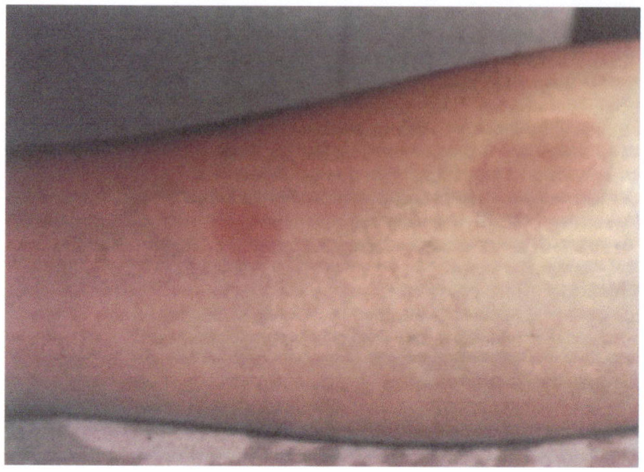

Fig. 4.43 An intradermal positive allergic probe. *Credit* Turyanov et al. (1998)

The pathogen is transmitted by a variety of mechanisms, usually via faecal–oral transmission; contact and airborne transmission are also possible. Raw dairy products, meat and materials from goats and sheep infected with brucellosis are the most dangerous in this regard (Fig. 4.44). Infection via aspiration is possible when dust is inhaled in animal enclosures or raw animal materials are processes. An infected person poses no epidemiological risk (Turyanov et al. 1998). Brucellosis is known for its clear occupational characteristics: it is most common in rural areas among people who work with livestock (Pokrovsky et al. 2007) as well as among veterinarians, veterinary technicians, insemination service employees, zoo technicians, employees of slaughterhouses and meat processing plants (Galińska and Zagórski 2013).

Brucellosis is known for having a variety of focus types (Voronov 1989). The disease is widespread among wild animals; however, although humans are generally infected through milk, it is believed that its natural foci are not very epidemiologically significant. Brucellosis has been found in bison, boars, red deer, reindeer, moose and other animals, including small mammals (Godfroid 2002; Zheludkov and Tsirelson 2010). *Brucella* circulation in natural ecosystems has not been sufficiently studied, although it has been established that the bacteria can exist in natural ecosystems without livestock participation. The ubiquitous nature of brucellosis distribution is conditional upon the pathogen's ability to adapt to a variety of ecological niches. The possibility of *Brucella's* capability to migrate from its biologically adapted host to other animals is currently under study. For instance, nearly nothing is known about the possibility of pathogen exchange between wild animals and livestock (Godfroid et al. 2013).

Fig. 4.44 Large cattle serve as a reservoir for the brucellosis pathogen. *Credit* N. G. Kadetov (permission obtained)

Foci linked to infected domesticated animals, especially goats and sheep, pose the greatest danger for people. Human morbidity is closely linked to epizootics among animals. Human infection rates in these foci can reach 70%, and the disease has a very specific set of clinical symptoms. The greatest number of infections with goat and sheep brucellosis occurs in the spring and summer. In cases of infection from cattle, the disease is less seasonal, which is explained by the longer lactation period, as most infections are the result of transmission via milk or dairy products (Turyanov et al. 1998).

Fig. 4.45 David Bruce (1855–1931). *Credit* Wellcome Collection https://wellcomecollection.org/works/whrsy6e8 Attribution 4.0 International (CC BY 4.0)

4.11.3 Historical Background

Brucellosis has been known to humans since the era of Hippocrates, but it was first scientifically examined only in the 1860s. At the time, it was known as Mediterranean or Maltese fever, because it was widespread on the island of Malta in the Mediterranean (Tarshis and Cherkassky 1997). David Bruce, the Scotch doctor and microbiologist, first isolated the pathogen in 1886 (Fig. 4.45). Later, the bacterial genus was named after him (Wilks et al. 2003). The discovery of the pathogen in the milk of local goats allowed a conclusion to be reached on the disease's zoonotic nature. The Russian scientists A. A. Kramnik, E. I. Martsinovsky, A. F. Bilibin, V. N. Beklemishev, P. A. Vershilova and I.

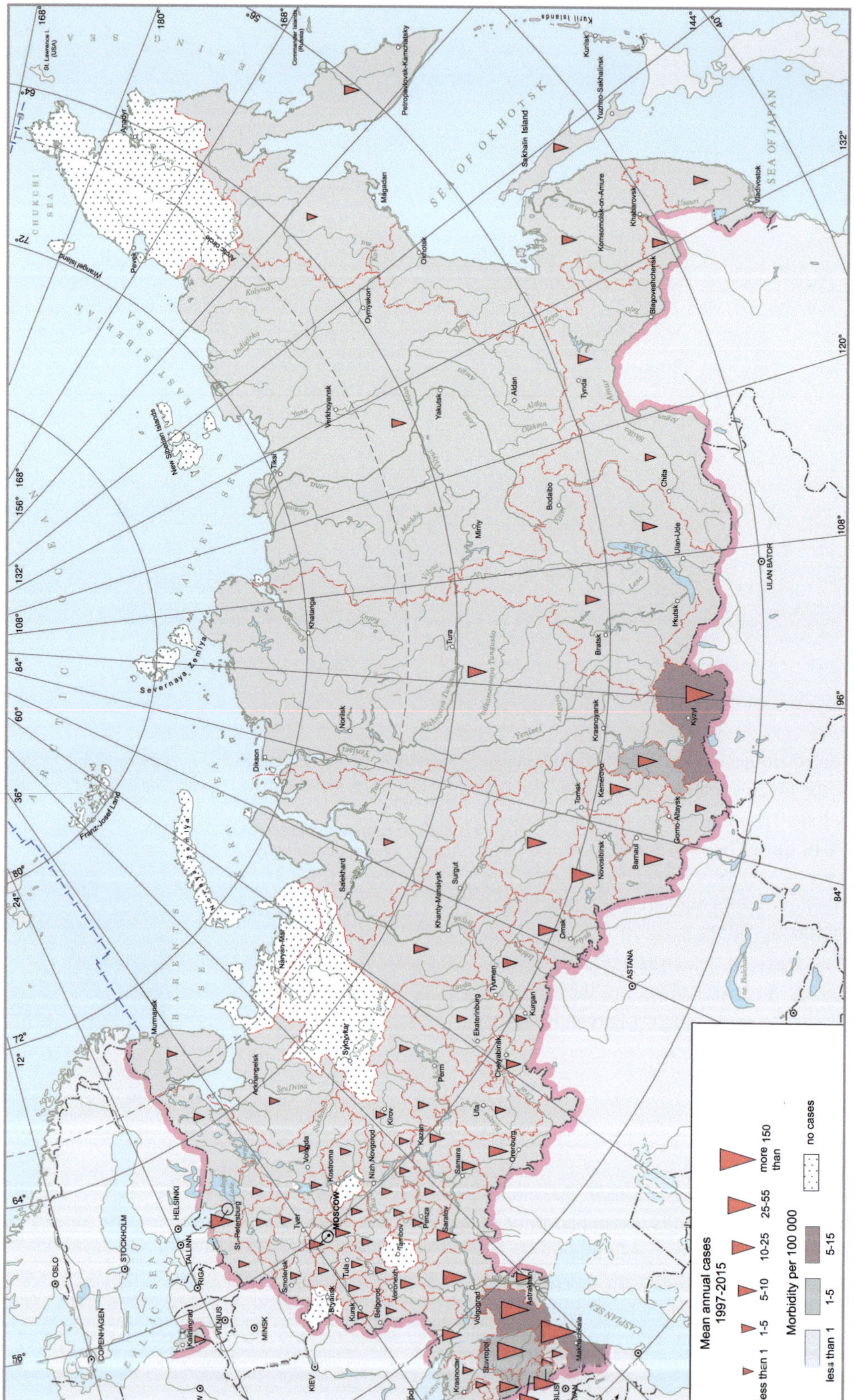

Map 4.20 Brucellosis morbidity, 1997–2015

S. Bezdenezhnykh made a large contribution to the study of the clinical symptoms and epidemiology of brucellosis, as well as the development of therapies for the disease.

4.11.4 Distribution in Russia and the World

Brucellosis foci have been found on every continent. Sporadic morbidity and periodic outbreaks are usually found in areas where people tend to livestock. In Europe brucellosis is most commonly reported in Greece, Spain, Italy, Portugal and France. In these countries, most cases of human disease are linked to infections among sheep and goats. 354 confirmed brucellosis cases were reported in 2014 in European Union countries (ECDC 2016). The majority of cases in the Americas are found in Mexico, Argentina and Peru (97%) and are related to infection from cattle (Pokrovskiy 1983; Pappas et al. 2006). There is not much data available on brucellosis in Africa and Asia. Epizootics and a high brucellosis morbidity rate remain in the Commonwealth of Independent States, mostly in Kazakhstan and Central Asia. Infected raw material may be brought to Russia from these countries (On the Status… 2014).

In the past decades in Russia, 400–550 new cases are diagnosed annually, while in 2015 the morbidity rate was 0.76 cases per 100,000 individuals. Brucellosis is non-uniformly distributed among regions. Higher morbidity rates are found in Dagestan, Kalmykia, Karachay-Cherkessia and other territories in the North Caucasus (Map 4.20). An increase in the number of epizootics has been tied to the uncontrolled introduction of large and small livestock from enzootic regions to adjacent territories. This introduction violates sanitary and veterinary regulations. Higher morbidity rates are also found in southwest Siberia, as well as Tyva and Khakassia.

4.11.5 Prevention

Prevention consists of a set of veterinary, facility and sanitary measures, including disinfection of livestock, disinfection of the livestock's environment and disinfection of animal products and materials. Vaccination is considered to be a temporary preventative measure.

4.12 Psittacosis

Psittacosis (also known as ornithosis or parrot's fever) is an acute infectious natural focal disease that affects birds, mammals and humans. It is characterized by overall toxicosis and malaise, pulmonary symptoms, neurologic symptoms and spleno- and hepatomegaly (Encyclopedia… 1998; Pokrovsky et al. 2007).

4.12.1 General Information

The disease is caused by *Chlamydia psittaci*, an obligate intracellular parasite in the *Chlamydia* genus of the Chlamydiaceae family. The pathogen can survive freezing and can survive in the outside environment for up to 2–3 weeks. People are natural highly susceptible to the pathogen (Pokrovsky et al. 2007; Prevention… 2010). In the vast majority of cases, disease onset is acute. Body temperature rises rapidly and reaches its peak within 24–48 h. Weakness and malaise are common symptoms soon after disease onset.

In addition, pulmonary symptoms soon appear, such as a mild cough (usually between the 2nd and 4th days of illness). By the end of the first week, most patients (70%) exhibit hepatomegaly, while splenomegaly is less common. After 7–14, the disease begins a decline, but full recovery may take longer. Furthermore, X-ray abnormalities may remain for a couple weeks. 10–12% of patients develop chronic disease, which usually takes the form of bronchitis or other system-wide attacks or symptoms and can last up to a few years (Turyanov et al. 1998 and Pokrovsky et al. 2007). Often the disease takes an asymptomatic or mild (flu-like) course and is misdiagnosed as pneumonia. A specific psittacosis study in a number of countries (Bulgaria, Holland, Germany, the USA, etc.) revealed that 10–20% of acute pneumonia are caused by psittacosis (Encyclopedia… 1998). Psittacosis was found in 18.4% of pneumonia patients in Moscow and 19.6% of pneumonia patients in St. Petersburg (Briko 2010).

4.12.2 Epidemiology and Natural Focus Characteristics

Wild, primarily aquatic, domesticated and pet birds serve as the natural reservoir for psittacosis (Fig. 4.43). Today, the psittacosis pathogen has been found in over 140 bird species in the world and 70 bird species in Russia. These species span from Russia's southern border to the far north. Domesticated and caged birds (ducks, turkeys, parrots (Fig. 4.47), canaries, etc.) have great epidemiologic significance, while feral pigeons (Fig. 4.46) pose the greatest risk (prevalence in this population wavers between 30 and 80%) (Sysuev et al. 2013). Bird exoparasites and rodents, as well as infected humans, can serve as additional pathogen reservoirs.

Fig. 4.46 Pigeons are the primary reservoir of psittacosis. *Credit* N. G.Kadetov (permission obtained)

Fig. 4.47 A parrot infected with psittacosis. *Credit* http://www.eastbundabergvet.com.au/wp-content/uploads/2016/06/P2050135.jpg

Pathogen transmission from birds can occur via direct contact if a sick or dead bird is touched and indirect contact via the air, if infected down particles, nasal mucus droplets or excrement particles are inhaled. Moreover, infection can occur through damaged mucous membranes or skin after a bite or scratch from an infected bird, as well as after consumption of contaminated food, including raw eggs. Airborne transmission is most common, 10% of transmission cases are foodborne. Humans transmit the infection via droplets (Prevention… 2010). Psittacosis is a gastrointestinal infection in birds, and the pathogen is first localized to the abdominal cavity. Feathers and down that are contaminated with discharge and dry excrement can be transported long distances.

Epidemic foci are usually occupational in nature and usually appear in poultry farms or plants, pet stores and zoos. Cases are most commonly reported in late summer and early autumn. Cases of home infection are typically sporadic, although familial psittacosis foci have been found. Such cases usually occur in the winter months and develop after new birds are purchased (Sysuev et al. 2013).

4.12.3 Historical Background

The disease was first described by T. Jurgensen in 1876 and termed 'atypical pneumonia'. In 1879, D. Ritter linked it to infections from parrots. A disease outbreak caused by parrot brought from Brazil occurred in 1892 in Paris, giving the disease the name 'psittacosis', which means parrot disease. For a long time, the pathogen was poorly studied and unclassified. It was first isolated by S. P. Bedson in 1930. Later, psittacosis was found in other bird species, and it was also named ornithosis in 1941 (Pokrovsky et al. 2007).

4.12.4 Distribution in Russia and the World

Psittacosis is widespread in every country, which is related to its circulation in birds and their migratory lifestyle and ability to travel long distances (Wilks et al. 2003). Sporadic infections and outbreaks among people are found everywhere in the world.

For instance, an average of 100 cases is reported annually in the USA, one of which is fatal. 935 cases were reported in the USA from 1998 to 2003 (Smith et al. 2005).

Psittacosis has been reported in 45 regions of the Russian Federation (Map 4.21). The long-term average annual morbidity rate over the course of 19 years (1997–2015) does not exceed 1 case per 100,000 individuals. The greatest annual average number of cases (23) was reported in St. Petersburg. In the majority of regions, in Russia, the annual average number of cases is less than 1. Leningrad Oblast, Komi Republic, Kurgan, Ulyanovsk, Voronezh and Lipetsk Oblasts, as well as Moscow, serve as the exception: the average annual number of cases ranges from 1 to 5 in these regions. Outbreaks are periodically reported on these territories. As such, two foci were recorded in Kurgan and Orenburg Oblasts from December 2008 to January 2009 with 21 and 15 cases respectively. Nine cases of psittacosis were reported in Moscow at the same time.

Psittacosis has been reported in the Russian Far East, but the morbidity rate and annual average number of cases are both low.

4.12.5 Prevention

Preventative measures are largely based on fighting the disease in birds. It is important to adhere to veterinary and sanitary norms when bringing birds from outside of Russia, transporting and maintaining birds in poultry plants, farms, zoos, etc. Personnel needs to be provided with personal protective equipment and disinfectants. Immunological preventative measures have not been developed.

4.13 Rabies

Rabies (hydrophobia) is a natural and anthropurgic infectious disease that affects warm-blooded animals and humans. It affects the central nervous system and leads to a fatal outcome. Rabies is a widespread disease among animals.

4.13.1 General information

Rabies is caused by a neurotropic virus in the *Lyssavirus* genus of the Rhabdoviridae family. Rabies pathogen strains that circulate in nature are pathogenic to all warm-blooded animals. The natural variability of the virus's biological traits has been established. Typical strains that cause the expected violent disease course are isolated most often. Intensive strains that lead to paralysis after a short incubation period are also known to researchers. Strains linked to rabies-like disease of Arctic foxes, 'wild dog disease' in West Africa and vampire bat transmission of rabies in cattle in Central and South America also have unique features. They cause disease with a prevalence of paralytic symptoms (Konopatkin et al. 1993). More than 99% of all rabies cases in humans are transmitted from dogs, predominantlly in low-income countries (WHO 2013; Hampson et al. 2015). Bats becoming an increasingly common source of human infection, especially in settings where rabies is controlled by vaccination in dogs and other carnivores (Begeman et al. 2017).

The virus can withstand low temperatures and can survive in the brains of buried animal cadavers (Konopatkin et al. 1993). It can remain in superficial soil layers for 2–3 months. However, it dies at high temperatures, after exposure to sunlight, drying and several founds of freezing and thawing (Rothschild 1992).

Transmission of the virus from an infected animal to a person occurs after a bite or saliva access through a wound or minor injury to the skin. More rarely, it can be transmitted through mucous membranes. Cases of human-to-human transmission have not been reported. The incubation period usually lasts 1–3 months, but can be as short as 1 week or over a year long. A high fever, pain and strange feelings of tingling or 'pins and needles' at the wound site are the earliest symptoms. As the virus spreads across the central nervous system, fatal inflammation of the brain and spinal cord occurs. People with violent rabies can be hyperactive, agitated, hydrophobic and periodically aerophobic. Death occurs after a few days (usually one week). 30% of rabies

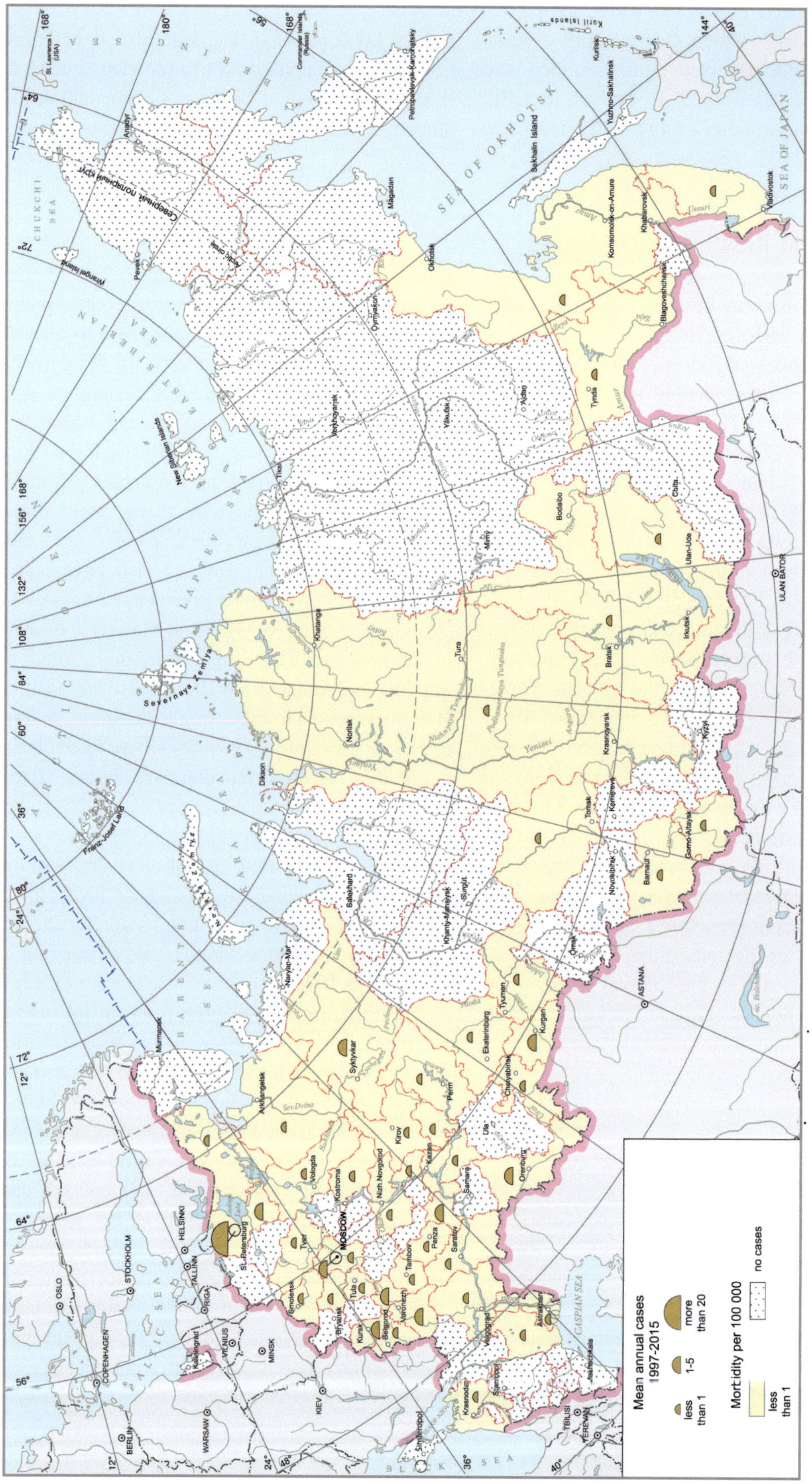

Map 4.21 Psittacosis morbidity, 1997–2015

cases in humans are paralytic. In this case, the disease course is less dramatic and usually longer. The muscles gradually become paralyzed, beginning from the wound or scratch site. A coma develops that ultimately ends in death. In this case, rabies is often misdiagnosed, artificially lowering disease data (Informational Bulletin 2014).

4.13.2 Epidemiology and Natural Focus Characteristics

Wild carnivores and dogs, as well as bats in some regions, serve as a natural reservoir for rabies. Based on the reservoir's unique characteristics, rabies epizootics are classified as urban or natural. In urban epizootics, feral dogs serve as the pathogen's main sources and distributors. The epizootic's scale depends on the size of their population. In wild epizootics, the pathogen circulates among wild predators, especially those in the Canidae family. They are highly sensitive to the virus and become aggressive when infected, often migrate long distances and excrete the virus intensively in their saliva. These conditions, along with the high population density of some predators, rapid generational turnover and the prolonged incubational period of rabies, allow for a continuous epizootic process despite the rapid death of each individual infected animal. In North America, grey and red foxes, skunks and raccoons play the main role in sustaining wild epizootics. This role is played by jackals in South Asia and North Africa, red foxes in Central and Western Europe, red foxes, corsac foxes, raccoon dogs and wolves in Russia and arctic foxes in the tundra (Fig. 4.48). Bat rabies in South America is seen as a unique and independent type of wild epizootic; however, cases of infection from bats were reported in 1985 and 2007 in Primorsky Krai (Leonova et al. 2010; Rabies… 2013). Earlier, two new species of rabies-related viruses were isolated from bats in Eastern Siberia near Baikal Lake and in the Western Caucasus Mountains (Botvinkin et al. 2003; Kuzmin et al. 2005).

Fig. 4.48 The red fox (*Vulpes vulpes*). *Credit* A. A. Kadetova (permission obtained)

In Russia, the rabies epizootic is characterized by its connection to steppes, mixed coniferous and deciduous forests, and the forest-tundra and tundra, while the vast Northern Taiga is relatively free of disease. Natural foci localization corresponds with the habitats of foxes, corsac foxes, raccoon dogs, jackals, wolves and arctic foxes. Epizootic intensity depends on the population density of these animals. High population density leads to the rapid spread of disease, which substantially reduces predator populations. Moderate population density leads to one-off cases in a wide range and does not lead to a marked decrease in predator populations. Low population density leads to epizootic abatement. Epizootic cycles are also related to changes in predator population sizes: rabies is characterized by increases every 2–3 years, which lead to an increase in the nosogenic territory. An increase in cases in the winter and summer, caused by the biological characteristics of the main carriers (mating season, which facilitates intense animal-to-animal contact), explains the seasonality of rabies. Small predators, such as martens, weasels and least weasels, as well as rodents and wild herbivores, are sometimes drawn into the epizootic chain. However, these animals cannot participate in disease spread based on their lifestyle and behaviour and thus become a biological dead end for the pathogen (Konopatkin et al. 1993).

Arctic rabies virus is reported in the tundra and is common among arctic foxes. In a number of countries in Central and South America, vampire bats, fruit bats and insectivorous bats play a major role in the transmission of rabies. In some cases, bats can infect people and animals without biting them. Experiments confirming the airborne transmission of rabies, which is linked to its excretion in urine and milk, leading to air contamination in caves (Cherkassky 1985).

Humans, dogs, cattle, sheep, goats and horses are moderately sensitive to rabies, while birds are mildly sensitive. Young animals and humans are more susceptible to the virus (Konopatkin et al. 1993).

4.13.3 Historical Background

Rabies as a disease affecting people and dogs was described by doctors in the Ancient Middle East (3000 BCE). The first official mention of rabies was found in the code of law in Ancient Babylon (2300 BCE). Democritus (circa 500 BCE) described the disease in animals in detail, while Celsus (First century CE) described hydrophobia in humans (Neville 2004). Rabies has been known of in Western Europe since 1271. At this time, data on the spread of infection among wolves in France were recorded. In the sixteenth century, the

Italian Girolamo Fracastoro classified rabies as a disease transmitted by 'spores' (Anina-Radchenko 1979; Konopatkin et al. 1993).

In 1780, Danilo Samoylovitch expressed the conviction that rabies was contagious and could not occur spontaneously. But incontrovertible evidence of rabies' contagiousness was only found in the nineteenth century. In 1804, George Zinke, a German doctor, experimentally proved that the saliva of dogs transmits the infection. The French veterinarian P.-V. Gaultier artificially produced rabies in rabbits from 1879 to 1881 and reported on the possibility of vaccinating sheep by intravenously injecting the saliva of infected animals. The results of these studies laid the foundation for the discovery of a rabies vaccine by Louis Pasteur (Fig. 4.49) and his students (1881–1889). People were first vaccinated in 1885 (Konopatkin et al. 1993).

Rabies was reported in wild animals before the seventeenth century. Wolves served as the main reservoir. The role of domesticated animals, especially dogs, in disease transmission was insignificant. But in the early eighteenth century, epizootics among domesticated dogs began to be reported in Italy. Soon after, in 1728, epizootics were reported in France, Germany and Hungary. In the nineteenth century, an epizootic of rabies in dogs seized the entire continent. England, Germany, France and Russia were most severely affected (Selimov 1963). Some countries (Norway, Sweden, Denmark) developed sanitary measures, such as mandatory extermination of sick animals and pet quarantines. A rabies law was passed in 1897 in England: it allowed for the capture and extermination of stray dogs and chaining up of pets. As a result of such measures, rabies morbidity in dogs decreased substantially. By the end of the nineteenth century, rabies in dogs was eradicated in Great Britain and Scandinavia, and its spread was sharply reduced in Central Europe by the start of World War II (Anina-Radchenko 1979).

Rabies epizootics recurred intensely in the war and post-war years, which is when foci among badgers and foxes were found in Europe, Asia and the Americas. An epizootic

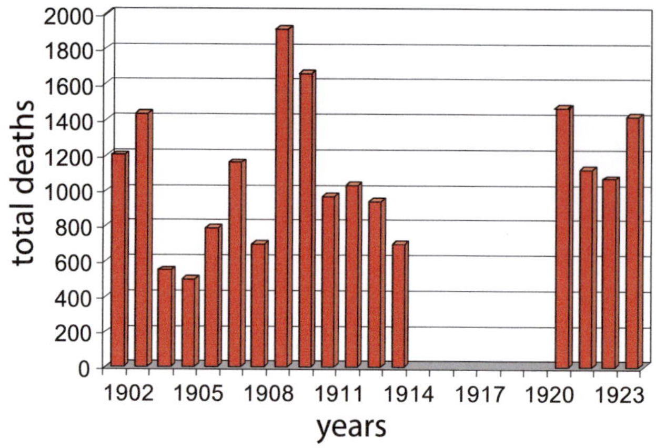

Fig. 4.50 Rabies Mortality Rate, 1902–1923 Pokrovsky et al. (2003)

of fox rabies infiltrated Austria, Switzerland, East and West Germany and other countries between 1950 and 1960 (Konopatkin et al. 1993).

Rabies cases have been officially reported in Russia since 1887, i.e. the year after the first Pasteur stations were opened. The first Pasteur station in Russia and the second in the world appeared in Odessa in 1886. By 1912, 28 such stations operated in Russia: they conducted population census and vaccination (Pokrovsky et al. 2003).

In the early twentieth century in pre-revolutionary Russia, 500–2000 people died of rabies annually (Fig. 4.50). After the revolution, morbidity rates gradually decreased, and almost no cases of rabies were reported during the Great Patriotic War (World War II).

Natural foci of fox rabies in Russia were found in the Volga Delta (Astrakhan Oblast) in the 1950s. In the following years, fox rabies foci appeared in a number of regions that used to be rabies free (Pokrovskiy 1983). In the 1950s, the number of rabies cases among humans in Russia reached 600 (200 of which were on the territory of modern Russia). By the late 1980s, the morbidity rate in Russia was successfully brought down to 50 cases. On average, about 10 cases of rabies were annually reported in Russia in the 1990s (Fig. 4.51).

Fig. 4.49 Louis Pasteur (1822–1895) Wellcome Collection https://wellcomecollection.org/works/gze65hju

4.13.4 Distribution in Russia and the World

All of North America and a substantial part of Europe have been overtaken by rabies epizootics fuelled by wild carnivores in the past 40–45 years. At the same time, dogs are the main transmitters of rabies in the majority of Asian, African and South American countries. According to the WHO's most recent data, rabies has been reported on every continent, in 160 countries (Fig. 4.52), however timely and quality control and preventative measures allows for the maintenance of a favourable epidemiological situation in

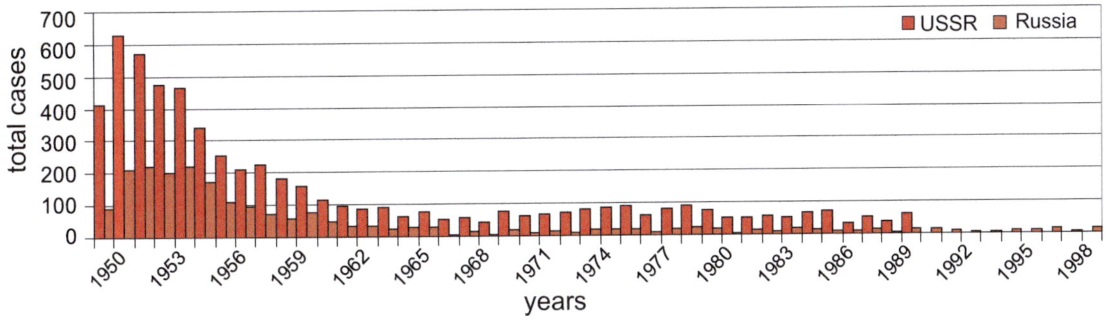

Fig. 4.51 Human rabies morbidity rate in Russia and the USSR, 1950–1998

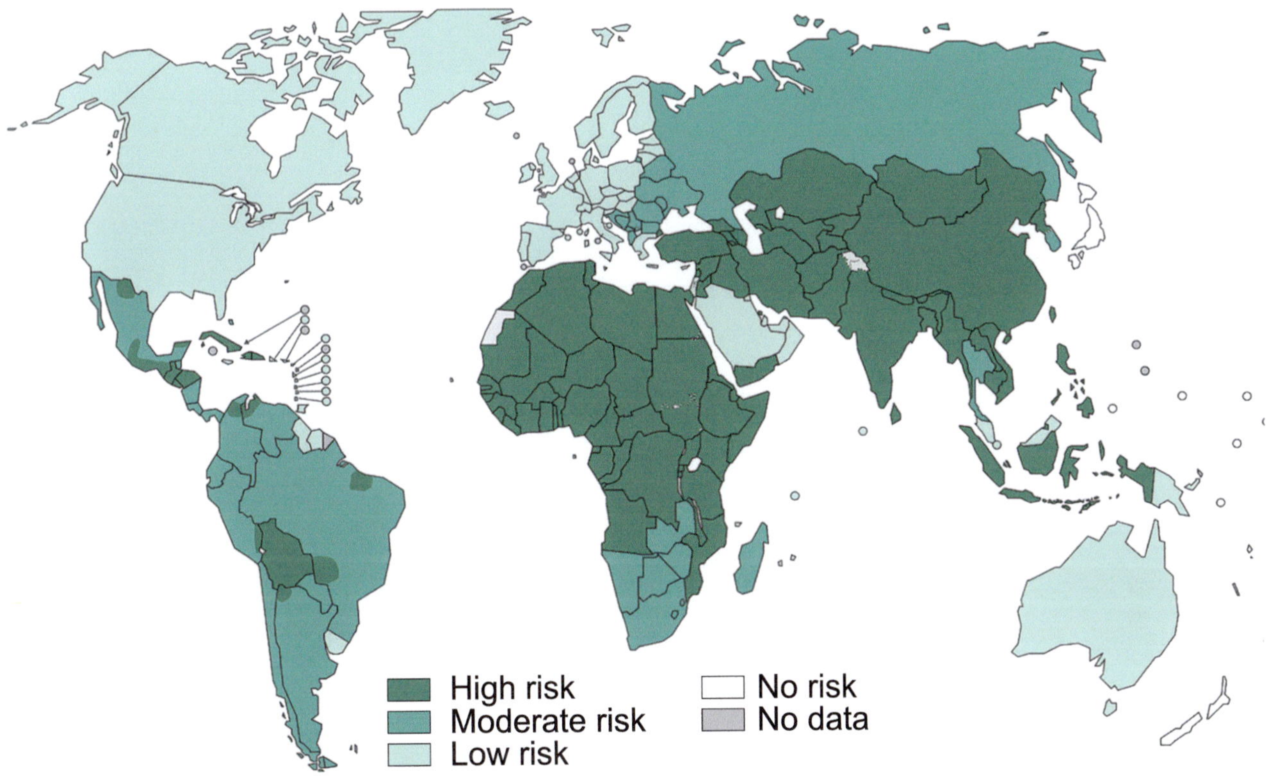

Fig. 4.52 A schematic map of rabies infection risk among humans in the world (WHO 2013). *Credit* http://www.who.int/neglected_diseases/NTD_A_statistical_update_latest_data_available.pdf?ua=1

Australia, Japan and the majority of developed European countries (Zavoloka and Zavoloka 2013; Informational Bulletin 2014).

Nine natural foci had been identified in Russia and the Commonwealth of Independent States by the early 1990s (Fig. 4.53). each of these regions is characterized by a unique qualitative and quantitative set of wild canines and varying morbidity rates in humans (Kuranshina et al. 2014).

An analysis of rabies cases in animals in Russia between 1960 and 2011 (Fig. 4.54) demonstrates a high level of cases among farm animals in the 1970s, low morbidity rates among animals between 1981 and 1995 with a small increase in 1985. A trend towards higher infection rates among wild and domesticated animals can be seen from 2000 to 2011. This may be related to changes in rabies control and prevention systems.

Data from 2010 to 2011 show that rabies cases in animals were reported in 58 and 62 administrative subjects of Russia respectively. 2751 unfavourable locations were reported, and 3188 animals fell sick and died, about half of which were wild. The most unfavourable regions for rabies are found in the Central, Volga and Ural Federal Districts, and 77% of all reported rabies cases in animals and 78% known epizootic foci are found there. In 2011, the greatest number of

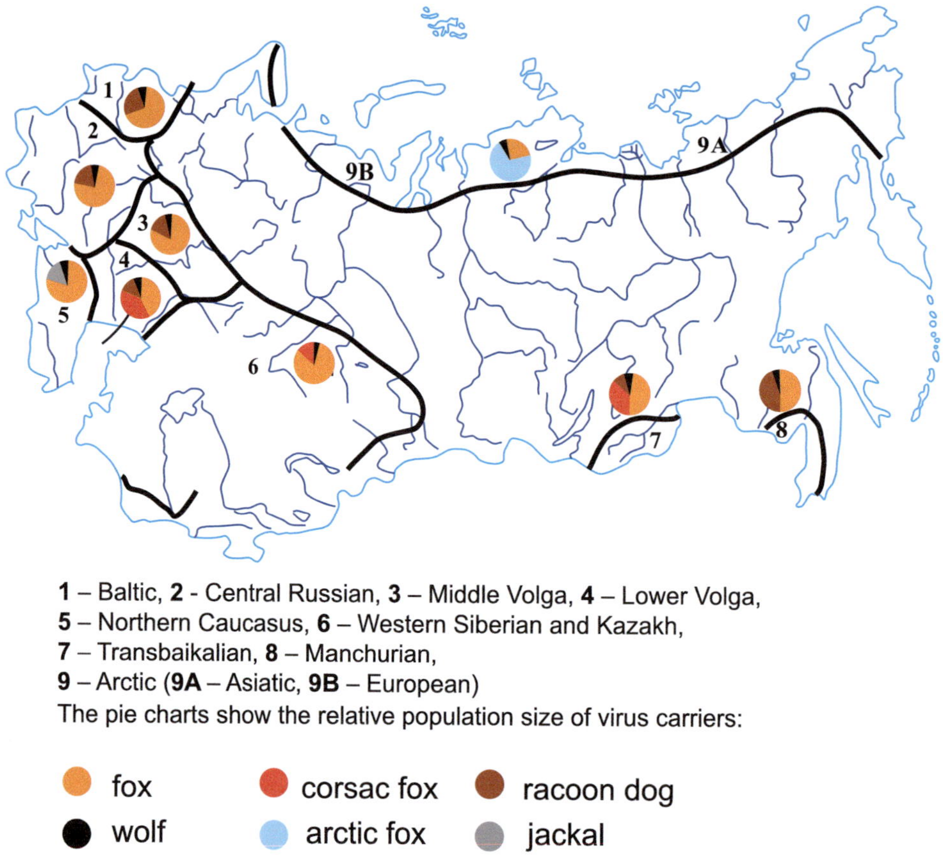

Fig. 4.53 Rabies natural foci in Russia and adjacent territories. *Credit* adapted from Sidorov et al. (1992)

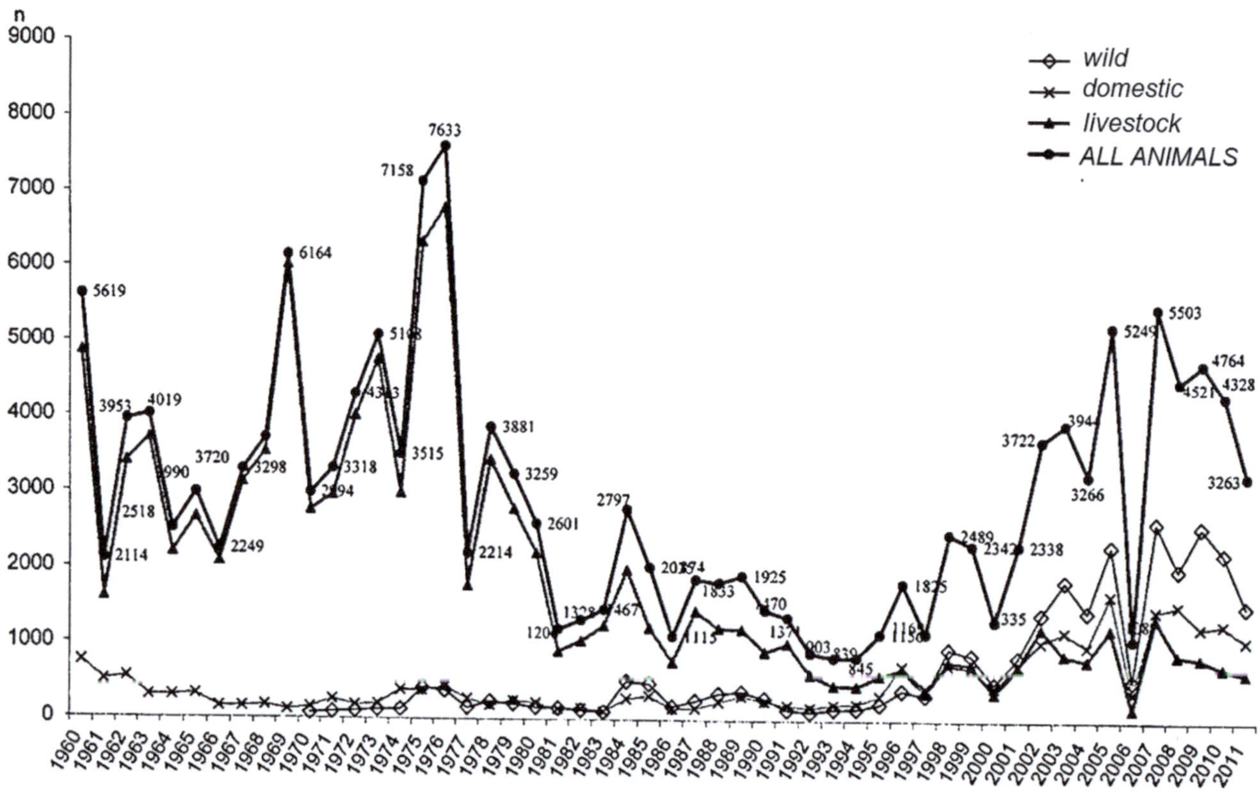

Fig. 4.54 Absolute rabies morbidity rates in animals in Russia, 1960–2011 (Rabies… 2013) [Wild, Domesticated, Farm, Total]

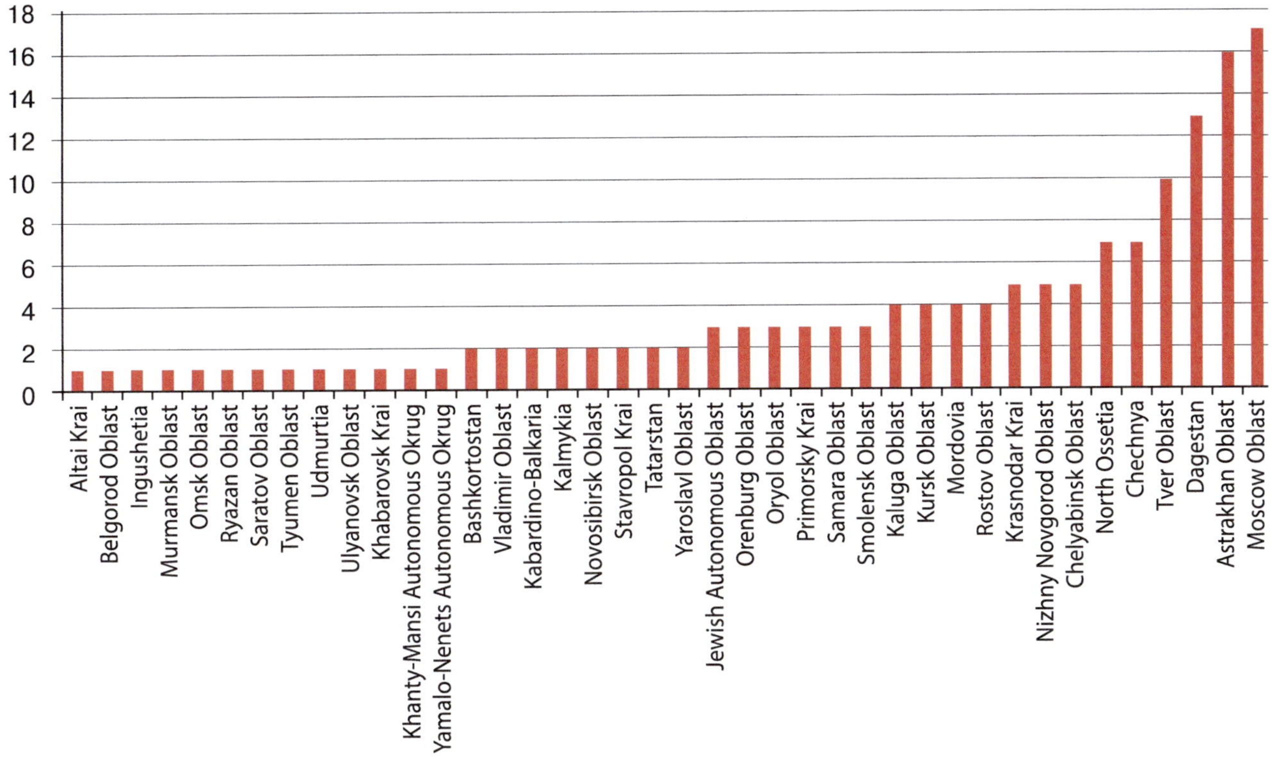

Fig. 4.55 Rabies cases in various administrative subjects of the Russian Federation, 1960–2011 (Rabies… 2013)

unfavourable stations was recorded in Moscow (235), Smolensk (200), Chelyabinsk (107) Oblasts and Tatarstan (118) (Epizootic… 2012).

148 rabies cases in people were reported in Russia in 2010 (Fig. 4.55). The majority is related to a refusal to vaccinate or patient non-compliance (alcohol consumption). A small number of cases was caused by dangerous bites to the face and neck, which cannot be helped by emergency vaccination.

4.13.5 Prevention

Annual vaccination is used to prevent rabies among animals. In any case of disease, and especially when rabies symptoms become evident (excessive salivation, difficulty swallowing, seizures), it is important to immediately seek help at the local veterinary station. Wild animals are vaccinated by way of bait delivery of rabies vaccines in common habitats.

In order to prevent rabies, anyone bit, scratched or covered with an animal's saliva, should immediately go to the nearest urgent care center to begin a course of rabies vaccines. Today's rabies vaccine has few side effects and leads to high immunity. The Russian rabies vaccine course consists of 6 injections: the vaccine is injected into the deltoid muscle (shoulder) at a dose of 1.0 mL.

Rabies in humans can only be prevented via preventative vaccines, the effectiveness of which depends on how quickly medical help was sought. An earlier course of rabies vaccines will lead to faster development of immunity and prevention of disease.

4.14 Anthrax

Anthrax is a natural focal infectious disease that affects animals and people. It is caused by a Category A priority pathogen, just like the plague and cholera (Act… 2004). One of anthrax's characteristics is its ability to take root in a location and pose the risk of repeat outbreaks for many decades after appearing there once (Bakulov et al. 2001).

For a long period of time, the anthrax pathogen was studied as a potential bioweapon. It has been officially reported that at least six countries weaponized anthrax, and some of them tested these bioweapons. In 2001, the anthrax pathogen was used in postal terrorist attacks in the USA that led to fatal outcomes (Marinin et al. 2008).

4.14.1 General Information

Anthrax is caused by the *Bacillus anthracis* bacterium, a relatively large spore-producing gram-positive bacillus 3–8, sometimes 10, nm long. It is found in vegetative and spore forms.

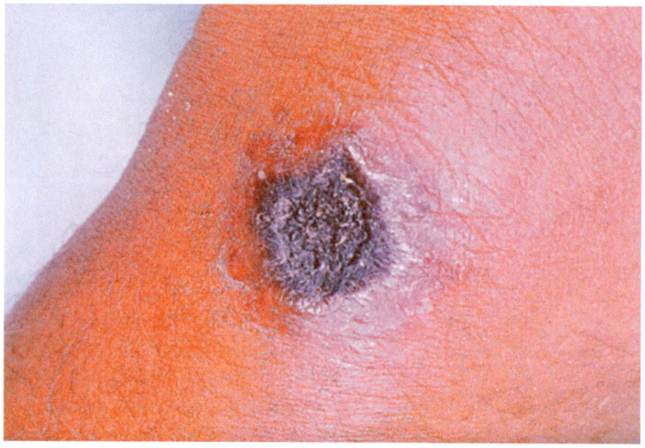

Fig. 4.56 Cutaneous anthrax lesion. *Credit* CDC—Division of Bioterrorism. Preparedness and Response (DBPR), National Center for Preparedness, Detection, and Control of Infectious Diseases (NCPDCID), Office of Infectious Diseases). **Public domain**

Vegetative *Bacillus anthracis* are relatively fragile; cells survive for 24–48 h in an animal organism, up to 1 week in the bone marrow, and up to 14 days on the skin. The anthrax pathogen is still viable after 1 successive freezes and thaws and can survive for up to 15 days in frozen meat at a temperature of −15 °C (Onischenko et al. 1999). In the outside environment, where oxygen is freely available, the vegetative cells turn into a spore. Unlike the vegetative form, spores are extremely resilient: after being boiled for 5–10 min they are still able to vegetate, and can survive in water or soil for over 10 years (Cola 1976; Cherkassky 1996).

Anthrax primarily attacks various wild animals and livestock, as well as people. Domesticated animals are the typical infection source for humans. A variety of epidemiologic classifications of human anthrax morbidity has been put forward (Onischenko et al. 1999). Under normal conditions, 95–97% of human cases are localized (cutaneous). Overall, three types of disease have been defined based on their origin: agricultural, industrial and accidental (or household) (Cherkassky and Zhanuzakov 1980). A red spot appears on the skin that gradually turns into an ulcer (Fig. 4.56). Fever begins at the end of the first 24 h and lasts for 5–7 days.

Both inhalation and gastrointestinal anthrax types are systemic. Fatality rates usually do not exceed 2–3% in cutaneous anthrax and reach 100% in systemic anthrax (patients die after 2–3 days after the onset of infectious and toxic shock) (Cherkassky 1996).

4.14.2 Epidemiology and Natural Focus Characteristics

Soil is the main habitat of *Bacillus anthracis*: it allows for stable population survival and the species' continued existence. Extensive data in the scientific literature serve as evidence for the fact that the primary physical and chemical conditions for the vegetation of anthrax in soil are: an environmental temperature no lower than 12–15 °C, humidity within the bounds of 29–85%, a neutral or slightly basic pH, and 3.1–7.3% humus (Ginsburg 1975; Cherkassky 2002).

Herbivores (wild hoofed mammals, small and large cattle, horses, deer, reindeer, camels, swine) are the main source of anthrax. Sheep, goats and cattle are most susceptible. An infected person poses no epidemiological risk (Burgasov et al. 1970; Cherkassky 1996). An animal is contagious for as long as it is sick, during which time it excretes the microbe in its urine, faeces and blood. The pathogen enters the organism through micro-injuries in the skin or external mucous membranes during contact with an infected animal or its cadaver. Infection occurs when people care for or bury diseased animals, slaughter livestock or process meat, as well as after contact with infected animal products sown with anthrax spores. Raw material from an infected animal (wool, pelts, etc.) and the products that are made from it pose an epidemiological risk for many years. It should be noted that lately infection through infected animal products has played the main role in spreading anthrax among people (Marinin et al. 2008). Moreover, infection can occur via contact with soil sown with spores, or via airborne transmission. Humans' natural susceptibility to the pathogen is relatively low (approximately 20% of exposed individuals contract the disease), but it is nearly universal in cases of airborne transmission (Cherkassky 1996).

A number of reports of anthrax in wild animals have accumulated in the scientific literature, especially among hoofed mammals, members of the Cervidae and Bovidae families, as well as the excretion of anthrax pathogens by rodents (ground squirrels, great gerbils, Ural field mice). Horseflies, flies, mosquitoes and black flies can serve as vectors for the transmission of anthrax between animals. Based on data collected by V. F. Nagorsky, N. G. Olsufiev and P. P. Lepnin, horseflies and other blood-sucking insects were a crucial factor in the transmission of anthrax infection in Czarist Russia among horses and large cattle in taiga and forest regions of Russia, while mosquitoes played a similar role in the tundra (Burgasov et al. 1970).

The stationary nature of anthrax foci is not only linked to the stability of anthrax spores, but also, more importantly, related to their accumulation in soil by way of vegetation. In certain conditions, they are capable of germinating and forming new spores, supporting the existence of a soil focus (Cherkassky 1996). Back in 1880, Louis Pasteur proved that the anthrax pathogen moves vertically through the soil thanks to earthworms (Schuch and Fischetti 2009). Pathogen transport from lower soil layers can also be conducted by other invertebrates, such as snails and beetles, and burrowing rodents. A link between the population size of field rodents in anthrax foci and infection rates among livestock has been established. The higher the rodent population, the greater the number of spores brought to the soil's surface and the higher the chance of infection among livestock. Long-term vegetation in the soil leads to changes in the pathogen's characteristics. For example, in unfavourable conditions, including the function of antagonists (bacteria and Actinomycetales, as well as antagonistic plants such as clover, Medicago, rye, Hordeum and garlic) or as the result of a prolonged saprophytic lifestyle, anthrax microbes partially or entirely lose their pathogenicity (Cola 1976). Direct sunlight, insufficient humidity, insufficient free oxygen, the presence of ferrous iron salts or aluminium salts, etc., all have bactericidal effects on anthrax (Ginsburg 1975). All of these factors can lead to changes in the soil focus' level of activity.

A number of perspectives exist with regards to the proper definition of an anthrax focus. Depending on the point of view, the defining span of the focus can vary from insubstantial (a burial ground for infected animals) to large pasture territories (Cola 1976). Many authors who studied the territorial distribution of stationary and high-risk locations noted their affinity for black earth and chestnut soils (Fig. 4.57) (Ginsburg 1975).

Climate plays a major role in the distribution of anthrax. Ample precipitation washes out soil foci and facilitates the transfer of spores to the surface of the earth and their distribution to adjacent areas. At the same time, increases in anthrax morbidity rates usually do not coincide with periods of ample precipitation and instead follow rapid decreases in precipitation combined with high atmospheric temperatures. An 1885 review of the spread of anthrax lists 'an extraordinarily long period of dry and hot weather' as one of the main reasons for an anthrax epizootic that occurred in the majority of Russia's provinces. In 1902, V. Nagorsky cross-referenced the morbidity dynamics of anthrax in animals in Novoladozhsky district of Petersburg province with atmospheric temperature, cloud cover and levels of precipitation and concluded that anthrax epizootics only occur under the following conditions: high atmospheric temperatures, scant cloud covers and low precipitation. In many cases, meteorologic changes have an effect on epizootic dynamics thanks to their compound influence on soil conditions (the anthrax pathogen's habitat), transmission mechanisms and animal susceptibility (Ginsburg 1975). One

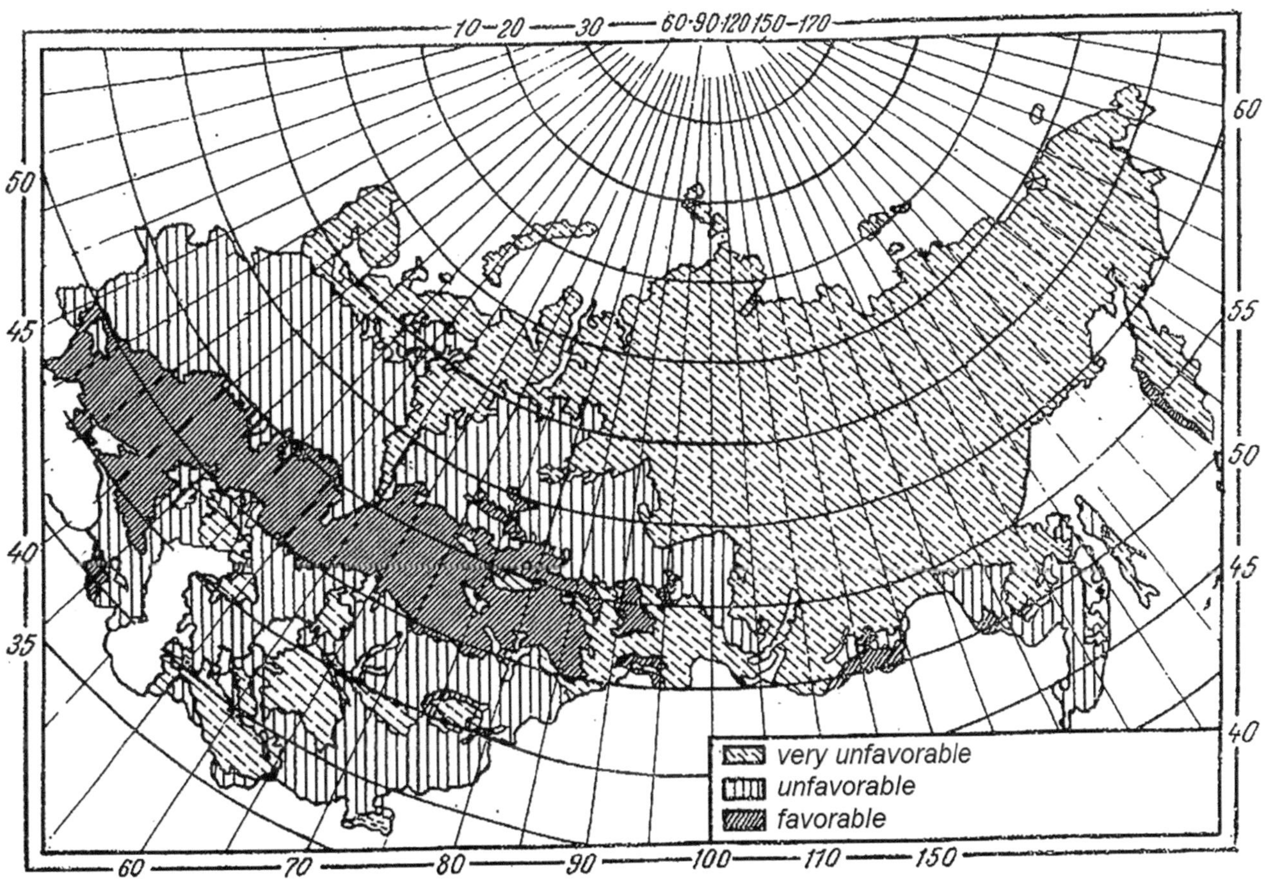

Fig. 4.57 Anthrax vegetation conditionals in the soil in the former USSR. *Credit* Anthrax (1975)

proposal argues that, in drought years, anthrax pathogens can move vast distances as the result of dust storms (Cola 1976).

The local terrain has a similarly complex effect on anthrax transmission. Numerous observations serve as evidence for the fact that anthrax foci are often found in river floodplains and are activated by river overflow during the spring thaw. Back in 1831, Stepan Khotovitsky, while describing the distribution of anthrax in Russia, noted that 'fire disease [anthrax] rarely visits tall mountains, and elevated, dry, and cold places in general', while 'places that are low-lying, wet, slightly salty, those that dry up in droughts, as well as river and lake valleys, those that lose water during droughts, are quite dangerous' (Ginsburg 1975). V.I. Adamovich and N. N. Nikonov pointed at the direct relationship between the quantitative distribution of anthrax foci and intensity of erosion on a specific territory. The stronger the erosion, the greater the number of anthrax foci (Cola 1976). However, this relationship does not apply to all regions. For example, in Bryansk Oblast epizootic indices in areas of high soil erosion could be greater than those in areas of moderate and weak erosion by a factor of 20, and yet this pattern does not prove to exist in Moldova (Cherkassky 2002).

As described in the literature, areas with a high anthrax risk have an affinity for specific regions in Russia and its adjacent countries. At least three landscape types of anthrax foci have been defined: Piedmont and low-mountain arid zones, floodplain zones in mixed coniferous forests of the East European Plain, and terraced river valley zones in the taiga. It must be noted that the territorial distribution of anthrax has undergone substantial changes in the past two centuries. Compared to pre-1917 (Czarist) period, between 1950 and 1979 anthrax's localization shifted from northern landscapes to southern open landscapes, especially to steppes, forest steppes, deserts and semi-deserts (Cherkassky and Zhanuzakov 1980). Analysis of the proportion of animal disease in various geographical zones of the USSR in the second half of the twentieth century revealed that 25% of anthrax cases occurred in steppes, 15% in forest steppes, over 17% in the desert, 15% in semi-deserts, 10% in forests and 4% in the subtropical zone. Approximately 10% of cases were reported in mountainous areas, including 3% in mountain steppes and 7% in mountainous forests. The tundra, which previously had high anthrax morbidity and mortality rates among reindeer, became essentially disease free (Baryshnikov et al. 2007). Overall, black earth regions in the steppe and forest steppe pose the greatest epidemiological risk today.

Anthropogenic activity has a significant influence on the function of anthrax foci. Soil foci can be activated by pasture expansion at the expense of unbroken or unused land or construction that involves soil excavation (the building of canals, roads, etc.) (Cherkassky 2002).

4.14.3 Historical Background

The anthrax bacillus was first discovered by A. Pollender, a German physician, in 1849. In 1850, two French researchers, Casimir Davaine and Pierre Rayer, found thread-like immobile bodies in the blood of sheep that were sick with or died of anthrax. Friedrich Brauell found the same bodies in animal cadavers in Russia at the same time, in 1857 (Fig. 4.58). However, the meaning of these bodies remained unknown, and only in 1863 did Davaine establish their role as the anthrax pathogen. In 1876, Robert Koch reported on his isolation of anthrax in cell culture and its ability to form spores.

Aloys Pollender (1799-1879) | Friedrich Brauell (1807-1882) | Casimir Davaine (1812-1882)

Fig. 4.58 Anthrax researchers. *Credit* http://commons.wikimedia.org/wiki/Category:Aloys_Pollender?uselang=pt#/media/File:Aloys_Pollender_1869.jpg **Public domain**; https://de.wikipedia.org/wiki/Friedrich_Brauell#/media/File:Friedrich_Brauell.jpg **Public domain**; https://fr.wikipedia.org/wiki/Casimir_Davaine#/media/ **Public domain**

Anthrax has been known to man since ancient times. Presumably, anthrax as an infection appeared before the development of man and was a natural focal disease that affected hoofed mammals. It likely appeared in the early Quaternary Period of the Cenozoic Era (Cola 1976).

Ancient Greek and Roman writers knew anthrax as the 'sacred fire', while ancient Arab doctors called it 'Persian fire'. In Russia, it was called 'snake rogue', 'fire danger' or 'fire disease'. The modern Russian-language term for anthrax, Siberian Ulcers, was coined by S. S. Andreevsky, who proved that the same disease affected humans and animals and named it after the location of study (Baryshnikov et al. 2007).

A disease clinically similar to anthrax was found in Ancient Greece (it was described in Homer's *Iliad* as the disease that overtook the Greek's camp during their siege of Troy in 1218 BCE) and Ancient Rome (the Roman poet Virgil depicted the disease in *The Georgics* in 37–30 BCE). It is also mentioned in Hippocrates' ancient writings (460–377 BCE) and in *Hippiatrica,* the ttenth-century veterinary texts (Turnbull 2002). The first mention of anthrax in Russia can be found in the Nikon Chronicle (979 CE). A large anthrax epidemic occurred in Moscow in 1649, leading the Tsar Alexei Mikhailovich to issue a decree banning people from leaving animal corpses in the streets of Moscow. The first scientific description of anthrax in Russia was written by Johann Gmelin, a scholar from St. Petersburg, in 1767: he observed the disease during a voyage to Siberia. However, the most thorough description of anthrax in Russia was S. S. Andreevksy's work "On the Siberian Ulcer", which was published in 1788: it contained a detailed map of the disease's distribution, a description of its symptoms, and a set of drawings based on the findings of a 30-year expedition in Chelyabinsk Okrug in the Ural Vicariate (Burgasov et al. 1970).

Anthrax was one of the most common and dangerous diseases in pre-1917 Russia. An enormous number of livestock died from it annually, and large epidemics also occurred in the human population. There was one human case for every 3–4 animal cases (Fig. 4.59).

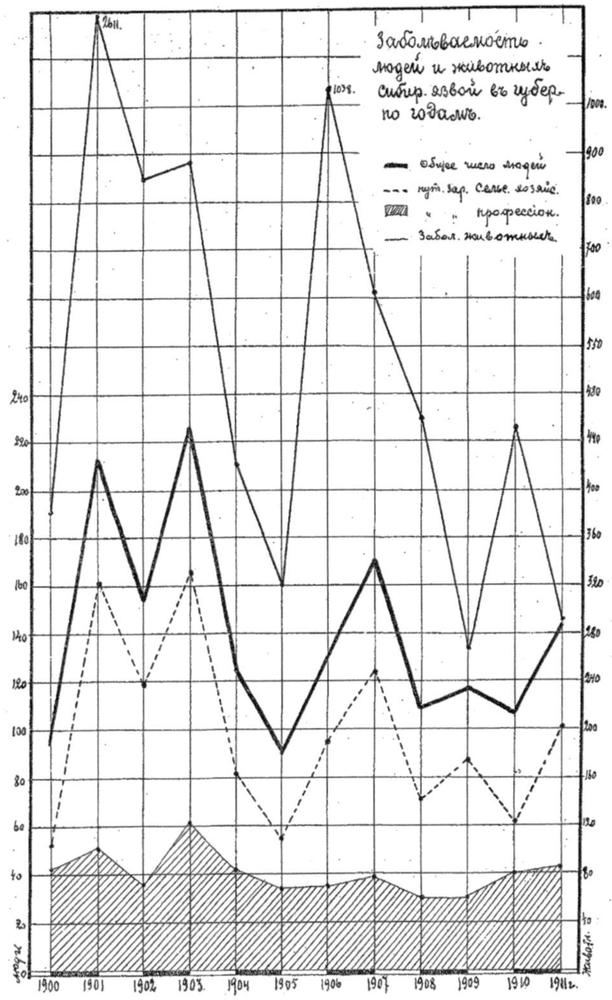

Fig. 4.59 Anthrax cases among people and animals in the Czarict Russia (Nizhny Novgorod Province) (Kalantarov 1912). Legend: Anthrax perennial human and animal cases in governorates Thick black line—total human cases; Dashed line—infected in agriculture Shaded area—infected in other professional groups Thin black line—animal cases. *Credit* Kalantarov (1912)

Louis Pasteur was involved in the development of the anthrax vaccine: he demonstrated its function in 1881 (Fig. 4.60). The introduction of animal anthrax vaccination in Russia in 1883 somewhat decreased the loss of livestock and human morbidity rates, however epidemics and epizootics continued to occur in following years. Anthrax in humans was reported in 57 out of 60 provinces of imperial Russia (Fig. 4.61). Approximately 268,000 people contracted anthrax between 1896 and 1913, and 25% of them died. Anti-anthrax measures were extremely insufficient. Thwarting of sanitary and hygienic norms in agricultural work and on animal material plants led to mass occupational infection with anthrax. In 1891, Vladimir Svyatlovsky wrote, "Anthrax is constantly present in Russia. It's our own form of national disease…" (Burgasov et al. 1970). It should be noted that European countries faced similar issues (Fig. 4.62).

In the late nineteenth and early twentieth centuries, 50% of infected people were peasants who did agricultural work, 40% were industrial labourers that worked with wool, sheepskin and leather, and 1.7% worked with hair, bristles, horns and hooves. For instance, in 1926 in Moscow 27 out of 47 cases of anthrax were related to occupational risks: 13 people in the wool industry, 9 people in the leather and fur industries and 5 people who worked with animal hair and bristles in artisanal conditions. Human infections had a high fatality rate, which averaged 15% (Cherkassky and Zhanuzakov 1980; Cherkassky 2002).

Fig. 4.60 Louis Pasteur vaccinates a sheep against anthrax. *Credit* http://wellcomeimages.org/indexplus/image/L0003758.html

A substantial number of other forms of anthrax has occurred in the past. Inhalation anthrax used to be very common and was known as "ragpickers' disease", caused by the collection of rags contaminated with infected animal waste. During the Russo–Japanese War in Manchuria in 1904, 956 cases of cutaneous anthrax occurred over the course of several months in the Russian army due to use of clothing and other items made out of contaminated animal materials.

That same year 2000 horses and 30 people died over 2 weeks during the construction of the Ussuriysk Railroad. In 1927, a batch of sausages was made out of contaminated beef in a small sausage plant in Yaroslavl, infecting 23 people with gastrointestinal anthrax. In 1953, anthrax appeared in a family in the Tatar ASSR, as a result of which 14 out of 15 infected people died. The people contracted anthrax from eating vegetables grown in a garden where an anthrax-infected lamb had been buried (Cherkassky and Zhanuzakov 1980; Cherkassky 2002).

Vaccination improvements in the USSR, annual mass vaccination of livestock, structural changes in agriculture,

Fig. 4.61 The distribution of anthrax in Moscow Governorate (Anthrax… 1910). *Legend* epizooty area in 1910 Anthrax locations in 1910 Anthrax locations in 20 years (no data for 1891, 1892, 1896 and 1898) Anthrax… 1910

Fig. 4.62 Newspaper illustration. In Britain, Bradford's phenomenal growth and prosperity were the result of the wool industry. Garments were made from alpaca and mohair brought from overseas. These bales of wool were often infected with the anthrax pathogen. *Credit* https://100objectsbradford.wordpress.com/2011/07/13/21-death-and-the-woolsorter-bradford-doctors-against-anthrax/

changes in land use and improvements in veterinary and sanitation control led to a 21.5-fold decrease in morbidity rates in 1966 compared to 1913 (this was 2.5-fold decrease compared to 1950). The late twentieth century was characterized by stabilization: an average of 40 cases per year, or 0.026 cases per 100,000 individuals (Cherkassky 2002).

The theory that risk of anthrax appearance in areas that were at risk many years ago is confirmed by the following evidence. In 1964, anthrax reappeared in a settlement in Kirov Oblast 76 years after the last case. This outbreak was preceded by the spreading and dispersal of soil due to geologic work that did not take into account the presence of anthrax soil foci. In 1997, animal anthrax outbreaks occurred in areas where the disease had not been seen since 1949 (Samara Oblast), 1953 (Altai Krai) or 1958 (Udmurtia) (Bakulov et al. 2001).

4.14.4 Distribution in Russia and the World

Analysis of anthrax surveillance has shown that the disease's world geography is fairly stable. The modern range of anthrax in livestock and wild animals spans every continent. The disease has not been reported in the northern part of North America or in a few island territories. Conditions for anthrax are optimal in West and Central Africa, Central and South Asia, South America and the Mediterranean (Bakulov et al. 2001). Anthrax is widespread in Afghanistan, Iraq, Pakistan, Turkey, Greece, India and other countries (Onischenko et al. 1999).

Based on the extensiveness of the anthrax epizootic process, the territory of former USSR was divided into three parts:

- Sporadic (north of 56 N and the eastern areas of the Transbaikal, where anthrax is limited).
- Periodic (between 56 and 52 N in Russia's European territory and Siberia (west of Lake Baikal), where anthrax levels are higher).
- Stable (from 53 N to the southern border of the former USSR, where the conditions for anthrax are most favourable).

The epizootic situation in different regions reflects the contrasts in anthrax's territorial distribution among people (Onischenko et al. 1999). For a long time, most anthrax morbidity was found in the Northern Caucasus, Transcaucasia, Kazakhstan and Central Asia (Burgasov et al. 1970). In the late twentieth century, anthrax was found in the form of sporadic cases or small outbreaks in the Northern Caucasus, Central Black Earth Region and the Volga Region. Cases were reported in Voronezh, Kursk, Saratov and Ulyanovsk Oblasts, Altai Krai and Bashkortostan (Bakulov et al. 2001). Anthrax is generally localized to rural areas (90% of all confirmed cases are there) (Cherkassky 2002).

According to the Inventory of Stationary At-Risk Areas in Russia, there are over 35,000 such areas that account for approximately 8,000 anthrax-infected animal burial grounds. Moreover, many burial grounds have likely not been accounted for.

In 2008, the greatest number of such areas was found in Russia's European Territory: in some regions, their number exceeds 2000 (Map 4.22).

Areas of animal burial are similarly distributed (Maps 4.23 and 4.24). 37% of these grounds are in accordance with veterinary and sanitary norms. The greatest number of these grounds is found in the North Caucasus Federal District (69.9%) and the smallest number is in the Ural Federal District (7.5%). The proportion of anthrax

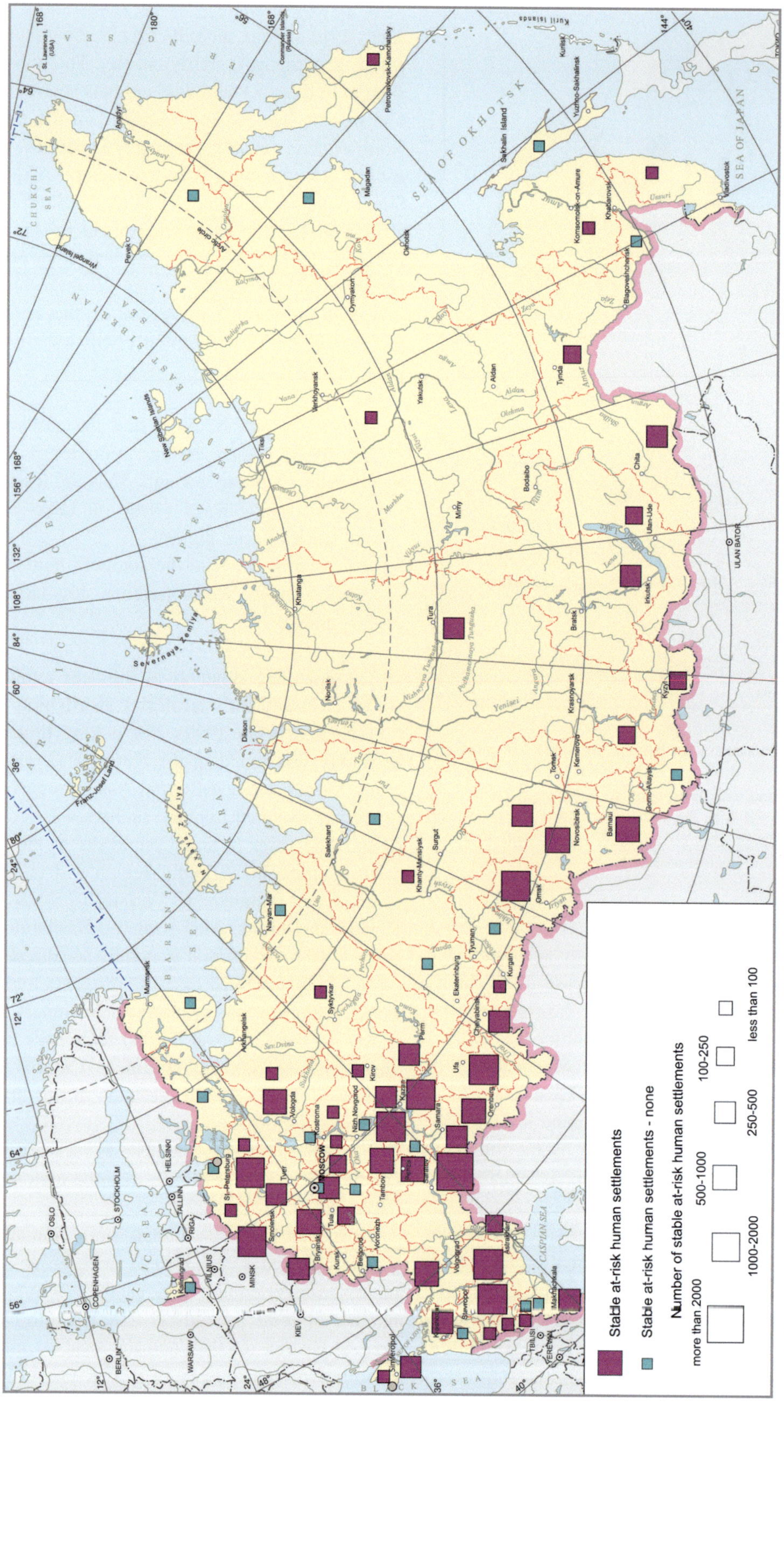

Map 4.22 Stable at-risk human settlements

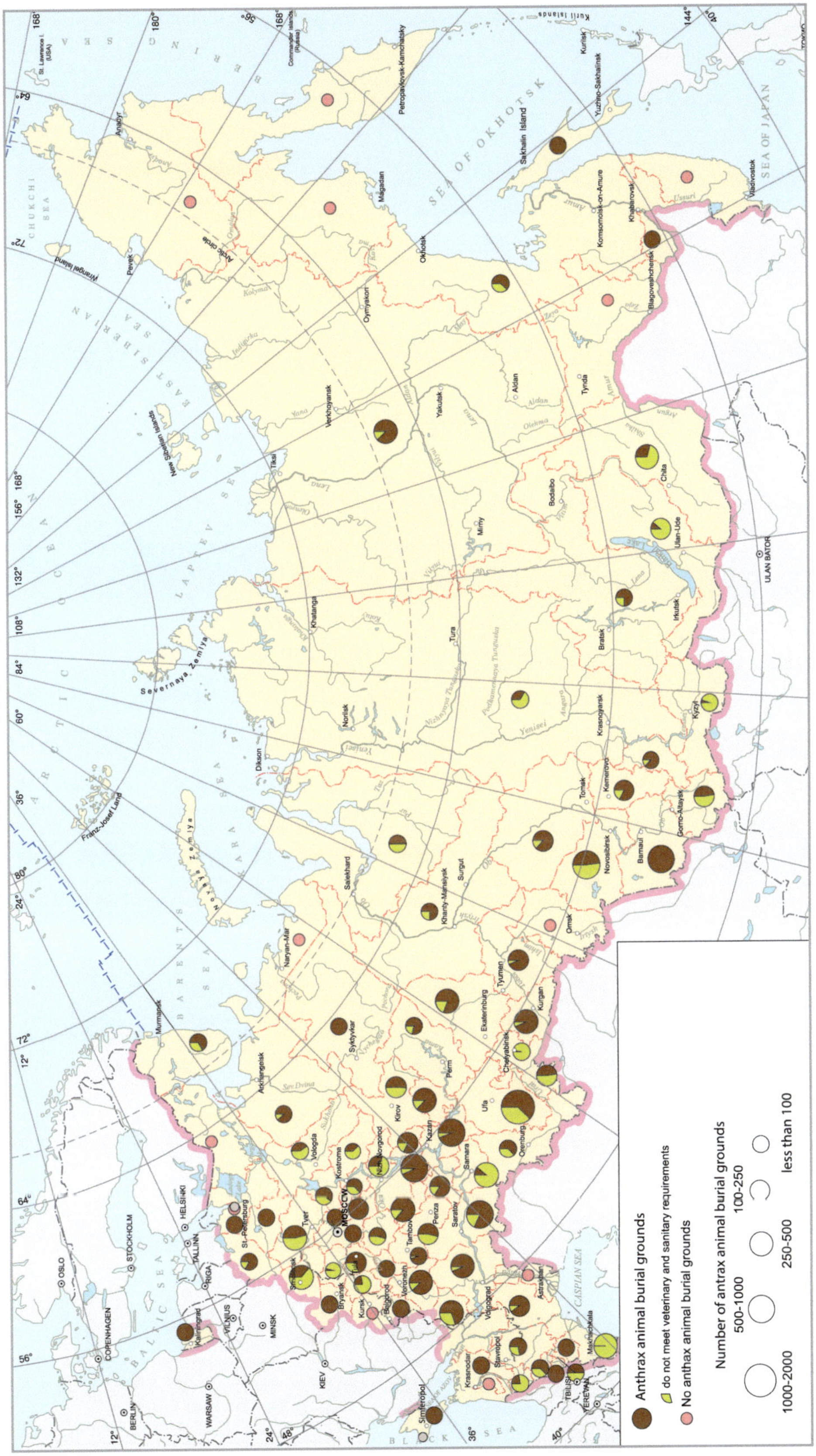

Map 4.23 Anthrax animal burial grounds

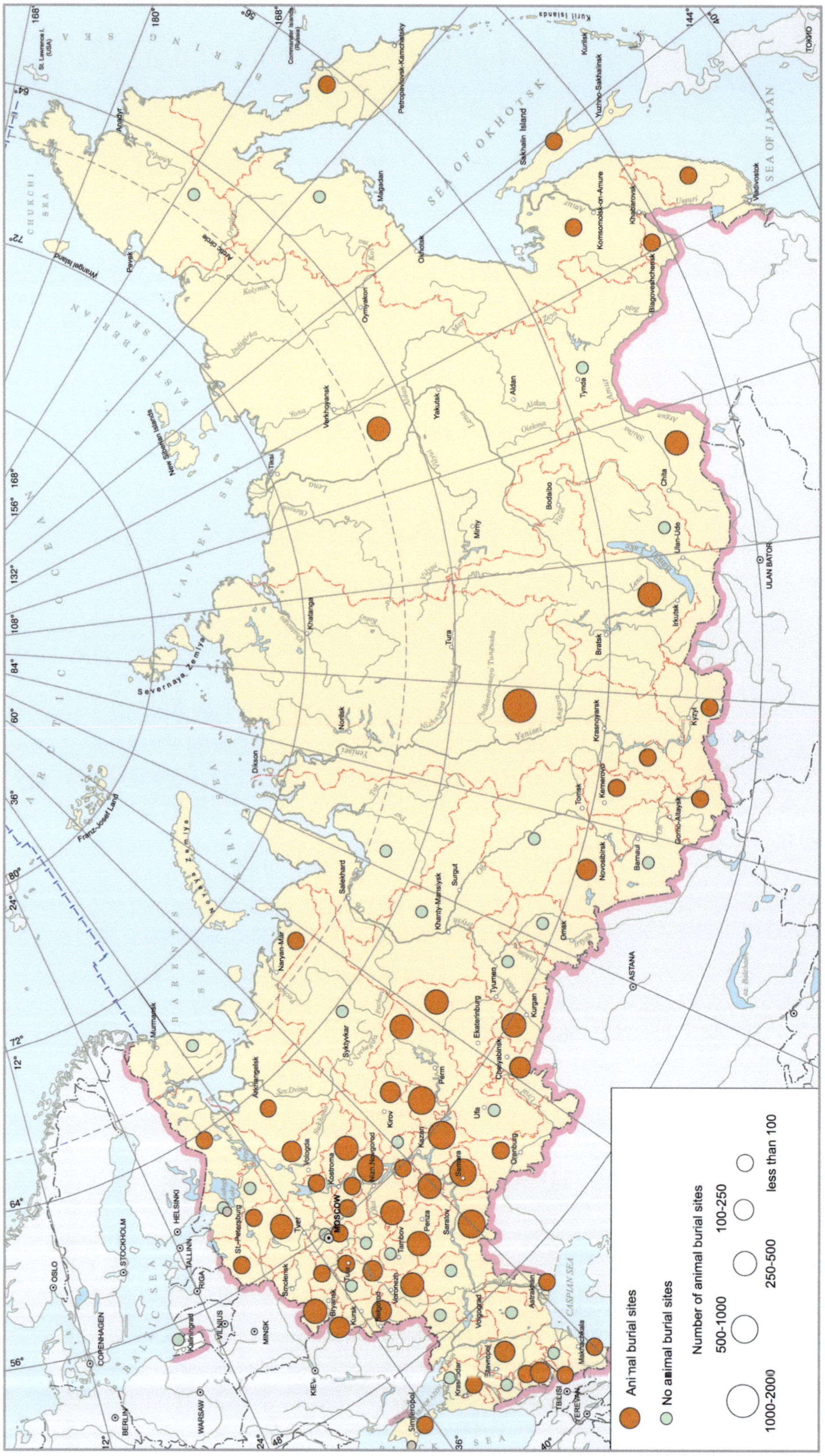

Map 4.24 Animal burial sites

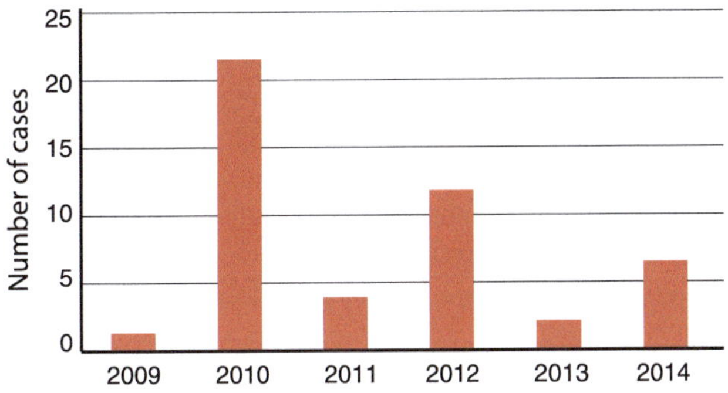

Fig. 4.63 Cases of anthrax in humans in Russia, 2009–2014. *Data source* www.rospotrebnadzor.ru

animal burial grounds which have clearly defined border coordinates is 39.3%, 94.1% in the Southern Federal District, 83.9% in the North Caucasus Federal District, 70.5% in the Volga Federal District, 50.7% in the Northwestern Federal District, 25.6% in the Siberian Federal District, 24.2% in the Far East Federal District, 21.8% in the Central Federal District and 8.15% in the Ural Federal District. Burial grounds in the path of spring floodwaters are especially dangerous. According to data collected by Rospotrebnadzor in administrative subjects of the Russian Federation, in the spring of 2014, 7 subjects (Buryatia, Crimea, Sakha (Yakutia), Stavropol Krai, Khabarovsk Krai, Ivanovo Oblast and Nizhny Novgorod Oblast) contained 66 anthrax animal burial grounds that fell into the flooding zone during the spring thaw (Act… 2014).

Forty cases of anthrax in people were reported in Russia from 2009 to 2014. This was a 43% increase compared to the previous 5-year period (Fig. 4.63). Two anthrax cases, in 2010 and 2012, led to a fatal outcome.

Anthrax cases have been reported in 3 federal districts: 20 cases in 4 subjects (14 (70%) of which were in Dagestan, 3 in Chechnya, 2 in Stavropol Krai and 1 in North Ossetia) of the North Caucasus Federal District, 9 cases in 3 subjects (4 each in Krasnodar Krai and Volgograd Oblast and 1 in Rostov Oblast) in the Southern Federal District and 11 cases in 2 subjects (6 in Omsk Oblast and 5 in Altai Krai) in the Siberian Federal District. Direct contact with diseased livestock during butchering and animal burial, cooking of infected meat, care for infected animals or sale of meat in farmer's markets. According to Rospotrebnadzor, a large outbreak of anthrax occurred among reindeer in the Yamalo-Nenets Autonomous Okrug in June and July or 2016. Over 2500 animals died, and 36 human cases were reported. This outbreak has evoked public concern, is a reflection of the critical situation in reindeer husbandry, a traditional economic activity for indigenous peoples living in the tundra (Bogdanov and Golovatin 2017).

4.14.5 Prevention

Preventative measures can be veterinary or sanitary. Veterinarians identify, catalogue and record areas that pose an anthrax risk, along with routine animal immunization. Sanitary measures include control over general sanitary measures in risk-prone areas, as well as control of the preparation, storage, transport and processing of raw animal-borne materials. People in risk zones are routinely vaccinated. Antibiotics are used as an emergency measure of prevention (Cherkassky 1994).

4.15 Tetanus

Tetanus is an acute infectious disease that affects humans and animals. It is an anaerobic infection that enters the body through a wound on the skin. It is characterized by neurological symptoms, specifically, muscle spasms that can lead to respiratory failure (Turyanov et al. 1998; Encyclopedia… 1998).

4.15.1 General Information

The tetanus pathogen (*Clostridium tetani*) is a rod-shaped, anaerobic, gram-positive bacterium 2–5 mm in length. It forms round spores on its ends. It enters the human body in spore form through broken skin or mucous membranes. Spores are so resilient to environmental factors that they can withstand being heated to 90 °C over the course of 2 hours, die only after being boiled for 1–3 h, in their dry form can be heated up to 150 °C, can survive in salt water for up to 6 months and on various objects for over 10 years (Turyanov et al. 1998; Tetanus 2015). In an anaerobic environment, at temperatures of 37 °C, and sufficient humidity, spores develop into a fragile bacterial form (Pokrovsky et al. 2007; Postovit 1997; Turyanov et al. 1998).

According to recent findings, *C. tetani* forms two exotoxins: *Tetanolysin* and *Tetanospasmin*. The former has not been sufficiently studied. *Tetanospasmin* is one of the most powerful toxins known to man, and it is responsible for causing the clinical symptoms of tetanus. Specialists have determined that the minimal lethal dose of this toxin is 175 nanograms per 70 kg adult weight (Tetanus 2015). The incubation period lasts from 24 h to 1 month. Disease onset occurs with the appearance of lockjaw, which is caused by jaw muscle spasms that can also spread to the facial musculature (Fig. 4.64). As a result, the patient's facial expression is distorted into a "sardonic smile". This is followed by muscle spasms in the back of the head, the back, the ribcage

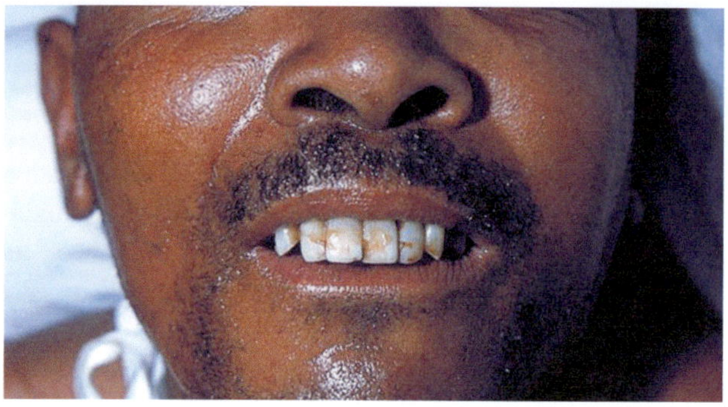

Fig. 4.64 Lockjaw. *Credit* Emond et al. (2003)

Fig. 4.65 Opisthotonos is the characteristic muscle spasm caused by tetanus. Painting by Charles Bell (1809). *Credit* upload.wikimedia.org/wikipedia/commons/d/d8/Opisthotonus_in_a_patient_suffering_from_tetanus_-_Painting_by_Sir_Charles_Bell_-_1809.jpg Public domain

and proximally in the limbs (Fig. 4.65) (Fusetti et al. 2013; Tetanus 2015). In 1–5 days seizures begin (they manifest as jerking or staring) and body temperature rises. The patient remains conscious. Other symptoms are tachycardia and excessive sweating. Fatal cases are the result of respiratory or cardiovascular distress or damage to the autonomic nervous system (Fusetti et al. 2013). Survivors usually begin to improve in 2–3 weeks. People who have had tetanus do not form adaptive immunity, as the dose of the toxin needed to cause symptoms is too small to trigger an adaptive immune response. A person who has received the full course of vaccinations preserves the ability to rapidly (in 2–3 days) produce an antidote to the toxin for about 10 years (Postovit 1997; Turyanov et al. 1998; Emond et al. 2003; Encyclopedia… 1998; Pokrovsky et al. 2007; www.who.int).

4.15.2 Epidemiology and Natural Focus Characteristics

Tetanus morbidity is reported everywhere in the form of sporadic cases. There are no tetanus epidemic foci (www.rospotrebnadzor.ru). Herbivores, rodents, birds and humans serve as the infection's reservoir and source—the pathogen lives in their digestive tracts. The tetanus spore is also widespread in the soil and other objects in the outside environment, where it can reproduce and survive for a long time. Black and red soil with high humus content, as well as soil rich in organic matter, provide optimal conditions for tetanus spores. Bacteria can enter any location from the soil with dust particles (including procedure and operating rooms). They can also land on surgical and medical materials (various medications, plaster, talc, cotton, clay, mud, etc.). The disease is more common in the spring and summer (Wilks et al. 2003) and people who work on farms, in orchards or gardens or in construction are at a higher risk of infection (Harding et al. 2014). Tetanus is transmitted via direct contact—it enters the body through broken skin or mucous membranes (wounds, burns, frostbite). Surgical wounds can also become entry points for tetanus spores. Infected umbilical wounds after birth can cause tetanus in newborns. Humans are naturally highly susceptible. Tetanus cannot be passed from one person to another.

4.15.3 Historical Background

Tetanus has been known to man since ancient times, and its onset has long been associated with wounds. Hippocrates gave tetanus its name and was the first to describe its clinical symptoms. Nestor Monastyrsky was the first to discover tetanus bacilli in a human cadaver in 1883, while Arthur Nicolaer was the first to isolate it from experiments on animals in 1884. Kitasato Shibasuro, a Japanese bacteriologist, was the first person to grow the tetanus bacillus in pure culture in 1887. Later he isolated the tetanus toxin, and worked with Emil von Behring to develop a serum antitoxin. The French immunologist Gaston Ramon developed a method of tetanus antitoxin production, and it was first released in 1924. Thanks to this antitoxin serum, the number of tetanus cases during World War II was only 0.44 cases per 100,000 individuals, compared to the peak of 13.4 cases per 100,000 individuals that was reached during World War I (Tetanus 2015).

4.15.4 Distribution in Russia and the World

Tetanus is found around the world, but it is most common in densely populated regions with a hot and humid climate (Fig. 4.66) and with soil rich in organic matter (WHO. Tetanus 2015). Tetanus prevention is a global concern: based on official data, over 250,000 people die of tetanus every year (www.who.int). In Russia, the tetanus morbidity

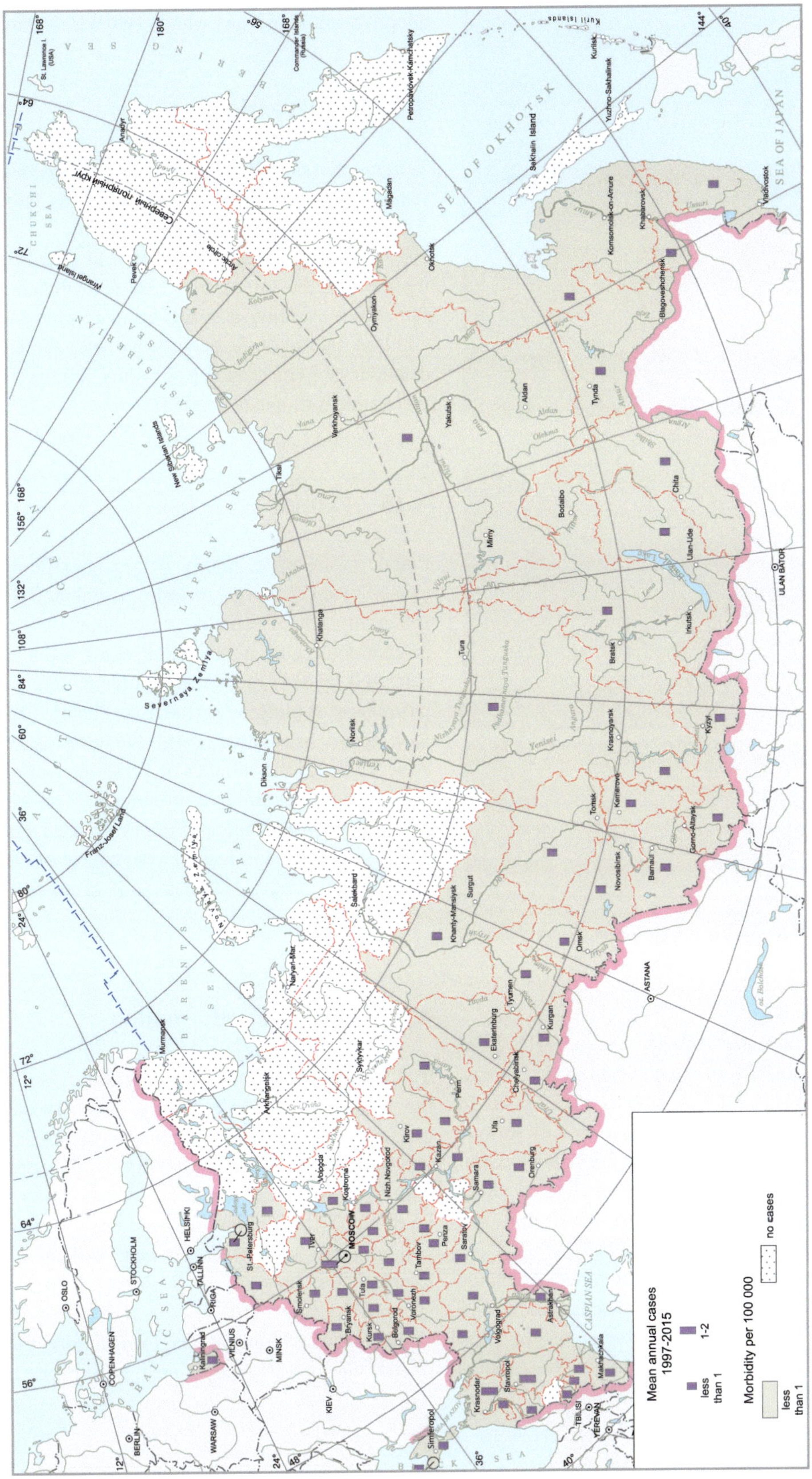

Map 4.25 Tetanus morbidity. 1997–2015

(a)

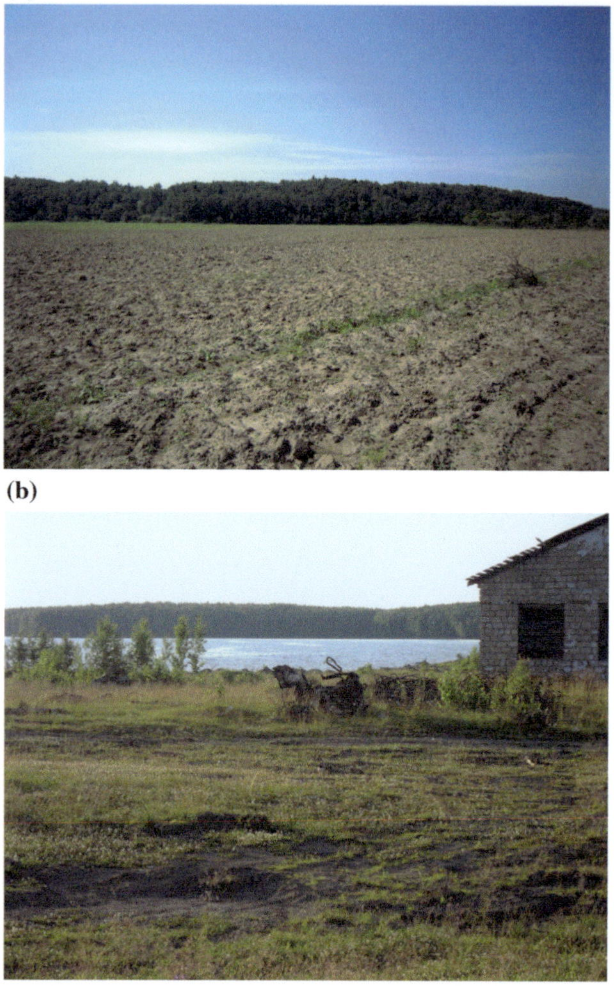

(b)

Fig. 4.66 Farmland (a) and bare lot soil (b) serves as an epidemiologically significant reservoir for *C. tetani*. *Credit* N. G. Kadetov (permission obtained)

rate is stable at 0.01 cases per 100,000 cases: 10–15 cases are reported annually, 30–45% of which occur in rural areas.

According to the WHO, 80% of tetanus cases occur in newborns (www.who.int). This problem is especially serious in developing countries (Pokrovsky et al. 2007, www.who.int). In developed countries, more than half of all cases affect people over 60. More than half of this age group does not have adaptive immunity (Aksoy et al. 2011).

Between 1997 and 2015, the highest number of tetanus cases occurred in Stavropol Krai, Krasnodar Krai and Moscow (Map 4.25). Thanks to targeted vaccination campaigns in Russia, which began in 1975, most cases occur in elderly people who were either never vaccinated or whose adaptive immunity has waned (www.rospotrebnadzor.ru).

4.15.5 Prevention

Tetanus is most effectively prevented by vaccination (Pokrovskiy 1983, www.rospotrebnadzor.ru). Non-specific and specific measures are employed in order to prevent tetanus. Non-specific prevention is geared at avoiding wounds and injuries (in the home, at work, in healthcare institutions). Specific prevention is made up of routine immunization in children and adults in accordance with the national vaccine calendar and emergency post-injury immunization (Sanitary and Epidemiological… 2013).

4.16 Plague

Plague is an acute, highly dangerous natural focal disease caused by a Category A priority pathogen (NIAID Emerging Infectious Diseases/Pathogens, https://www.niaid.nih.gov/research/emerging-infectious-diseases-pathogens). It is characterized by skin-related symptoms, enlarged lymph nodes, pulmonary symptoms and sepsis. Plague's range of natural foci spans substantial parts of Eurasia, Africa and the Americas, and the number of vertebrates affected by plague epizootics is extremely high compared to that of other diseases. In the past, plague epidemics caused a great number of deaths, sometimes influencing the course of history (Neronov et al. 1991). Local outbreaks still occur today and pose a serious risk.

4.16.1 General Information

The plague is caused by the *Yersinia pestis* bacterium in the *Enterobacteriaceae* family (Domaradsky 1998). The plague bacillus is fairly sensitive to outside factors. The length of its survival in the corpses of people and animals who died from the plague is directly related to the speed with which the corpse decomposes. The lower the temperature, the longer the bacteria will survive. At glacial temperatures, bacteria can survive for up to a few months (Fedorov et al. 1955).

Plague is a highly contagious disease that can be contracted via vector-borne, contact or airborne transmission. It has various clinical forms, which is why the degree of infection risk a plague-infected person poses to others can vary. Bubonic and pneumonic plague are most common, while septicemic and cutaneous plague are more rare, with gastrointestinal plague being the most uncommon.

The incubation period lasts 3–6 days. Disease onset is sudden and acute with chills and a rapid rise in body temperature to 39–40 °C. Consciousness may be impaired, and delirium or agitation may occur. General intoxication, fevers and lymphatic symptoms are common. The disease has a high fatality rate.

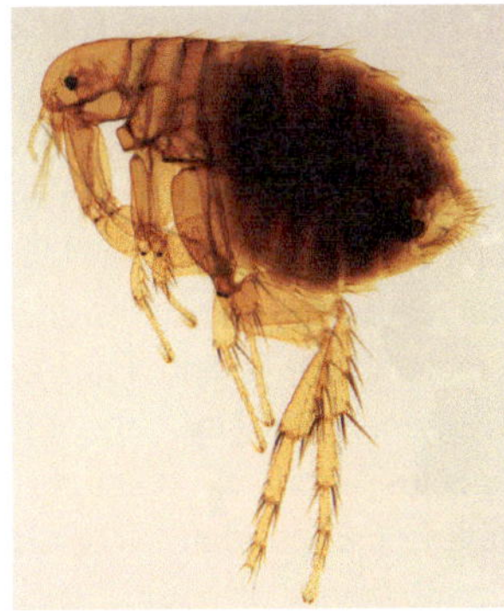

Fig. 4.67 The *Xenopsylla cheopis* flea

4.16.2 Epidemiology and Natural Focus Characteristics

A few classification systems of plague natural foci exist (Kucheruk 1959; Kozakevich et al. 1981; Suntsov and Suntsova 2006). Based on their history of formation, foci are divided into 3 types:

- Primary natural foci in arid areas, inhabited by the plague pathogen regardless of human activity as a native member of the biocenosis (Asia and the far southeast of Europe).
- Synanthropic (anthropogenic, anthropurgic) plague foci in port cities, whose existence is supported by human acivity. Synanthropic rodents are the main carriers.
- Secondary natural foci, which sprang from synanthropic foci in Africa, the New World and on large oceanic islands, such as Madagascar, Hawaii and Java.

Today 341 mammal species that can carry the plague are known (Sludsky 2014). Rodents (273 species) are the pathogen's main carriers. In Russia and its adjacent countries, this role is filled by the little ground squirrel (*Spermophilus pygmaeus*), the Daurian ground squirrel (*S. dauricus*), the long-tailed ground squirrel (*S. undulates*), the Tarbagan marmot (*Marmota sibirica*), the grey marmot (*M. baibacina*), the great gerbil (*Rhombomys opimus*), the mid-day jird (*Meriones meridianus*), the tamarisk jird (*M. tamariscinus*), the common vole (*Microtus arvalis*), etc. Members of other orders are accidentally brought into epizootics. Birds have been known to be infected in the wild. The plague bacterium may be able to survive in soil (Neronov et al. 1991; Sludsky 2014).

The pathogen circulates among wild animals in natural foci as the result of vector-borne transmission. Fleas in the *Siphonaptera* order, especially the Oriental rat flea (*Xenopsylla cheopis*), play the main role in transmission, and they have acquired a variety of adaptive qualities throughout their evolution in order to serve as vectors. Plague bacteria stay in the stomach and gizzard (proventriculus) of the vector (Fig. 4.67). Under favourable conditions (temperature, food, etc.), the bacteria divide rapidly, forming a gelatinous mass called a "plague plug" which clogs the digestive tract (Bacot and Martin 1914; Fedorov et al. 1955; Bibikova and Klassovsky 1974; Tarasov 2002). A plugged flea does not lose the necessity for food, but thanks to the blocked digestive tract the sucked blood is washed with the bacterial mass in the stomach and regurgitated into the bite wound along with a large number of microbes. The plague plug, in particular, allows for the flea to become the plague's main natural vector.

People become infected when they enter a natural focus, either via a bite by an infected flea, or from direct contact with an infected rodent (during skinning). People are highly susceptible to the disease. Adaptive immunity does not prevent the recurrence of infection.

Rats and mice—the most common plague carriers in anthropurgic foci—may become infected in a natural focus. In such a focus, the pathogen begins to cycle between synanthropic rodents. Epizootics in grey rat and house mouse populations can lead to vector-borne infections in humans. The disease can be transmitted through fleas (bubonic and cutaneous plagues), or via infected airborne droplets from one person to another (pneumonic plague). This form of disease, in particular, along with septicemic plague, is the most dangerous. Plague is a rare example of a disease where contracted by humans from animals where humans also act as a significant source of infection.

4.16.3 Historical Background

The *Y. pestis* bacillum, which causes plague, was discovered in 1894 by Alexandre Yersin, a French scientist during a plague epidemic in Hong Kong. The pathogen's vector-borne transmission method was described by M. Ogata in 1897 and Paul-Louis Simond in 1898. In 1900, Danylo Zabolotny (Fig. 4.68) proposed that various rodent species serve as plague's natural carriers, thus not only laying the foundation for the study of plague, but also for the entire theory of natural foci, which was later generalized by Yevgeny Pavlovsky (Rall 1958; Neronov et al. 1991).

Plague has always been thought of as a terrifying disease. At times, it had a substantial influence on the history of humankind. This disease is mentioned in ancient Indian and Chinese source. Plague is also described in the bible in fairly clear terms (presumably, it was plague that accompanied the Jews' exodus from Egypt in 1320 BCE). The first recorded pandemic, the Justinian Plague, spread across the people of the Mediterranean in the sixth entury. This pandemic killed approximately 1 million people in 50 years and undoubtedly facilitated the fall of the Byzantine empire. The second pandemic, also known as the "Black Death", engulfed Europe, Asia and North Africa in the fourteenth entury. In China alone, it led to 13 million deaths, while it caused 24 million deaths in other countries in Asia and 25 million deaths in Europe. The third pandemic occurred in the late nineteenth century, after a prolonged lull, when the plague had been all but forgotten in many countries (Naumov et al. 1972).

Plague epidemics were fairly common in Russia. Old Russian chronicles contain mentions of various pestilences, some of which are thought to be mentions of plague epidemics. There is more or less clear evidence of the "Black Death" in the fourteenth century. The disease was first recorded in Pskov in the early 1350s and soon spread across the entire country. The epidemic had a high fatality rate. Chronicles quite precisely describe the clinical forms of pneumonic and bubonic plague (Vasiliev and Segal 1960).

Plague epidemics often occurred in the eighteenth entury. In the 1700s and 1710s, they followed one another at short intervals. The most intense epidemic raged in 1770 during the Russo–Turkish War and affected a large part of Russia, including Moscow (Fig. 4.69). This is when the first anti-epidemic campaigns were developed: they consisted of quarantine, patient and survivor surveillance, disinfection (smoking out rooms, etc.), public health education campaigns and the collection of statistical data on morbidity and mortality rates (Vasiliev and Segal 1960).

Short-term disease outbreaks in Russian Empire occurred in Odessa in 1901, 1902 and 1910, and in Batumi, Georgia in 1901 and 1916. This leads to the assumption that plague was brought in by ship rats to various Russian seaport cities on the Baltic and White seas and the in the Russian Far East, but it could not spread under the local conditions in those areas (Fedorov et al. 1955). Outbreaks also occurred in 1918 in the Caspian region, in 1920 in Batumi (this outbreak was large, reaching 100 cases), in the Russian Far East in 1921, and in the Transbaikal region in 1922, where cases were reported for the following 6 years (Supotnitsky and Supotnitskaya, 2006). Lev Zilber (1966) reports a plague outbreak in the Hadrut region of Azerbaijan in 1930. 365 cases of plague infection were reported in the USSR from 1945 to 1967 (Artemenkov 2000).

By the early 1960s, plague was understood to have a natural focal nature (Kucheruk 1972). A substantial number of studies was geared at studying the biocenotic and spatial structure of specific foci. However, researchers in the last third of the twentieth century posed new questions and discovered that the established theoretical groundwork for the natural focal nature of plague did not have sufficient scientific evidence (Rothschild 1978; Soldatkin and Rudenchik 1988; Popov 2002). Today, scientific discussions

Проф. Д. К. Заболотный на чумѣ.

Fig. 4.68 "Prof. Danylo Zabolotny on the Plague". Newspaper cutout from *The Spark*, issue 44, 10 November 1913

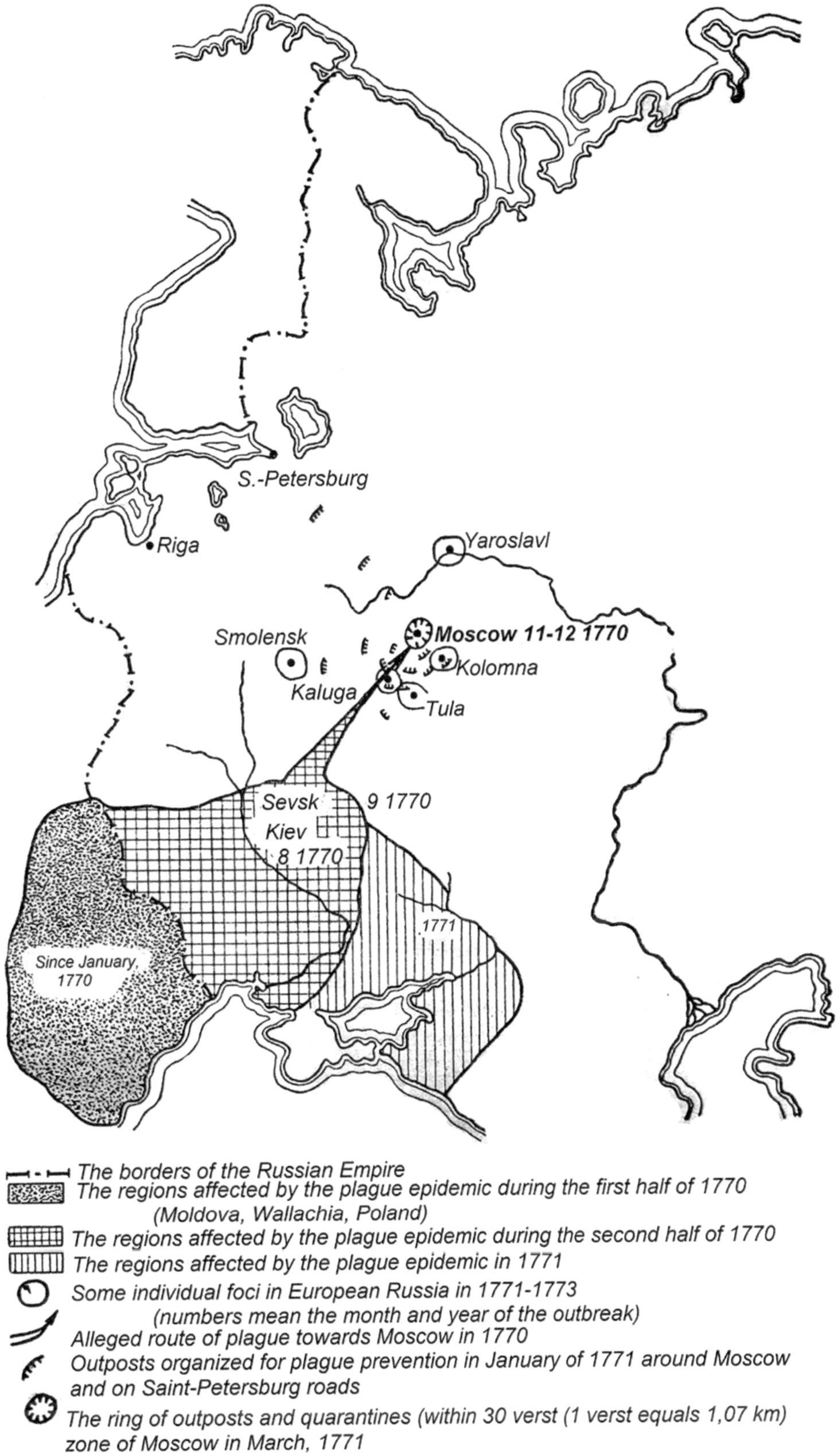

Fig. 4.69 The 1770–1773 Plague Epidemic. *Credit* Vasiliev and Segal (1960)

on the time of *Y. pestis*' emergence and the obligatory nature of its circulation between fleas and warm-blooded animals, the microbe's forms in nature, and the mechanism, development and evolution of the "interepizootic" period (Suntsov and Suntsvoa 2006; Supotnitsky and Supotnitskaya 2006; Sludsky 2014).

4.16.4 Distribution in Russia and the World

According to WHO data, plague cases have been recorded on all continents except Australia and Antarctica. Natural foci lie between 48–49°N (rarely reaching 50°N) and 40–41°S. This primarily depends on the climatic and geographic conditions that define the ecology of local rodents and their exoparasites (Fedorov et al. 1955). Anthropurgic foci are generally located between 35°N and 35°S.

Between 1989 and 2003, 38,310 cases, 2845 of which had a fatal outcome, were reported in 25 countries (WHO 2004). Over 5000 thousand cases, 90% of which occurred in African countries (mostly Madagascar and the Democratic Republic of the Congo, where local epidemics occurred almost annually), were reported from 2005 to 2010. Outbreaks occurred in Peru in 2009 and 2010 (17 and 25 cases respectively) and in China in 2009 (12 cases). The latest large outbreak was reported in Madagascar (263 cases, 71 of which had a fatal outcome) (World Health Statistic 2009, 2010, 2011). Cases are annually reported in China, Mongolia and Kazakhstan. Sporadic cases were recorded in the Altai Mountains in Russia between 2014 and 2016.

There are 11 plague natural foci in Russia (Map 4.26), 4 of which are located in the Caucasus, 4 in the Caspian Depression and 3 in the Altai and Transbaikal regions (Onishchenko and Kutyrev 2004; Grizhebovsky 2006).

The largest natural focus is the Caspian Sandy Focus, spanning an area of 63,276 km^2. The focus is located within a couple nature zones in the eastern part of the northwest Caspian Depression, from the lower reaches of the Volga to the lower reaches of the Terek. It is bordered by the coast of the Caspian Sea from the east. The north-eastern part of the

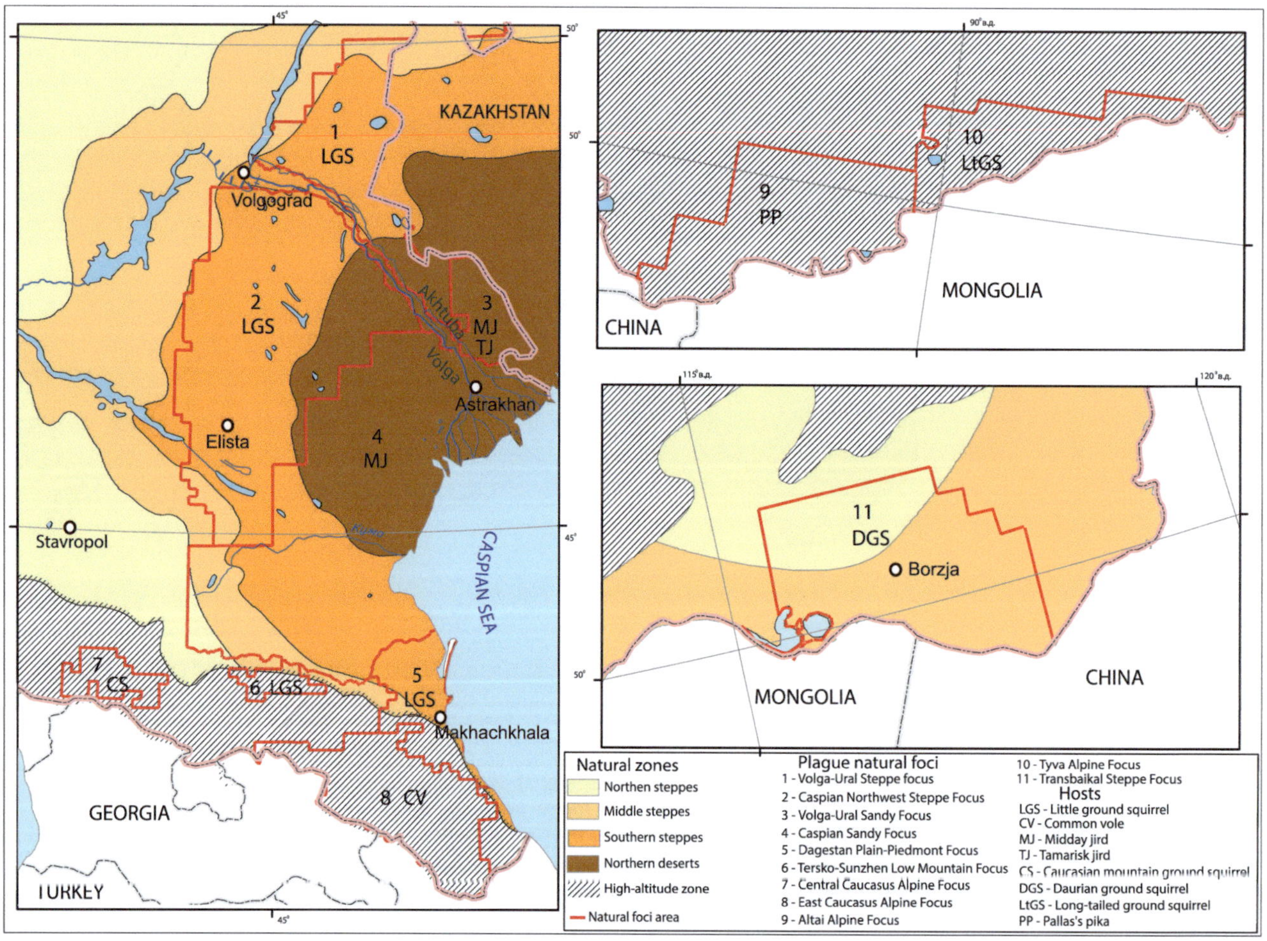

Map 4.26 Plague natural foci

focal territory is in the subzone of the northern Caspian deserts in the desert zone; the middle part is in the southern steppe subzone; the western part is in the mid-steppe subzone, while the western margins reach the northern steppe subzone. The midday jird (*Meriones meridianus*) and the tamarisk jird (*M. tamariscinus*) are the main carriers, while their specific fleas (*Nosopsyllus laeviceps* and *Xenopsylla conformis*) are the main vectors. The focus is characterized by moderate epizootic activity. Epidemic events were seen in 1923–1924, 1935–1936, 1947–1948 and 1979. The majority of outbreaks was reported in the central and eastern parts of the focus: 82 people in 13 settlements contracted the disease (Onishchenko and Kutyrev 2004; Kutyrev et al. 2015).

The next largest natural focus is the Caspian Northwest Steppe Focus, spanning 51,152 km^2. Today it occupies the southwestern regions of the East European Plain, mostly the Ergeni hills, the northwestern part of the Caspian Depression, the Davan valley and the adjacent Western brim of black earth land. A substantial part of the focus is located in the southern steppe subzone; the southwestern part lies in the northern Caspian steppe subzone. The little ground squirrel (*S. pygmaeus*) serves as the main carrier in this focus, and its fleas (*Neopsylla setosa* and *Citellophilus tesquorum*) are the vectors. This focus is not very epizootically active. It used to be one of the most dangerous plague foci: epidemic outbreaks were reported here from 1878 to 1935, and the total number of cases was 1615 people in 212 settlements (Onishchenko and Kutyrev 2004).

The Volga-Ural Sandy Focus has been known of since 1913. Its main portion is located in Kazakhstan. Its total area is 51,620 km^2. Only the southwestern edge of the focal territory (8625 km^2) in the northern desert subzone overlaps with Russian land. The focal territory's borders coincide with the outline of the desert landscape in the southern interfluvial area between the Volga and the Urals, including the Volga-Akhtubinsk Flood Plain and part of the Caspian region. Two main carriers have been found in this territory: the midday and tamarisk jirds (*M. meridianus* and *M. tamariscinus*). *Nos. laeviceps* and *X. conformis* are the plague vectors. This focus has moderate epizootic activity. Plague outbreaks were recorded in 1899, 1904–1907, 1909–1918, 1922–1930, 1932–1933, 1935–1941, 1945–1946, 1951, 1975 and 1997. 117 outbreaks have been recorded overall, totaling 250 cases in 214 settlements (Onishchenko and Kutyrev 2004).

The Volga-Ural Steppe focus is located in Russia and Kazakhstan. Its overall area is 85,000, 20,873 km^2 of which are on Russian territory. It is located on clay-rich semi-deserts. The little ground squirrel (*S. pygmaeus*) is the main carrier, and its fleas (*N. setosa* and *Cit. tesquorum*) are the main carriers. This focus has a moderate level of epizootic activity. Epidemic plague outbreaks have been recorded from 1878 to 1933: 1082 cases occurred in 73 settlements (Onishchenko and Kutyrev 2004).

Four plague natural foci are found in the North Caucasus. The largest of these is the East Caucasus Alpine Focus, located on the northern mountainsides of the Eastern Caucasus. It was discovered in 1977. Its area spans 23,420 km^2. The common vole (*Microtus arvalis*) is the main carrier, with the *Callopsylla caspia* flea as the main vector. This focus has high epizootic activity, but the stable epizootic territory is small. Regular appearances of the plague have been recorded in a local region in the heads of the Kulinka and Chirakhchai rivers on the slopes of Mt. Kokmadag. There are no verifiable data on epidemic plague outbreaks in this focus (Onishchenko and Kutyrev 2004).

The Dagestan Plain-Piedmont Focus is next in size and is located on the coast of the Caspian Sea. It occupies the southern part of the North Dagestan Depression, the Caspian Depression and the Piedmont of the east Caucasus. It was first recorded in 1951, and its overall area is 11,150 km^2. It mostly has a plain landscape, although the eastern part of the focus, which is adjacent to the Caspian coast, is in the steppe zone. The little ground squirrel (*Spermophilus pygmaeus*) is the main carrier, and the *N. setosa* and *Cit. tesquorum* fleas are the main vectors. The focus is characterized by low epizootic activity (Onishchenko and Kutyrev 2004).

The Central Caucasus Alpine Focus is located on the northern slopes of the central part of the main Caucasian mountain range. It was discovered in 1971 and spans 4309 km^2. The Caucasian Mountain ground squirrel (*S. musicus*) is the main carrier and the *N. setosa* and *Cit. tesquorum* fleas are the main vectors. This focus has high epizootic activity, although beginning in 2007, given the low numbers of carriers and vectors, plague cultures have not been isolated in this focus. There are no verifiable data on plague cases among the local population (Popov et al. 2014).

The Tersko-Sunzhen Low Mountain Focus is located between the Tersk and Sunzhen ridges in the Alkhanchurt Valley in the Piedmont of the Northern Caucasus. The focus was discovered in 1970. It spans 2336 km^2. The little ground squirrel (*S. pygmaeus*) is the main carrier, and the *N. setosa* and *Cit. tesquorum* fleas are the main vectors.

Three foci are located in Siberia. The Altai Alpine Focus includes the territory between the Sailug, Chikhachev, Kuray, North Chuya and South Chuya mountain ridges, surrounding the Chuya steppe and the steppe part of the Ukok Plateau. Its overall surface area is 11,597 km^2 and it was discovered in 1961. It makes up the northern part of the Sailugem Natural Focus, which is shared between Russia and Mongolia. Both a less virulent strain of the Altai plague subtype and a highly virulent strain of the main plague subtype circulate in this focus. Pallas's pika (*Ochotona pricei*) is the main carrier of the Altai plague subtype, and its fleas (*Paradoxopsyllus scorodumovi, Amfalius runatus* and

Ctenophyllus hirticus) are the main vectors. Epizootics of the main plague pathogen subtype develop in grey marmot habitats. *Oropsylla silantiewi* and *P. scorodumovi* fleas serve as vectors for this strain. This focus exhibits constant epizootic activity. Three cases of bubonic plague in humans, all linked to butchering marmots, were reported between 2014 and 2016 (Kutyrev et al. 2014; Popov et al. 2016).

The Tyva Alpine Focus is located in Tyva. It is also the northern part of a focus that is shared with Mongolia and located in the offsets of the eastern Altai Mountains. The focus was discovered in 1961, and plague epizootics have been reported in this area with small intervals and nearly annually since 1964. The focus spans 7849 km^2 on Russian territory. The long-tailed ground squirrel (*S. undulatus*) is the main carrier, while *Cit. tesquorum altaicus* is the main vector. This focus has high epizootic activity. No epidemic events have occurred in this focus (Onishchenko and Kutyrev 2004).

The Transbaikal Steppe Focus is located in the southern part of the Zabaykalsky Krai. It was discovered in 1911 and is part of the Manchurian (East Mongolian) plague natural focus, which is shared between Russia and Mongolia. On Russia's territory, it spannedd the arid mountain steppes of Dauria and the interfluvial region between the Onon and Argun rivers. Smooth-sloped mountains and river valleys make-up most of its landscape. The area of the Russian part of the focus is 18,150 km^2. In the past, the Tarbagan marmot (*M. sibirica*) was the main carrier in this region and the *Or. Silantiewi* flea was the main vector. Due to a decrease in the marmot's range and population size, they have been replaced by the Daurian ground squirrel (*S. dauricus*) and *Cit. tesquorum sungaris*, its flea. This focus has low epizootic activity. Verifiable data on plague epidemics are available beginning from 1868. Plague outbreaks were reported in this focus until 1930. 1052 human cases in 39 settlements occurred throughout this period. The focus has been in a dormant state since 1971 and animal cases of the plague have not been reported (Popov et al. 2014).

4.16.5 Prevention

Plague prevention consists of sanitary and epidemiological control in natural foci conducted by specialized anti-plague institutions within Rospotrebnadzor, sanitary control and prevention of plague introduction with shipments, infected animals or people in large international ports and on the country's borders. Plague natural foci are constantly monitored for their epizootic status; epidemiological studies supervise population health; laboratory diagnosis is conducted; a group of emergency and routine, specific and non-specific measure for plague prevention is conducted: patient treatment, population vaccination, camel vaccination, sanitary measures to prevent rat and flea infestations. Healthcare and veterinary specialists undergo training in order to be able to diagnose, treat and prevent plague. Public educational campaigns are conducted, along with insect and rodent extermination (Matrosov 2007).

4.17 Opisthorchiasis

Opisthorchiasis is a group of natural focal helminthiases cases by trematodes in the *Opithorchis* genus characterized by a chronic course with hepatic and bile symptoms. Two trematode species cause opisthorchiasis: *O. felineus* (the cat liver fluke) (Fig. 4.70) and *O. viverrini* (the Southeast Asian liver fluke). The latter has a limited range in Southeast Asia, while the range of the former is substantially greater. A substantial part of its range (up to 70%) is on Russian territory (Be'er 2005). *Opisthorchiasis felinea* foci are affiliated with river basins that have favourable conditions for gastropods of the Bithyniidae family (primarily in the *Bithynia* and *Codiella* genera) and fish in the *Cyprinidae* family.

4.17.1 General Information

The life cycle of *O. felineus* (Fig. 4.71) consists of a few ordered stages, part of which must occur in water (Lysenko et al. 2002). *O. felineus* eggs containing mature larvae (*miracidia*) enter the environment with faeces. After coming into contact with water, the larvae exit the eggs and are swallowed by their first intermediate hosts—freshwater mollusks in the *Bithyniidae* family. Larvae advance to the next stage—*cercariae*—inside the mollusks' bodies. *Cercariae* leave the mollusks and enter fish in the Cyprinidae family, which serve as the second intermediate hosts.

Metacercariae, which can infect humans, form in the fish skin or flesh. People or warm-blooded mammals become infected after ingesting undercooked, raw or poorly cured fish in the Cyprinidae family.

Opisthorchiasis has a severe clinical course, affecting not just the organs in which the helminths localize, but also other organ and tissue systems. Opisthorchis organisms parasitize in the bile ducts, gallbladder and pancreatic ducts in humans and a number of carnivorous mammals. In humans, the infection affects the immune system and manifests in various symptoms characteristics of gastrointestinal infections, fevers, weakness, fatigue, muscle and joint aches. Severe cases can also lead to jaundice, nausea, vomiting, bloating, diarrhoea and severe allergic reactions. People living in endemic regions are typically chronically infected.

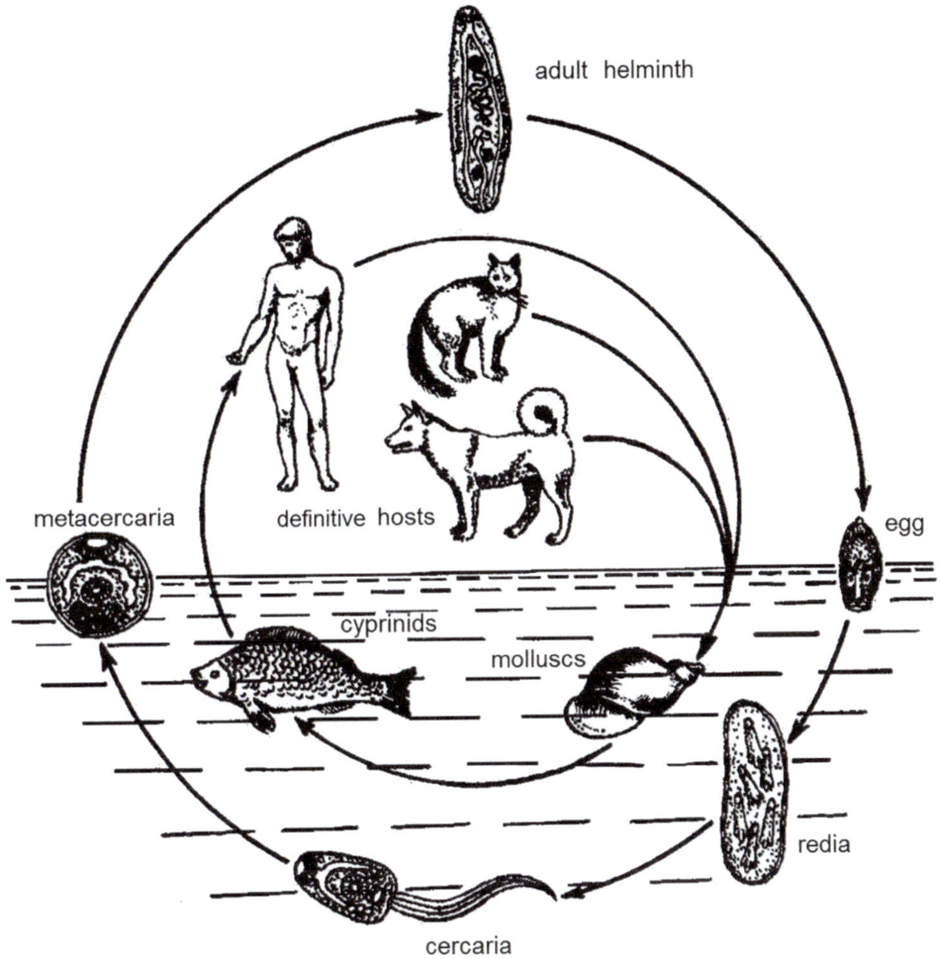

Fig. 4.70 The *Opisthorchis felineus* life cycle. *Credit* Lysenko et al. (2002)

Fig. 4.71 Bithynia leachi. *Credit* Фото Michal Maňas) https://upload.wikimedia.org/wikipedia/commons/5/5f/Bithynia_leachii.png This file is licensed under the Creative Commons Attribution 4.0 International license

4.17.2 Epidemiology and Natural Focus Characteristics

Opisthorchiasis is a natural focal disease. European water voles, muskrats, weasels and wild boars are its main carriers in natural foci (Sidorov 1983). Foci that involve wild animals can exist in locations that are quite far from human settlements. However, anthropurgic foci play the main role. In these foci, wild animals are minimally involved, while humans and domesticated animals (cats, swine, dogs) serve as the parasite's definitive hosts. Humans are most intensively involved in opisthorchiasis circulation in the Ob–Irtysh River Basin, especially in its central part, where up to 80% of a settlement's population can be infected (Be'er 2005).

The opisthorchiasis focal structure consists of three subsequent links in the biological chain: warm-blooded host—molluscs—fish. However, given the mobility of one of these links—the fish—it is quite hard to characterize the spatial

Fig. 4.72 Ides are one of the most commonly *Opisthorchis*-infested fish in the Cyprinidae family. *Credit* фото Kristian Kostov https://ru.wikipedia.org/wiki/%D0%AF%D0%B7%D1%8C#/media/File:Mazdruga_danube_vidin_bulgaria.jpg Attribution-ShareAlike 3.0 Unported (CC BY-SA 3.0)

structure of these natural foci, and nearly impossible to draw the boundaries of a natural focus within a single body of water (Sidorov 1983). According to contemporary understanding, river basins are the focal territory of opisthorchiasis, while mollusc habitats are their epicentres, given the relative inertia of molluscs in the parasitic system.

The first intermediate hosts—Bithyniidae—are a typical component of freshwater biocenoses. They generally live in coastal zones or small rivers with slow currents, channels, bays and bayous. These are small bodies of water that are easily warmed and have a wealth of embryophytes and middy riverbeds. The systematic position of the first intermediate hosts is not quite clear, because the formerly polytypic species *Bithynia leachii* (Fig. 4.72), which was considered to be widespread, was reclassified as a couple independent species, some of which were later classified in different genera. The majority of sources indicate *Codiella inflata* as the main host, however today it is thought that molluscs in at least five different closely related genera can serve as intermediate hosts in the nosogenic territory of opisthorchiasis (Be'er 2005). These molluscs vary in their susceptibility to opisthorchiasis and have vastly differing ecological needs, which ultimately has an effect on the parasite's circulation in natural and synanthropic foci. Nonetheless, *C. inflata* evidently plays the key role.

Fig. 4.73 Konstantin Vinogradov (1847–1907), the Russian pathologist who first discovered the opisthorchiasis pathogen in humans. *Credit* https://commons.wikimedia.org/wiki/File%3AVinogradov_Konstantin_Nikolaevich.jpg **Public domain**

Fish in the Cyprinidae family are the second intermediate hosts of Opisthorchis. They are infected by metacercariae and are the final factor in the epidemiological process. About 23 species in this family inhabit the disease's nosogenic territory. Ides (Fig. 4.73), roaches, common daces, common breams, common rudds, white breams and common bleaks are most commonly infected. *Metacercariae* are fairly resilient and remain infectious if the fish is not cooked through or if cold smoked. They die after 3–10 days if the fish is cured.

Cyprinidae infection rates can be quite significant, reaching 100% in certain foci (Lysenko et al. 2002), depending on the type and characteristics of bodies of water in the floodplain (Map 4.27). Usually, fish are most infected near human settlements and industrial lakes and bayous.

4.17.3 Historical Background

The opisthorchiasis pathogen was discovered in a cat by Karl Rudolphi, the German naturalist, back in 1822, and was considered a domesticated animal parasite for a long period of time. Only in 1891 did Konstantin Vinogradov, a professor at Tomsk University (Fig. 4.74), discover a

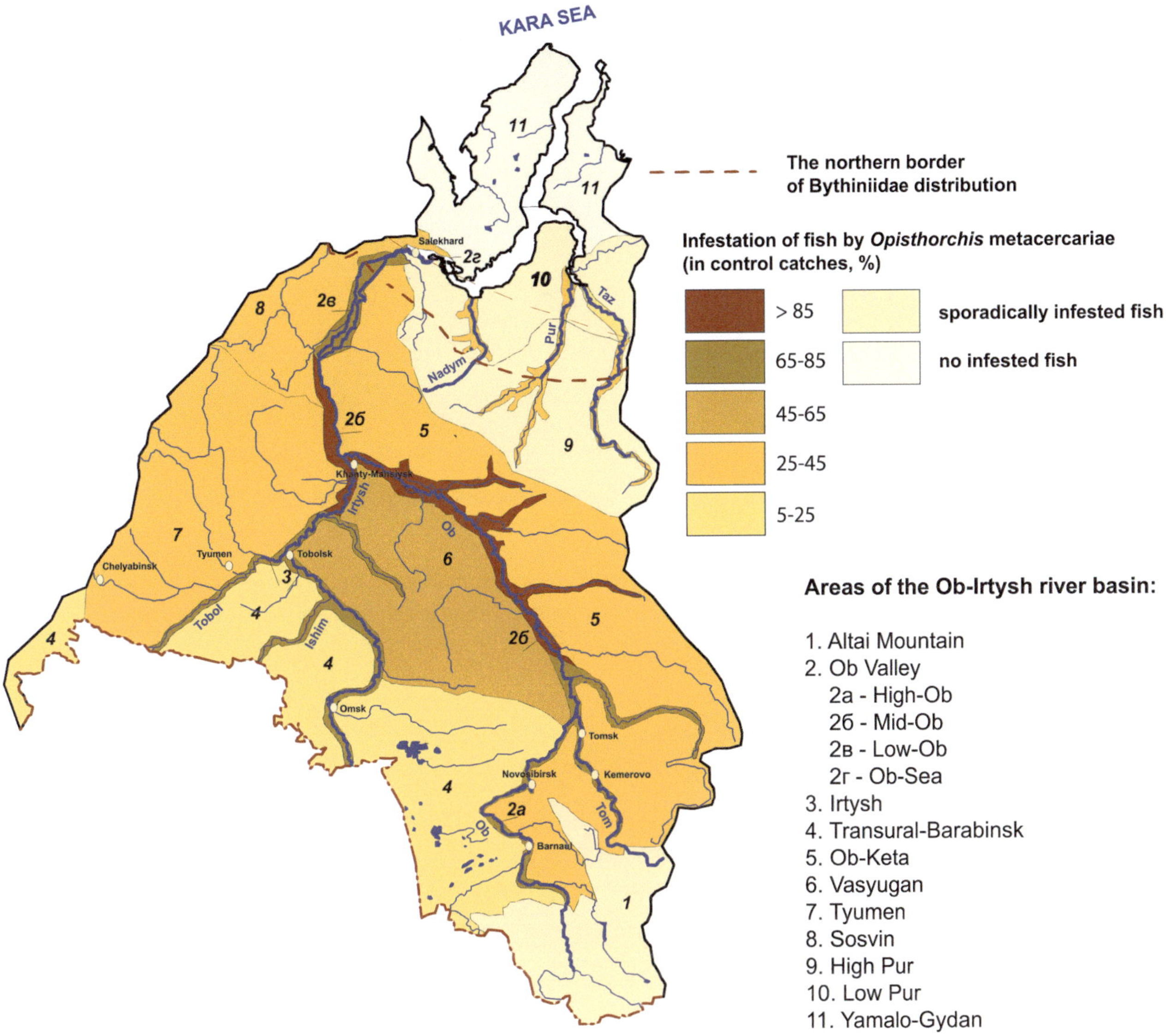

Map 4.27 The northern border of Bithiniidae distribution. Infestation of fish by *Opisthorchis*

previously unknown trematode while conducting an autopsy on someone who had presumably died of "Ob disease", long familiar to the West Siberian population. Vinogradov called this parasite the Siberian fluke.

Over the course of the following century, the spread of opisthorchiasis was studied intensively. In 1930, Konstantin Scriabin and Nikolay Plotnikov, the founders of Soviet helminthology, underscored the grave sanitary and socio-economic issues posed by opisthorchiasis. Under their leadership, the 70th helminthological expedition established that North Tobolsk (the lower reaches of the Ob River) is a massive opisthorchiasis focus that affects people, domesticated animals and fur game that feed on fish. In 1946, Evgeny Pavlovsky first proposed the disease's natural focal nature (Be'er 2005).

4.17.4 Distribution in Russia and the World

Opisthorchiasis felinea is found in a number of European countries: France, Bulgaria, Holland, Italy, Switzerland, Sweden and Romania. However, human morbidity is sporadic in all these locations. The majority of opisthorchiasis' nosogenic territory is on the territory of the former USSR, and the most intensive foci are concentrated in the Ob and Irtysh River Basin, the Kama River basin and, to a lesser degree, the Volga, Dnieper, Don, Donets and Northern Dvina river basins (Stepanova 2002; Be'er 2005).

Opisthorchiasis leads parasite morbidity rankings in Russia: in 2010, its proportion among other helminthiases nearly reached 40% (Guzeeva 2011a, b). In the biohelminthiasis morbidity structure, it was in first place (74.4%). The average annual

Fig. 4.74 The Ob River. Opisthorchiasis is most widespread in the Ob–Irtysh River Basin. *Credit* P. I. Konstantinov (permission obtained)

number of cases in Russia from 2006 to 2015 is 33,466, which corresponds to 23.43 cases per 100,000 individuals. However, the true number of opisthorchiasis cases in Russia is much higher. According to V. P. Sergiev and his co-authors (1993), expert estimates did not correspond to the number of reported cases by a factor of 15.6. It is thought that this discrepancy remains up to this day (Stepanova 2002; Guzeeva, 2011a).

Opisthorchiasis is the most common helminthiasis transmitted through contaminated fish. The issue has been complicated thanks to an increase in the number of private and privatized fish plants, which flagrantly violate the technical code of fish disinfection and providing the population with epidemiologically dangerous products (Guzeeva 2011a, b).

The Ob–Irtysh River Basin is home to the highest risk situation. The nosogenic territory is continuous here, and the human population is affected everywhere. Maximum prevalence is found in the mid-Ob region (Map 4.28). As demonstrated by numerous field studies (Be'er 2005), human infestation rates decrease as one moves upstream along the Ob and Irtysh, as well as their tributaries, but they never fall below 10%. In the Dnieper, Northern Dvina, Kama, Volga and Don River Basins, opisthorchiasis has a pronounced focal quality and is widespread in animal populations.

Certain trends can be noted based on the review of opisthorchiasis morbidity rate dynamics in Russia from 2006 to 2010. They vary based on non-uniform long-term oscillation (decrease, increase, oscillations with amplitudes ranging from significant to insignificant) (Map 4.29). Areas with very high and high morbidity rates (the central part of Western Siberia) demonstrate a gradual decrease in morbidity in the 5-year period. This decrease is most pronounced in Khanty-Mansi Autonomous Okrug, which is the most at-risk region in Russia. However, a few regions in the southern Ob–Irtysh River Basin have shown an insignificant growth in morbidity. In the rest of the country's regions, the curves' dynamics, with varying (from moderate to low) morbidity rates, are relatively stable throughout the period in question. Comparing morbidity rates between three individual consecutive years (2011, 2012, 2013) demonstrates substantial stability in focal territories.

Today the issue of opisthorchiasis as a natural focal helminthiasis has become extremely serious as a result of anthropogenic changes to ecosystems. In recent years, natural foci have been turning into mixed foci, i.e. humans become actively involved in the epidemic process: this is even happening in systems where humans' roles used to be insignificant (Be'er 2005). Therefore, practically any parasitic system needs to be viewed in the context of social factors and technogenic contamination, as human activity disrupts many crucial functional connections that have formed in natural foci. This has led to a number of consequences that should be taken into account when sanitary and epidemiologic monitoring organizations decide on their priorities.

4.17.5 Prevention

Based on Rospotrebnadzor's recommendations, fish preparation guidelines must be followed in areas with a high risk of infection with opisthorchiasis. Fish must be boiled or fried for no less than 15–20 min. Fish curing needs to use a large amount of salt and a long time period: no shorter than 2 weeks (2–3 kg of salt should be used per 10 kg of fish). Raw fish or ground fish meat should not be tasted for seasoning or doneness. Well-cooked or well-cured fish will not be a source of infection. Cold smoking does not guarantee the safety of the final product if it is not first treated in a saline solution.

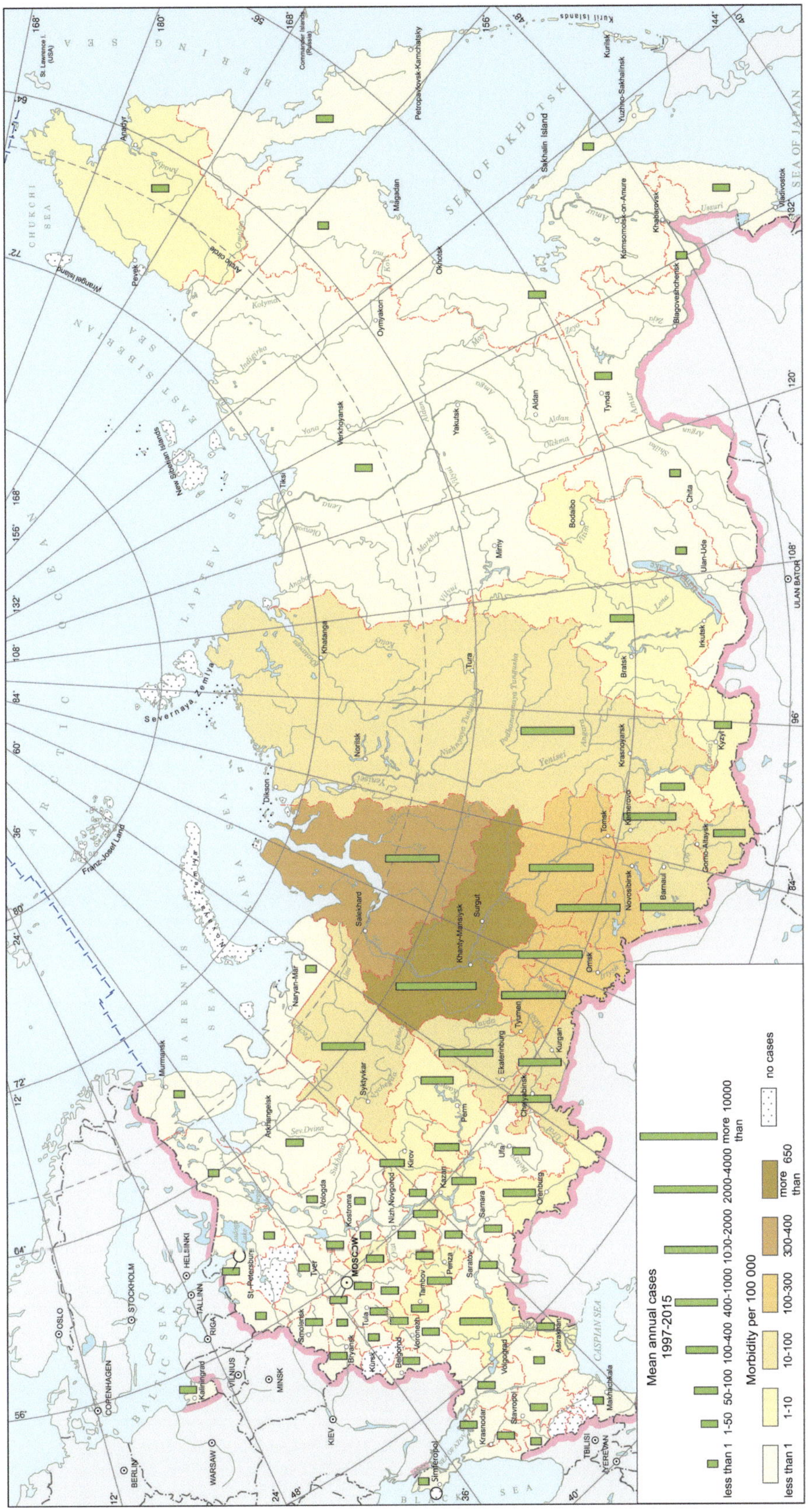

Map 4.28 Opisthorchiasis morbidity, 2006–2015

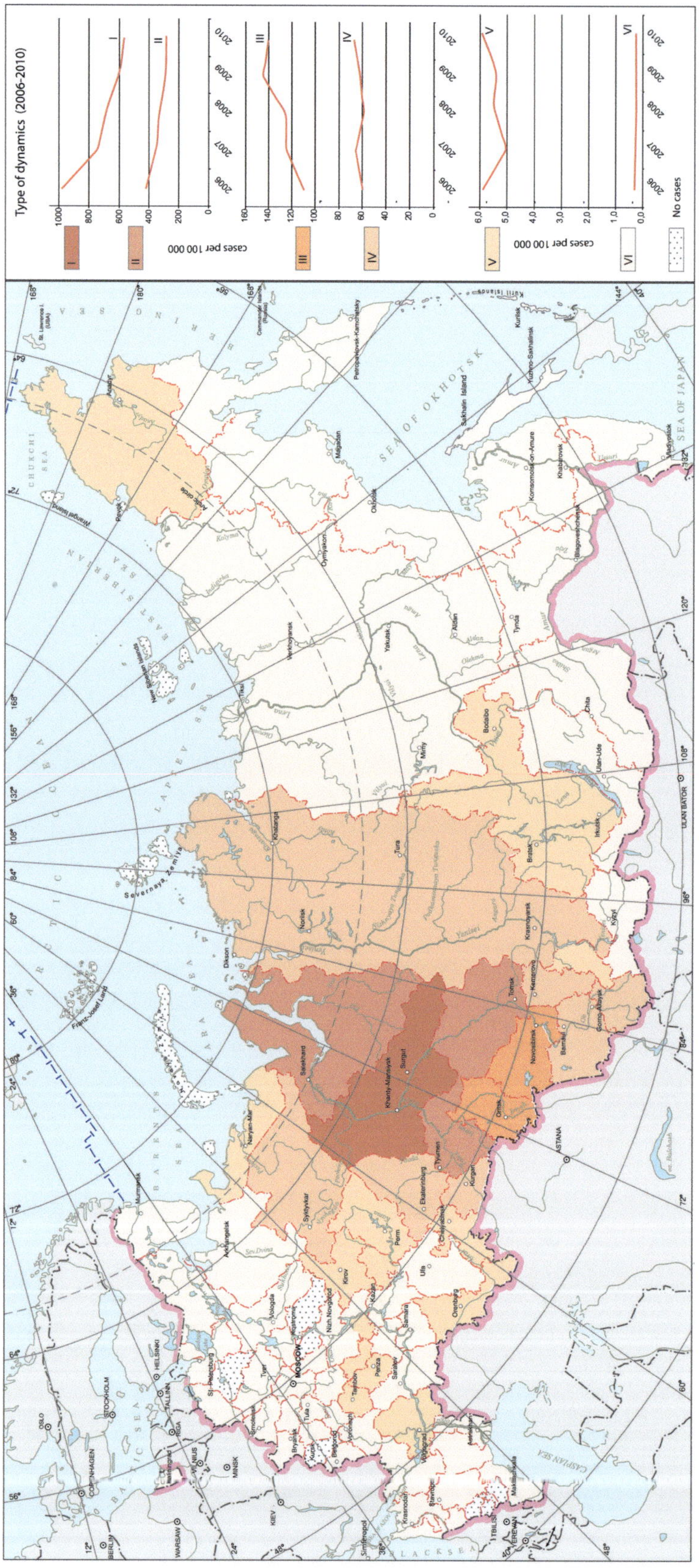

Map 4.29 Types of opisthorchiasis morbidity dynamics, 2006–2010

Opisthorchiasis *metacercariae* are resistant to freezing, so long-term freezing is necessary for parasite destruction: larvae die in 72 h at a temperature of −23 to 25 °C and in 6 h at temperatures below −30 °C. Lightly cured fish, caviar and fresh stroganina (a dish made out of raw lightly frozen fish that is extraordinarily popular in Western Siberia) are thought to pose the greatest risk of infection.

If the aforementioned guidelines cannot be met, it is best to avoid consuming fish in the Cyprinidae family altogether.

The following recommendations for opisthorchiasis prevention are also important: pets (dogs and cats) should not be fed raw fish as they can also serve as hosts for Opisthorchis, and human, cat, dog, swine or fur game faeces should not be allowed to contaminate bodies of water.

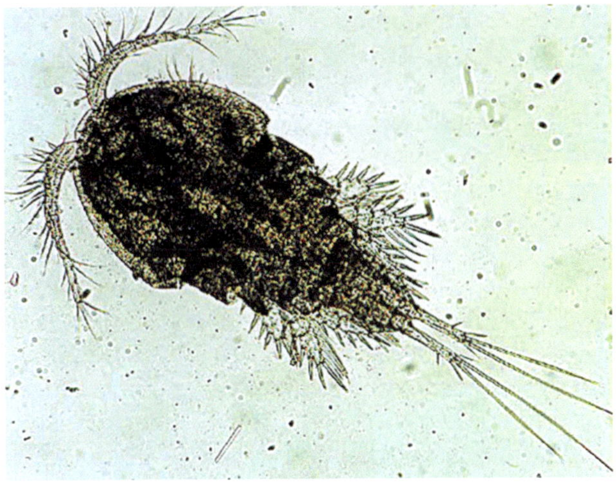

Fig. 4.75 The copepod is an intermediate host for tapeworms. *Credit* https://en.wikipedia.org/wiki/File:Cyclops.jpg#file

4.18 Diphyllobothriasis

Diphyllobothriasis refers to a group of chronic natural focal gastrointestinal helminthiases. Today, it is thought that 20 million people are infected with diphyllobothriasis worldwide, although morbidity and prevalence have been decreasing in developed countries (Scholz et al. 2009). However, given that this tapeworm infection often takes a light or subclinical course, data on diphyllobothriasis infection rates are probably inaccurate. Diphyllobothriasis is especially common in countries, where consumption of raw or undercooked fish or raw caviar is popular.

4.18.1 General Information

Diphyllobothriasis is caused by tapeworms (Cestoda) in the *Diphyllobothrium* genus. Of the 14 *Diphyllobothrium* species that affect humans, 5 have been found in Russia, of which *D. latum*—the broad tapeworm—is the most common (Lysenko et al. 2002).

The broad tapeworm is the largest parasite that affects humans. Its body, which is segmented, can be as long as 28 m.

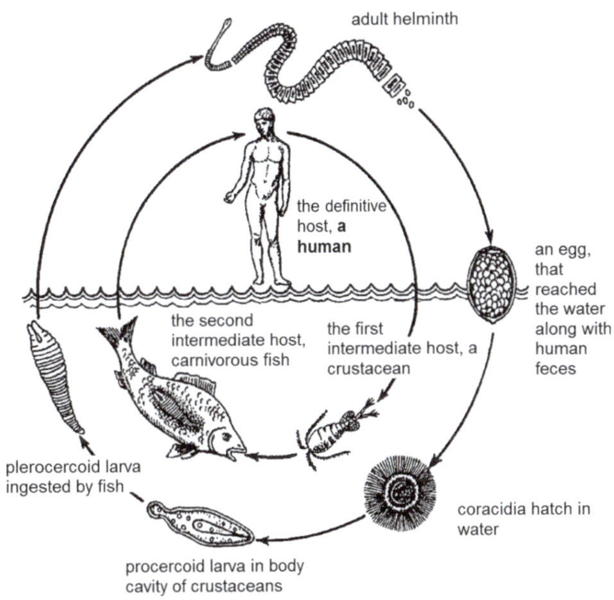

Fig. 4.76 The *Diphyllobothrium latum* life cycle. *Credit* Lysenko et al. (2002)

Humans and piscivorous animals are the tapeworm's definitive hosts, while copepods (Fig. 4.75) and fish are its intermediate hosts. The tapeworm's life cycle consists of multiple stages (Fig. 4.76). The mature helminth produces unembryonated eggs in the definitive host's body at a rate of up to 1 million per day. The eggs enter a body of water with faeces, where they embryonate and hatch into *coracidia*, which are swallowed by copepods. In 2–3 weeks, they develop into a new stage—*procercoid* larvae. When the copepods are consumed by planktivorous fish, the *procercoid* enters into their musculature and develops into a *plerocercoid* larva, which can infect humans. If predator fish (such as pikes) eat the small fish, *plerocercoids* form in their internal organs and roe as well. It is thought that about 20 copepod species can serve as intermediate hosts for *D. latum* (Semyonova et al. 2011).

Diphyllobothriasis is chronic and generally affects the small intestine, causing problems in the upper gastrointestinal tract as well as anaemia. An allergic response is also common during the first year of infection. Because the broad tapeworm, like all Cestoda, does not have a digestive system of its own and feeds by absorbing nutrients across its entire surface, the host's organism often experiences an imbalance of essential and nonessential amino acids and vitamins or has a lower albumin content. Diphyllobothriasis is asymptomatic in 10% of cases (Lysenko et al. 2002).

4.18.2 Epidemiology and Natural Focus Characteristics

Fish play the main role in transmitting diphyllobothriasis to humans. People serve as a crucial infection source and reservoir for *D. latum* because they infect bodies of water with sewage. The helminth can survive for much longer in a human's body than in that of an animal. All diphyllobothriases are zoonotic and natural focal in nature. Aside from people, cats, dogs, swine, bears and other piscivores may be infected, but their epidemiological significance is low.

The pike and the burbot are the main fish species that become infected with *D. latum.* The parasite is found not only in their musculature, but also in their internal organs and roe. The parasites only inhabit the musculature of perches (Fig. 4.77), ruffes and some salmon species.

D. latum diphyllobothriasis natural foci are found in freshwater river and lake ecosystems. Water with a weak current and standing water are most epidemiologically significant, as copepod concentrations in rivers with fast currents are low, reducing the likelihood of an encounter between a coracidium and its intermediate host.

Aside from the broad tapeworm, *D. dendriticum* and *D. ditremum,* who are primarily carried by piscivorous birds, are also found in Russia. For a long period of time, it was thought that *D. dendriticum*, which causes so-called "gull diphyllobothriasis", did not play a major role in human pathology. However, a number of recent studies have demonstrated that the role of *D. dendriticum* was underestimated (Kuchta et al. 2013). This parasite is characterized by the wide variety of fish (over 50 species) that can serve as its intermediate hosts.

D. nihonkaiense and the morphologically similar *D. klebanovskii* play the main role in foci in the Russian Far East.

Fig. 4.77 The European perch is one of the main fish hosts of the broad tapeworm. *Credit* Фото Karelj https://commons.wikimedia.org/wiki/Perca_fluviatilis#mediaviewer/File:Perca_fluviatilis_Prague_Vltava_2.jpg

In the past, these parasites were thought to be broad tapeworms given their extreme similarity (Arizono et al. 2009). Unlike *D. dendriticum,* these species can parasitize on both freshwater fish and migratory species. They have certain structural qualities that have been accepted as taxonomically significant (Muratov and Posokhov 1988). Piscivorous predatory mammals, primarily bears, serve as the main carriers for these tapeworms (Muratov 1993).

Diphyllobothriasis mortality is primarily related to the ancient tradition of consuming raw or undercooked fish: lightly cured, raw smoked, frozen, raw and fish jerky, as well as lightly salted caviar. The rising popularity of restaurants that serve these dishes, predominantly Japanese restaurants that prepare sushi and sashimi using raw salmon, plays a major role.

4.18.3 Historical Background

Evidently, tapeworms are extremely ancient human parasites. *Diphyllobothrium* sp. egg fossils have been found in ancient mummies (some are 6000 years old) in Peru and Chile and in Neolithic settlement excavation sites in Germany, Switzerland and France. The first recognizable tapeworm description dates from 1592. In 1758, Carl Linnaeus named the broad tapeworm *Taenia lata.* In 1747, Herman Spöring Sr., a Finnish doctor and amateur naturalist, first proposed a link between fish and tapeworms (Scholz et al. 2009), but only in the late nineteenth century was it proven that humans become infected after consuming contaminated fish (Fig. 4.78).

The study of the broad tapeworm's early life stages began in 1917, when Constantine Janicki and Felix Rosen established that copepods serve as the parasite's first intermediate hosts (Semyonova et al. 2011).

In 1916, Nikolay Kholodkovsky described a clinical case of gull diphyllobothriasis in an individual who lived on the coast of Lake Baikal, beginning research into the epidemiology of this tapeworm species. Later, a number of researchers gradually discovered the significance of other tapeworm species in human pathology.

4.18.4 Distribution in Russia and the World

Diphyllobothriasis is widespread around the world. Tapeworms' range is circumboreal. Broad tapeworm diphyllobothriasis foci are found in freshwater in Europe, North Asia and North America. In Europe, Sweden and Finland are most affected, but sporadic cases are annually reported in Switzerland, Austria, Greece, Spain, Italy, Norway and Romania.

In Russia, individual tapeworm species infections are not tracked, but the available statistical data reflect the overall

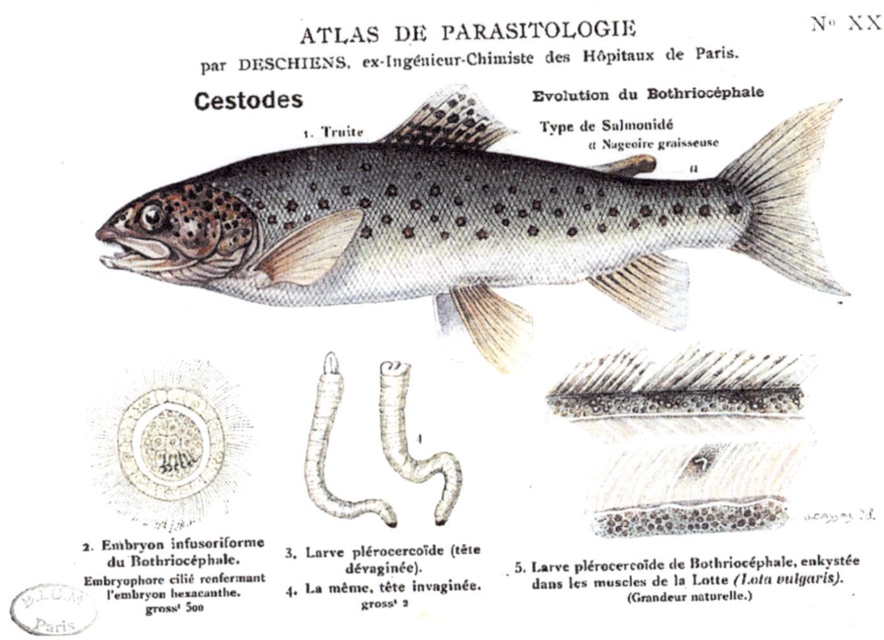

Fig. 4.78 A page from a parasitology atlas published in Paris in 1901 (Deschiens V. Atlas de parasitologie. Paris: 1901). The role of copepods in the tapeworm's life cycle was unknown at the time. *Credit* (Deschiens V. Atlas de parasitologie. Paris: 1901).https://sv.wikipedia.org/wiki/Diphyllobothrium#/media/File:Evolution_du_Bothrioc%C3%A9phale_from_Atlas_de_parasitologie_(1901).jpg

number of diphyllobothriasis cases. It can be assumed that the majority of cases are caused by *D. latum,* which one of the few endemic helminths in the Holarctic. In Russia, most diphyllobothriasis foci are concentrated in the basins of eastern and northern rivers. Six river and lake regions of Russia are thought to pose a high risk of diphyllobothriasis infections: North-Western, Volga-Kama, Ob–Irtysh, Yenisei–Lena, Far East and Black Sea-Azov (Lysenko et al. 2002).

Gull diphyllobothriasis (caused by *D. dendriticum*) foci are most widespread in Siberia, especially in the Selenga river basin and on Lake Baikal, where its distribution is linked to seagulls and the Arctic cisco (Timoshenko et al. 1995). In the Russian Far East, diphyllobothriasis morbidity is linked to *D. klebanovskii*, which is carried by bears and salmon.

The highest diphyllobothriasis morbidity rates are found in northern Siberia: Sakha (Yakutia), Krasnoyarsk Krai, Yamalo-Nenets Autonomous Okrug, as well as in Khakassia and Buryatia. In some regions, the average annual number of detected cases is over 1000, and the morbidity rate is over 150 cases per 100,000 individuals. The Nenets Autonomous Okrug is the most at-risk region in Russia's European territory, which is characterized by high morbidity rates (over 150 cases per 100,000 individuals), as well as Perm Krai, which lies within the Volga-Kama Focal Region (Map 4.30). However, it is important to keep in mind that, as mentioned above, reported morbidity rates may not be accurate given the large number of asymptomatic and subclinical cases.

Between 2006 and 2010, diphyllobothriasis morbidity rate dynamics, which vary based on their type of long-term oscillation (decrease, increase, oscillation with amplitudes varying from significant to insignificant levels), have similar tendencies across various regions in Russia (Map 4.31). Regardless of the morbidity rate in a specific region, the long-term morbidity dynamics have shown a tendency to decrease insignificantly in the 5-year period in question. In general, this corresponds with the tendency towards lower diphyllobothriasis morbidity rates in developed countries.

However, it must be noted that this tendency to decrease is quite weak. It is clear that the disease demonstrates pronounced consistency in the epidemiological behaviour of focal territories.

Many researchers point out that the nosogenic territories of diphyllobothriasis caused by *D. latum* and opisthorchiasis overlap in the number of territories, especially in Western Siberia. This congruence is the result of the overlap between the natural foci of these two infections. Ecological-faunistic associations of the diseases' first and second intermediate hosts are affiliated with aquatic biotopes and both diseases share definitive hosts (humans, cats, dogs, swine). This leads to the formation of combined opisthorchiasis and diphyllobothriasis natural foci. The influence of social factors, such as increased migration in the human population, leads to the expansion of nosogenic territories and the formation of combined foci in new places.

Anthropogenic factors also have a substantial influence on the formation of new opisthorchiasis and diphyllobothriasis foci. Thanks to the construction and use of water

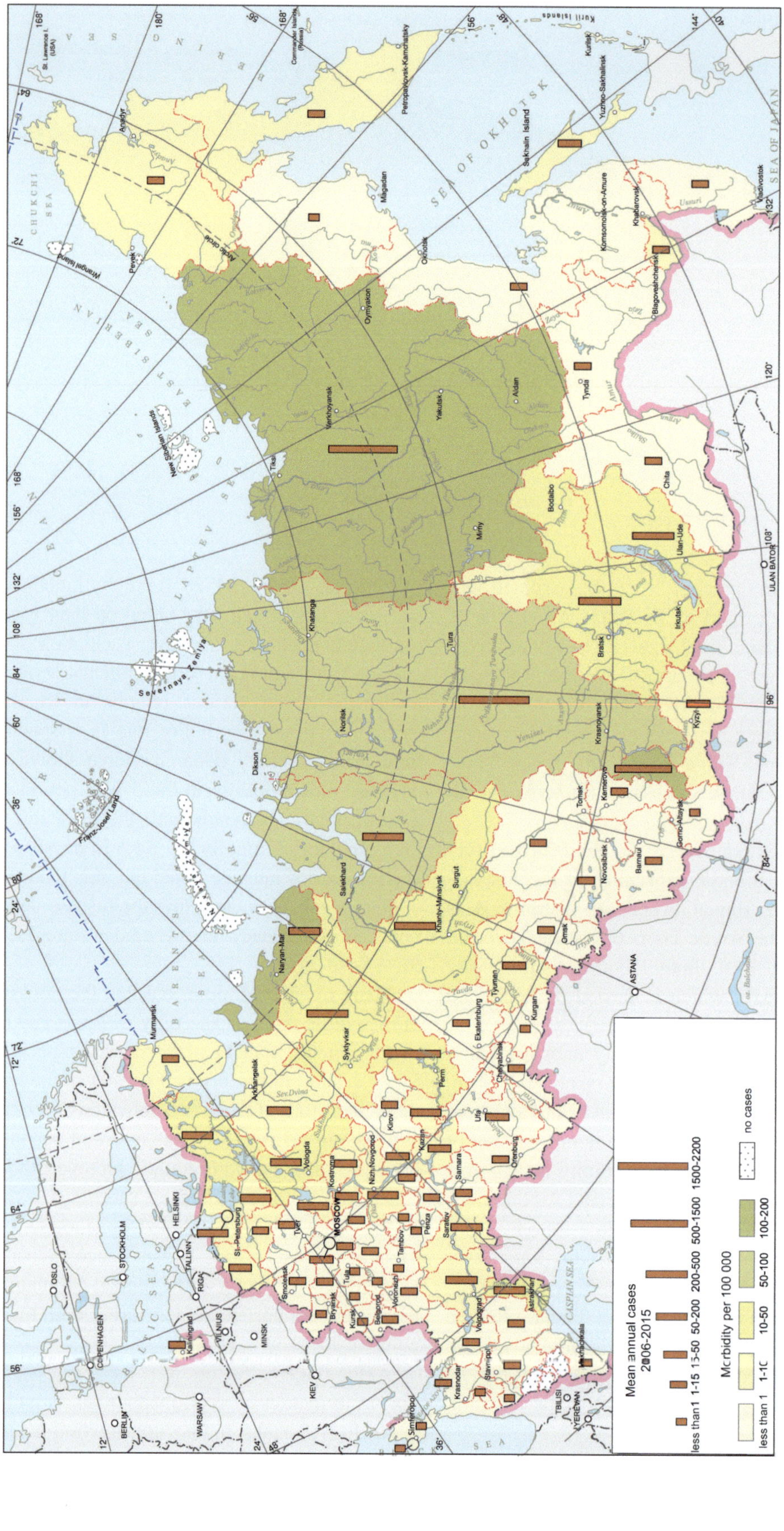

Map 4.30 Diphyllobothriasis morbidity, 2006–2015

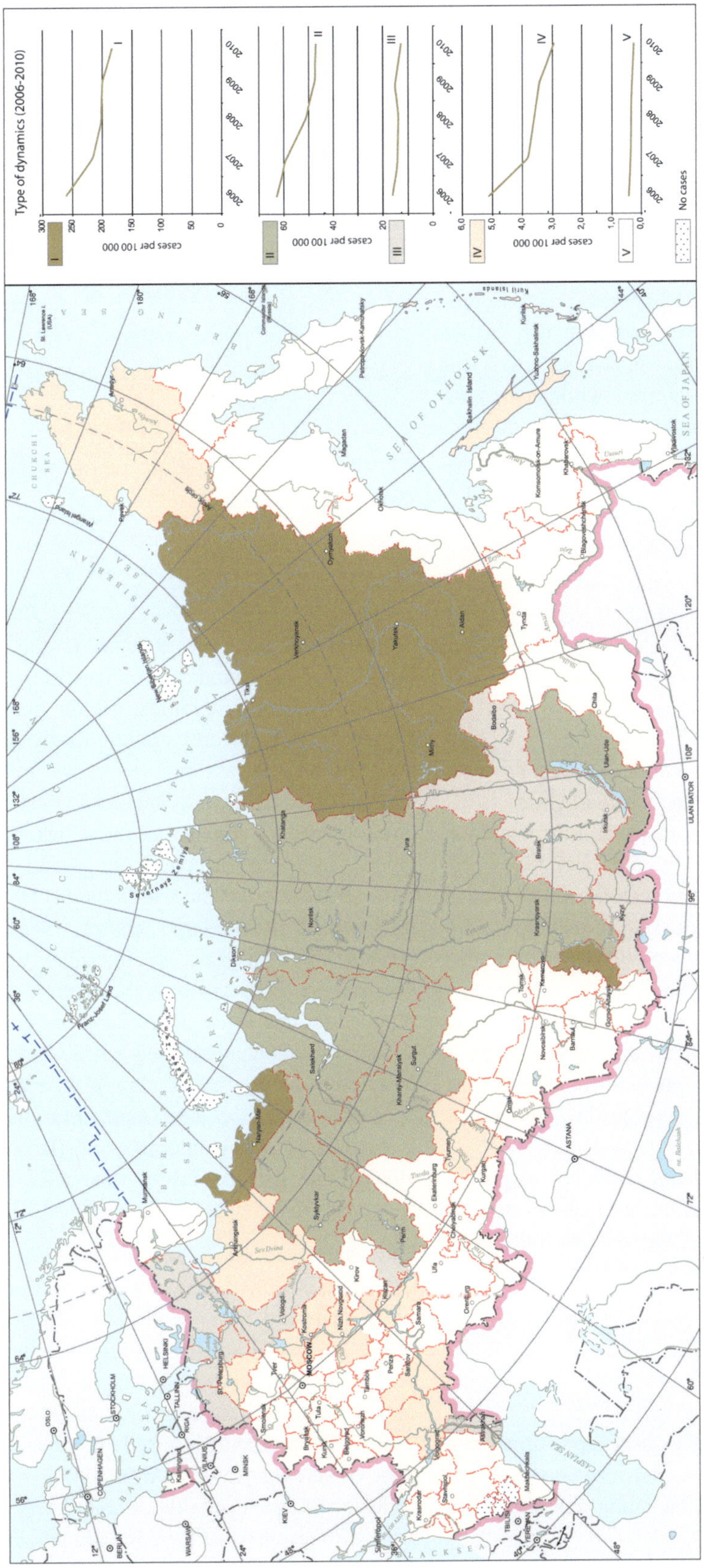

Map 4.31 Types of diphyllobothriasis morbidity dynamics, 2006–2010

reservoirs, new foci have formed on the Gorky, Kuybyshev, Volgograd, Krasnoyarsk and Bratsk water reservoirs (Stepanova 2001). Cases of gull diphyllobothriasis focus formation have occurred in lakes in Western Siberia after the introduction of certain predatory fish species.

4.18.5 Prevention

In accordance with Rospotrebnadzor's recommendations (www.rospotrebnadzor.ru), diphyllobothriasis can be prevented by thoroughly cooking river and lake fish before consumption. Fish filets, groundfish, cutlets and small fish should be fried for no less than 15–20 min, while non-sliced cuts of large fish should be fried for no less than 30–40 min. When fish is cured in a high saline solution, tapeworm larvae died after 8–10 days. Fish can be disinfected if frozen for 2 weeks at a temperature of −12 °C or lower. Raw fish should not be tasted while cooking. Moreover, consumption of lightly salted fish jerky and lightly salted homemade pike caviar should be avoided.

4.19 Taeniasis

Taeniasis refers to two nosogenic entities caused by tapeworms in the *Taeniidae* family of the *Cestoda* class.

One common type of taeniasis (one of the most common helminthiases overall) is caused by the beef tapeworm (*Taenia saginata*). It is found in many regions of the Russian Federation, usually in areas with extensive animal husbandry or among individuals who ate undercooked contaminated beef or veal (Lysenko et al. 2002). Another type is caused by the pork tapeworm (*Taenia solium*). It used to be very common in areas with extensive swine farming, but today is it quite rare. Taeniasis is an anthroponosis: humans are the definitive hosts of these parasites. As a result, these infections cannot exist without them.

Today taeniasis due to *T. saginata* continues to pose a serious medical and veterinary threat, while taeniasis due to *T. solium* is seen as a disease associated with regions with a great deal of swine farming, where it is not even that common.

4.19.1 General Information

Taeniasis due to *T. saginata* The beef tapeworm's life cycle includes two hosts (Fig. 4.79). Humans are the definitive hosts, while large cattle, including reindeer, serve is its intermediate hosts.

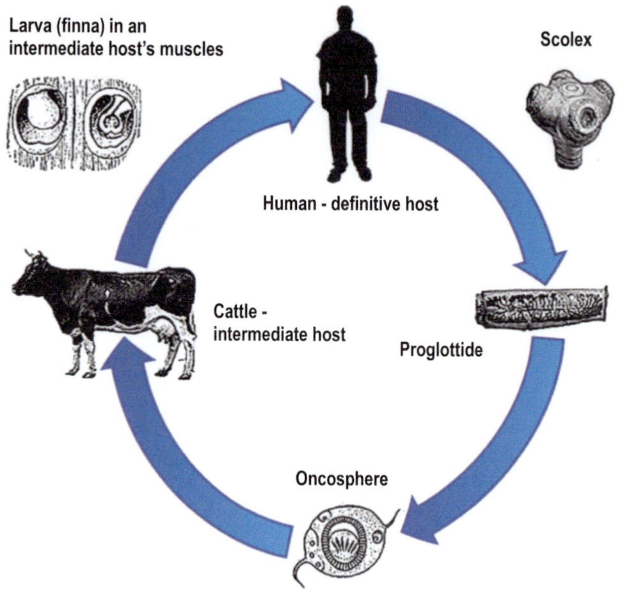

Fig. 4.79 Life cycle of the beef tapeworm

A mature *T. saginata* attaches to the wall of a human intestine using suction cups on its scolex (head). After this, it begins to form proglottids, which gradually mature, are shifted to the tapeworm's tail by newly appearing segments, and fall off. As a result, a human infected with a beef tapeworm excretes oncospheres, which enter the outside environment as contagious units that contain an embryo. Oncospheres can survive in the outside environment for a prolonged period of time, until they enter the organism of an intermediate host (large cattle), where the embryo is released. It penetrates the intestinal wall and enters the vasculature. They are brought to the musculature with the blood flow and settle there. People are infected when they consume meat that contains viable larvae: cysticerci. The helminths continue to develop in the human's intestine.

Taeniasis due to beef tapeworms usually causes mild symptoms or no symptoms at all (Beljaev and Alekseeva 1980). The parasite has a traumatic, irritating and psychogenic effect on its host.

Taeniasis due to *T. solium* The life cycle of the pork tapeworm is similar to that of the beef tapeworm, except that swine and wild boars serve as its intermediate hosts. Pork tapeworm proglottids are very similar to *T. saginata* proglottids, but have a number of unique anatomical qualities. Humans are the definitive hosts in taeniasis due to *T. solium*. They become infected after consuming undercooked pork containing pork tapeworm cysticerci. In this case, the parasite matures inside the human host. The clinical symptoms of both types of taeniasis are similar.

Pork tapeworms can lead to a more serious pathology—cysticercosis, which occurs when humans accidentally ingest worm proglottids or eggs. The disease that develops, as a result, poses a serious risk for the host's life. Cysticerci may develop in the host's muscles, eyes or brain. Neurocysticercosis, when the eggs develop in the brain, is the most dangerous, as it can lead to seizures similar to those caused by epilepsy. Some attacks do not lead to a loss of consciousness and manifest as the jerking of limbs on one side of the body or muscles on one side of the face, as well as extreme headaches or vision loss.

Fig. 4.80 A combined urban–rural *T. saginata* focus. *Credit* Zoonotic Diseases (1981)

4.19.2 Epidemiology and Natural Focus Characteristics

Taeniasis is a biohelminthic anthroponosis, meaning that the helminths cannot complete their life cycle without human participation but also require an animal intermediate host (large cattle or swine, depending on the type of tapeworm). Distinct stages parasitize on the intermediate hosts and on humans (Lysenko and Beljaev 1987). People infected with the tapeworm are the source of infection. Oncosphere excretion begins 2.5–3.5 months after infection and lasts many years.

People infected with *T. saginata* cannot directly infect others, but can serve as an effective source of infection for the intermediate hosts (Lysenko et al. 2002). Large cattle can be infected from humans, directly via hands that are contaminated with oncospheres or indirectly via contaminated water or feed. Usually, livestock ingests the parasite in pastures, while grazing near human settlements, and even in a barn, if it receives moist feed contaminated with oncospheres.

A taeniasis due to *T. saginata* focus consists of a settlement with a pasture or a group of settlements that shares a pasture, combined with infected humans and cattle. Urban *T. saginata* foci pose a special risk: they form as the result of consumption by city dwellers of beef brought from a rural area (Fig. 4.80). Subsequently, the infected urban population spreads the parasite outside the city, in recreational areas and along major roads. It has been established that livestock that grazes near roads or camping grounds has a higher *T. saginata* infestation rate than livestock that grazes far away from roads (Pawlowski and Schulz 1972).

Aside from social factors, natural factors also play a role in the formation of *T. saginata* foci. Temperature and humidity have some effect. Oncospheres are very sensitive to desiccation, as well as high (over 30 °C) and low (under −30 °C) atmospheric temperatures.

Unlike taeniasis due to *T. saginata,* taeniasis due to *T. solium*, as previously mentioned, is not a serious public health issue. This is due to a major shift in people's understanding of hygiene and sanitation, noticeable even in poor countries, and the qualitative improvement in sanitary and veterinary prevention measures. However, the parasite's biological characteristics are the main reason. Unlike the beef tapeworm, which is characterized by the active release and passing of segments, pork tapeworm segments are passively removed. A person infected with *T. saginata* is the main source of contagion, because active segment removal allows for the eggs to be spread across the patient's hands, clothing and everyday objects. As a result, one infected person can infect an entire herd of cows if he or she tends to them. Pork tapeworm infestation, however, does not have the same effect: swine are coprophages and can swallow the proglottids by itself. As a result, the parasite did not have to evolve in order to spread its contagious stage (Beljaev and Alekseeva 1980).

A second reason for the decrease in *T. solium* infection rates is the high number of eggs present in pig cadavers compared to that of *T. saginata* eggs in cattle cadavers, making it easier to identify the parasite in slaughterhouses and lowering the likelihood of human consumption of contaminated meat.

Taeniasis due to *T. solium* has always had a non-uniform distribution due to religious bans against pork consumption—as a rule, taeniasis due to *T. solium* is not found in human populations who practice Islam or Judaism and live in close proximity of one another. Traditions banning the consumption of undercooked pork also play an important role (beef, on the other hand, is often prepared rare). As a result, taeniasis due to *T. solium* is most common in areas where pork and pork product (such as lard striped with meat). Levels of sanitary and hygienic practices and sewage conditions also play a major role.

Cases of cysticercosis infection as a result of unhygienic practices in food production. One particularly famous case is that of a group cysticercosis infection among members of an orthodox Jewish community in New York in 1990 (these individuals did not eat pork). It turned out that the patients employed Latina maids, who had recently arrived from countries with high taeniasis due to *T. solium* rates and did not follow hygienic protocols while working with food (Schantz et al. 1992).

4.19.3 Historical Background

Cysticerci in pork were known of back in Ancient Greece. For example, they are mentioned in Aristotle's *History of Animals* as "hail-like bubbles in meat" (Nieto 1982). Although ancient works do not mention that these "bubbles" cause a disease in humans, the existence of bans on pork consumption in some cultures points to the ancient roots of the association between pork and human disease.

Scientific research on taenia began in the seventeenth century, when Edward Tyson (Fig. 4.81) first described the tapeworm. His later studies on tapeworm anatomy and physiology served as the foundation for the contemporary understanding of these parasites. In 1872, Johann Goeze, an eminent zoologist and German pastor (Fig. 4.82), proposed that beef and pork tapeworms belong to two distinct species. He also proposed that intermediate hosts participate in the parasites' life cycles. However, up until the mid-nineteenth century, this difference was unclear to the majority of zoologists, leading to confusion in defining the parasites' species. Only after Friedrich Küchenmeister thoroughly described the scolex structure of both tapeworms in 1857 was this taenia classification accepted. Küchenmeister also conducted experiments that are questionable from the standpoint of modern ethics: he fed a few individuals on death row meat containing beef tapeworm cysticerci and found mature parasites in their bodies post mortem, after they had been executed. He also noted in one of his works that these parasites are never found in people who practice Islam or Judaism. Soon after that, J. H. Oliver proved that people who consumed meat containing beef tapeworm larvae contracted taeniasis. As a result, by the end of the nineteenth century, an understanding of the life cycles of and differences between taenia had been reached, which had a large influence on the development of preventative measures against taeniasis (Cox 2002).

Fig. 4.81 Edward Tyson (1650–1708). Painted by Edward Lilly, 1965. The Royal Medical College, London. *Credit* http://www.bbc.co.uk/arts/yourpaintings/paintings/edward-tyson-16501708-192402 **Public domain**

Fig. 4.82 Johann August Ephraim Goeze (1731–1793). Unknown artist. *Credit* http://en.wikipedia.org/wiki/Johann_August_Ephraim_Goeze#/media/File:Johann_August_Ephraim_Goeze1.jpg **Public domain**

4.19.4 Distribution in Russia and the World

Taeniasis is found in all countries that raise large cattle and is especially widespread in Africa, South America, the Middle East and Southeast Asia (WHO/FAO/OIE 2005). In the USSR, taeniasis due to *T. saginata* used to be widespread in the Transcaucasia and Central Asia, and posed a serious epidemiological threat in the early 1950s. Over 14 million people were affected. In Russia, populations in Dagestan, the Altai Mountains and Buryatia were most affected. The USSR conducted a mass beef tapeworm eradication campaign in the 1960s which resulted in a tenfold decrease in prevalence between 1965 and 1973. However, taeniasis due to *T. saginata* was not completely eradicated.

Taeniasis due to *T. saginata* morbidity rates in Russia decreased from 49.06 to 0.16 per 100,000 individuals from 1960 to 2010. A 2.7-fold decrease in the population of cattle (57 million in 1991 to 21.4 million in 2006), deer (2.3 million in 1991 to 0.7 million in 2006), as well as an increase of deep-frozen beef imports to Russia, helped improve the epidemiological situation (Guzeeva 2011a, b).

The majority of beef tapeworm cases in Russia are concentrated in a few administrative subjects: Yamalo-Nenets Autonomous Okrug, Dagestan, Komi, Rostov Oblast and Kabardino-Balkaria. The morbidity rates are ten times higher than the national average in Yamalo-Nenets Autonomous Okrug and republics in the Northern Caucasus. Only sporadic cases are registered in Russia's remaining territory (Map 4.32).

Beef tapeworm morbidity rates for individual years are surprisingly variable, which is likely connected to the appearance of local outbreaks in individual regions. In addition, territories with high long-term average annual morbidity rates are quite stable, indicating that the issue is persistent in these regions.

Beef (raw ground meat and undercooked grilled meat) (83%) and reindeer brains (17%) are the two major causes of human beef tapeworm infections. Populations in Russia's northernmost regions traditionally consume freshly slaughtered, frozen and dried reindeer meat and raw reindeer brains, leading to high taeniasis prevalence among the Nenets, Komi and Khanty peoples and other indigenous populations in the Russian North. The historically developed lifestyles and popularity of animal husbandry among people of the Northern Caucasus lead to high taeniasis prevalence in that population. Limited pastures create favourable conditions for the spread of *T. saginata* in that region.

As mentioned above, taeniasis due to *T. solium* does not have high morbidity rates in any region. The maximum reported taeniasis due to *T. solium* rate has never exceeded 2% in past years (Lysenko et al. 2002).

In the USSR, relatively high taeniasis due to *T. solium* morbidity rates was primarily reported in Belarus and Ukraine, where pork lard is traditionally consumed. Periodic outbreaks occurred in Kazakhstan and Armenia. The Russian Federation is characterized by low morbidity rates, not exceeding 1 case per 100,000 individuals in most regions (Map 4.33).

Relatively high long-term average annual morbidity rates are only found in the Nenets Autonomous Okrug, Adygea and Khakassia, but even there they do not exceed 2 cases per 100,000 individuals. It should be noted that taeniasis due to *T. solium* is more prone to outbreaks linked to simultaneous infection among individuals who consumed pork from the same pig. Outbreaks of this type have been recently reported in Khanty-Mansi Autonomous Okrug and Krasnoyarsk and Perm Krais. Maps of morbidity rates for individual years clearly demonstrate the tendency of taeniasis due to *T. solium* to appear in the form of outbreaks.

Urban dwellers make-up the majority of individuals infected with taeniasis. According to Rospotrebnadzor (On the Morbidity Rates… 2014), in 2013, 58% of beef tapeworm and 64% of pork tapeworm cases occurred in the urban population. The majority of affected individuals were young men.

4.19.5 Prevention

In order to prevent infection with taeniasis, only meat that has undergone a veterinary inspection and sold by licensed retailers should be purchased. The meat must have a veterinarian's seal on it.

Meat should be cooked thoroughly. Raw ground meat or rare meat, especially pork, should not be tasted or sampled. Stroganina (thinly sliced frozen meat) is only safe if the carcass was thoroughly frozen at temperatures below −12 °C.

It is important to maintain hygiene while working with livestock on farms and to keep a clean toilet, barn and plot of land.

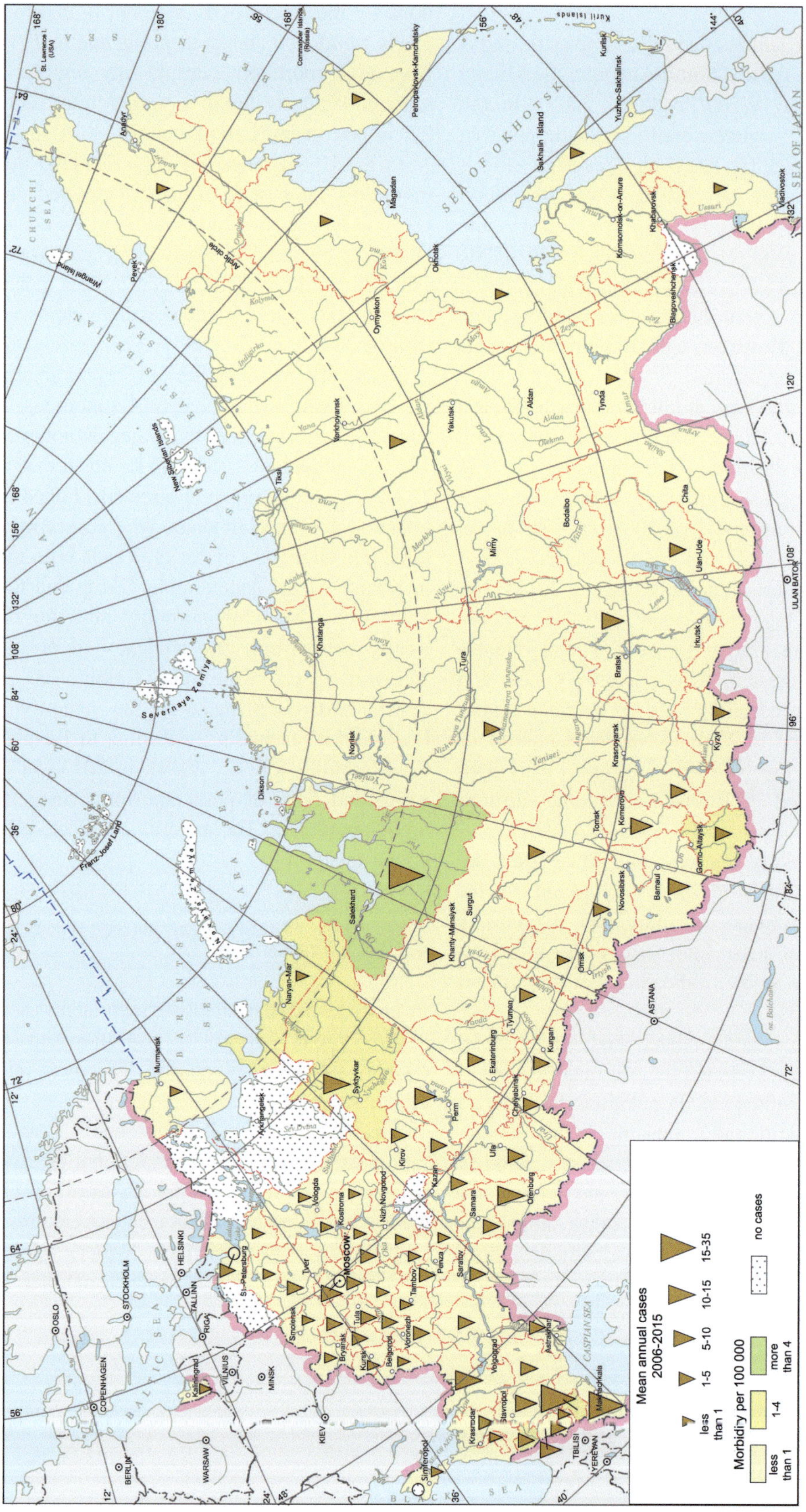

Map 4.32 Taeniasis due to *T. saginata* morbidity, 2006–2015

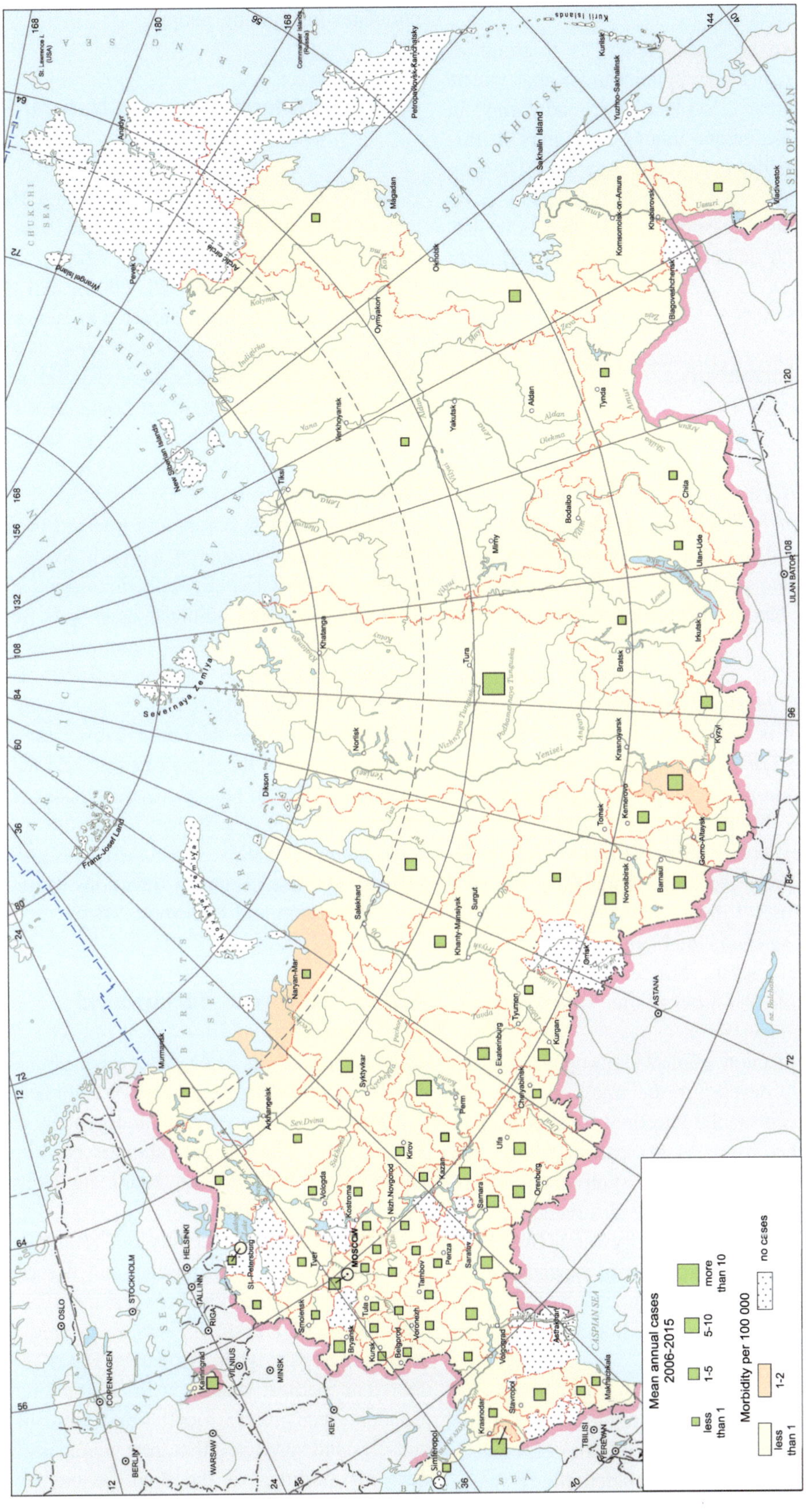

Map 4.33 Taeniasis due to *T. solium* morbidity, 2006–2015

4.20 Trichinosis

Trichinosis (also known as trichinellosis) is a group of extremely widespread nematodes helminthiases that affects mammals, including humans. Nematodes (roundworms) in the *Trichinella* genus are unique among helminths in that they undergo their entire life cycle in one organism. They are also unique in that they have no stage in the outside environment. Natural and anthropurgic foci exist thanks to parasitic circulation in wild animals and livestock (primarily swine).

4.20.1 General Information

Trichinella is small viviparous nematodes. *T. spiralis, T. native, T. nelson* and *T. pseudospiralis* are pathogenic (this is likely not an exhaustive list) (Bruschi and Murrell 2002). The first three species are morphologically identical but have a distinct group of hosts. The last species is unique in that its larvae do not encapsulate in host tissue. This is the only known species of *Trichinella* that can parasitize on both mammals and birds.

Until the 1970s, it was thought that the entire genus consisted only of the *Trichinella spiralis* species, but today it has become clear that it is made up of a group of morphologically close species. *Trichinella* circulates along trophic networks that include non-predatory animals, predators and scavengers. A predator becomes infected upon ingesting encapsulated (encysted) larva in the victim's striated musculature. The larvae are released in the stomach as the meat is digested, and mature in 3–4 days. They mate immediately, after which their offspring enters the bloodstream and travels to the muscles, where it encysts. The females give birth to larva over the course of 10–45 days, after which they die (Konstantinova and Beljaev 1997).

In the case of mass infection, clinical signs appear as soon as the parasite begins to develop in the small intestine and are gastrointestinal. However, the intestinal stage is usually subclinical. On the other hand, as the parasite enters the muscles, it can cause classic trichinosis symptoms (akin to myositis symptoms), sometimes affecting all striated muscles and even the myocardium. Disease severity varies from mild to severe and even lethal. Symptom severity depends on the ingested dose: trichinosis can be entirely asymptomatic if a small number of parasites is ingested and nonetheless lead to adaptive immunity.

Because there is no way to identify the parasite species while the patient is alive, trichinosis is reported as a single disease. However, an understanding of the differences between species is important from an epidemiology standpoint. In part, the species make-up of the pathogen can help identify if a specific case is natural focal in origin or if it circulates just among people.

4.20.2 Epidemiology and Natural Focus Characteristics

Based on today's understanding, around 100 mammal species serve as hosts to *Trichinella*: predators, omnivores and even some herbivores (horses, sheep, goats) (Bessonov 1979; Bruschi and Murrell 2002). All species become infected while consuming infected animals or their cadavers.

A typical characteristic of trichinosis is its ability to form two independent types of foci—natural and anthropurgic. However, there are no sharp boundaries between the two foci, and the pathogen is actively exchanged between wild and domesticated animals on many territories (Bronstein and Luchshev 2014).

People are most commonly infected in anthropurgic foci upon consuming contaminated pork. Swine are infected in a number of ways. Usually, infection occurs after consumption of infected or dead animals: dogs, cats, rodents and other small mammals, as well as contaminated meat and other slaughterhouse waste (Bessonov 1979). Evidently, horses and dogs become infected in the same way.

In the past, most infections came from wild animals, as the prevalence of these parasites in wild animal populations was fairly high. In recent decades, thanks to rising numbers of private farms and neglect of sanitary norms, the prevalence among privately owned swine has grown substantially, leading to a rise in human infections caused by unscreened pork (Bronstein and Luchshev 2014).

4.20.3 Historical Background

Evidently, *Trichinella* parasites were obtained by humans a long time ago, soon after the pig was domesticated. Encysted *T. spiralis* has been found in Egyptian mummies (Millet et al. 1998). The ancient religious bans against pork consumption in Judaism and Islam are likely related to the danger posed by pork.

Long before the discovery of *Trichinella,* doctors would find calcified cysts in muscles and describe them without drawing the connection between cysts and parasites. The larval stage of *Trichinella* was discovered in 1835 during a routine autopsy, when Sir James Paget (Fig. 4.83), then a first-year medical student at St. Bartholomew's Hospital in London, accidentally found small thorn-like bodies in the muscles and suspected that they were parasitic in nature. He sent his findings to the then famous zoologist Richard Owen (Fig. 4.84), who described the parasite and termed it

Fig. 4.83 Sir James Paget (1814–1899. Lithograph by T. H. Maguire, 1849. *Credit* Wellcome Collection https://wellcomecollection.org/works/jd2b8fpt?query=james+paget **Attribution 4.0 International (CC BY 4.0)**

Fig. 4.84 Sir Richard Owen (1804–1882). Lithograph by R. Hoffmann, 1858, after Maull and Polyblank. *Credit* Wellcome Collection https://wellcomecollection.org/works/t5sxf9zp **Attribution 4.0 International (CC BY 4.0)**

T. spiralis. Owen sent his study to the Zoological Society with a fleeting mention of Paget. As a result, the discovery of the trichinosis pathogen was somewhat unfairly attributed to Owen (Blumer 1939).

After thorough experiments, Friedrich Albert von Zenker described the *Trichinella* life cycle in 1860. Zenker conducted his experiments after his friend and colleague Rudolf Virchow, a famous German doctor, fed *Trichinella*-contaminated meat to a dog in search of the parasite's transmission method. Virchow is responsible for the proposition that the parasite can also cause a severe disease in humans. He also established that the larva die within 10 min when heated above 50 °C.

Trichinosis caught the attention of the broader public in the second half of the nineteenth century, before other parasitic disease, thanks to mass outbreaks. In 1865 in the village of Hedersleben, Germany (today's Saxony-Anhalt) about 347 people fell ill after the consumption of pork from one contaminated pig. 100 people died as a result (Brantz 2005) (Fig. 4.85). This event, along with a few other smaller scale outbreaks, caused consternation in Germany and called the issue of trichinosis to attention, serving as the cause for the introduction of sanitary inspections in many European countries.

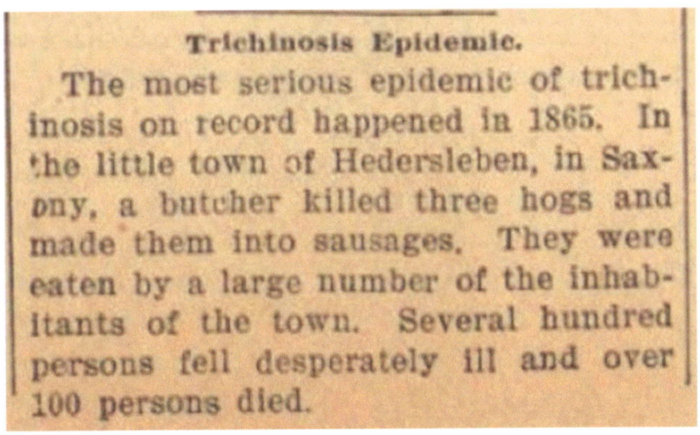

Trichinosis Epidemic.

The most serious epidemic of trichinosis on record happened in 1865. In the little town of Hedersleben, in Saxony, a butcher killed three hogs and made them into sausages. They were eaten by a large number of the inhabitants of the town. Several hundred persons fell desperately ill and over 100 persons died.

Fig. 4.85 The epidemic in Hedersleben in Germany had such an enormous influence that news of it reached the New World. A note on the epidemic in the *Ann Arbor Argus-Democrat*, 21 June 1899. *Credit* http://oldnews.aadl.org/node/162052 **Public domain**

4.20.4 Distribution in Russia and the World

Today, trichinosis is classified as a re-emerging disease (Bruschi and Murrell 2002). It is found around the world, although its prevalence among both humans in animals varies

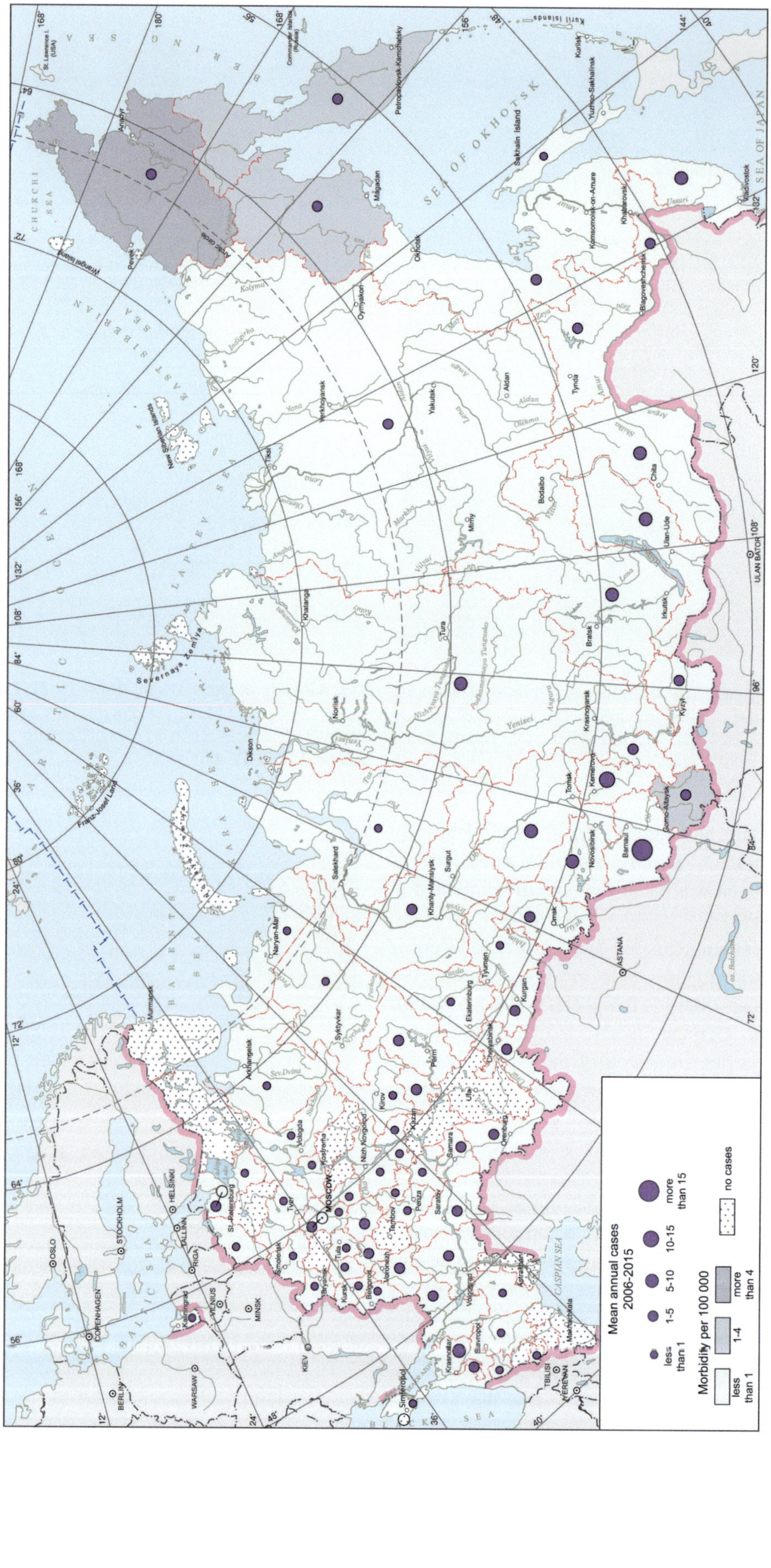

Map 4.34 Trichinosis morbidity, 2006–2015

greatly from one region to another. Overall, the number of reported trichinosis cases worldwide is not that great. WHO reported approximately 57,000 cases between 1986 and 2009, the majority of which occurred in Europe (Murrell and Pozio 2011). In Muslim countries, trichinosis is only found among non-Muslims or migrants. This case distribution, however, is likely not quite accurate, as trichinosis is not well diagnosed in many Asian and African countries and only cases that led to hospitalization are reported.

In Russia, 0.3% of all biohelminthiasis morbidity can be attributed to trichinosis. Trichinosis morbidity is wave like in nature. 84 cases were reported in 2011 and 118 in 2012 (On the Status… 2013). As a rule, morbidity is characterized by outbreaks linked to the consumption of both wild game (usually wild boar, bear or badger) and pork.

The disease was not found in a number of Russian regions between 1997 and 2015 (Map 4.34). The greatest number of trichinosis cases was reported in Altai, Zabaykalsky and Krasnoyarsk Krais and Kemerovo Oblast. Intensive pathogen exchange between natural and anthropurgic foci occurs in the most high-risk regions. According to the results of monitoring, wild animals and domesticated pigs are highly infested in Altai Krai (Ponomarev et al. 2011). The situation in Khakassia and a few other regions is similar. It can also be assumed that high trichinosis morbidity rates in humans (Chukotka Autonomous Okrug, Magadan Oblast, Kamchatka Krai) is linked to the popularity of amateur hunting in a number of regions, some of which pose a high risk of infection.

4.20.5 Prevention

Veterinary meat inspection is the main measure of trichinosis prevention in humans. Whole carcasses of omnivorous and carnivorous animals used for human consumption—pigs, wild boars, badgers, bears, beavers, etc.—must be inspected. In order to prevent infections in pigs, it is important to limit their contact with rodents, as that is a method of parasite transmission. Public health education campaigns that include information on modes of transmission are another important element of trichinosis prevention.

4.21 Echinococcosis

Echinococcosis is a group of helminthic zoonoses in which the larval and not the imaginal stage of parasite invades humans. A key distinction of such infections is that there are no mature helminths in the patient's body, meaning that there is no way that the person can affect other people. Echinococcosis has a severe effect on organ systems, because of which it is thought to be one of the most dangerous human parasitic diseases. Today WHO considers echinococcosis to be a neglected disease (Torgerson and Macpherson 2011, www.who.int).

Echinococcosis has two main clinical forms: cystic (or hydatid) and alveolar (Lysenko et al. 2002).

4.21.1 General Information

Echinococcosis is caused by cestodes in the *Echinococcus* genus. Species identification in this genus remains an actively discussed topic. Still recently, only two species, one of which was thought to be polytypic, were identified (Eckert 1997). Modern genetic studies led to a substantial revision of the genus's taxonomy (Romig et al. 2006). Today the majority of researchers identify 4–7 species (Romig 2003; Thompson 2008; Otero-Abad and Torgerson 2013).

Cystic echinococcosis is caused by the *Echinococcus granulocytes* cestode. Canines (dogs, wolves, jackals, arctic foxes, etc.) and, rarely, cats are the parasite's definitive hosts. Mature echinococci—small cestodes with a head and a few segments with eggs up to 5 mm in length—inhabit their small intestine. Infected animals excrete the eggs with their faeces into the environment, where they are ingested by their first intermediate hosts (various hoofed herbivorous mammals) along with contaminated grass or soil. These are predominantly sheep, goats, cattle and swine. After ingestion, the embryos are released from the oncospheres and carried everywhere in the body, ending up in the liver, lungs and other organs, where they form hydatid cysts, inside which protoscolices and daughter cysts form. Very small scolices form in the daughter cysts—they will only be able to fully mature once inside the definitive host.

The larval stage becomes contagious for its definitive hosts 4–5 months after the intermediate hosts are infected. Dogs and other carnivores ingest the parasite while eating the meat and intestines of infected animals (Beljaev and Konstantinova 1980).

People are accidentally infected when they ingest parasite eggs that are released into the environment in the definitive host's faeces. Humans are an ecological dead end for the parasites, as theoretically they could serve as sources of infection along with other intermediate hosts, but that situation is practically impossible (Lysenko et al. 2002).

Alveolar echinococcosis is caused by the *E. multilocularis* cestode and is one of the most severe parasitic diseases found in Arctic and temperate climates. It often leads to a fatal outcome. The parasite circulates along the "canines (usually foxes and arctic foxes, sometimes dogs)—rodents" chain. Cats can also become infected. Alveolar echinococci eggs are excreted with faeces by their definitive hosts and are ingested by rodents and small members of the Lagomorpha order (Vuitton et al. 2003).

4.21.2 Epidemiology and Natural Focus Characteristics

Echinococci can circulate both in humans' immediate environment and in the wild, meaning that both natural and anthropurgic foci have been identified. In natural foci, *E. granulosus* is found in jackals, wolves, foxes and raccoon dogs. Other species in the *Echinococcus* genus infect other mammals (Romig et al. 2006).

Anthropurgic foci consist of the combination of dog, sheep and other hoofed mammal populations, within which *E. granulosus* circulates. These foci are more powerful than natural ones, given the higher population density of domesticated hoofed mammals and dogs and closer contact between members of the parasitic system (Beljaev and Konstantinova 1980). They form in rural areas, in areas with highly developed sheep farming. Infection through dogs is the most often and significant method of echinococcosis in humans. Dogs receive the parasite after eating the contaminated insides of hoofed mammals. Sheep are the most effective source of infection for dogs, because echinococcal cysts in sheep specifically form scolices, which may not occur in other hoofed mammals (Beljaev and Konstantinova 1980). Overall, echinococcosis in dogs heavily depends on the availability of intermediate hosts (ultimately on the type of farm) and on their living conditions. In cities, echinococcosis is common in feral and stray dogs.

Hoofed mammals ingest the parasite both in pastures and through feed contaminated with echinococci eggs. It is thought that parasites associated with sheep and cattle are most pathogenic to humans, while the parasites that circulate in horses are either subclinical or very mild in people (Eckert 1997).

In Eurasia's temperate zone, alveolar echinococcosis primarily circulates between foxes and arctic foxes and rodents (Fig. 4.86), especially voles and lemmings, although domesticated cats and dogs are periodically included in the cycle. The pathogen has been found in raccoon dogs, wolves, Eurasian lynxes, wildcats and jackals, but they do not play a major epizootic or epidemiological role. Despite dogs' sporadic infestation, they are often the ones to allow for human infections, due to close contact between the two. Dogs are very efficient *A. multiocularis* hosts, and their relatively low infection rate is simply explained by their low level of contact with anthropurgic foci (Romig 2003). Today, there are no exact data on dog and cat infection rates in the temperate zone, although they are thought to be low (Romig et al. 2006). In Northern Europe and Siberia, the arctic fox (*Alopex lagopus*) is the main definitive host of *A. multiocularis,* and its infection rate can be very high.

Among rodents, the common vole and water vole (*Microtus arvalis* and *Arvicola terrestris*) are most commonly infected. Rodent infections depend on various ecological factors, rodent population sizes, habitat types and even average daily temperatures (Otero-Abad and Torgerson 2013). Reports exist on the widening of alveolar echinococcosis' nosogenic range in Europe, but they are not supported by convincing data.

Fig. 4.86 The red fox (*Vulpes vulpes*) is the main definitive host of *E. multilocularis* in Europe. *Credit* Dave Small—http://www.flickr.com/photos/usfwshq/6709465857/CC BY 2.0

4.21.3 Historical Background

Massive cysts in the liver and other organs were known to man back in ancient times. Ancient Babylonian texts mention cysts in the bodies of sacrificial animals. Hippocrates' (Fourth century BCE), Areteus' (First century CE), and Galen's (Second century CE) texts all include descriptions of the cysts.

The Italian Francesco Redi was the first to suspect the parasitic nature of these cysts in the seventeenth century, but the hypothesis that these cysts are larval stages of a tapeworm was only voiced in 1766 by Peter Simon Pallas, the German clinician and naturalist. In 1853, Karl Von Siebold demonstrated that cysts from a sheep's liver that are consumed by a dog give rise to mature parasites in its intestinal tract. In 1864 Bernhard Naunyn, a German clinician, discovered mature helminths in the dog that had been fed a cyst excreted by a person. This led to the discovery of the life cycle of echinococci (Cox 2002).

4.21.4 Distribution in Russia and the World

Echinococcosis is extremely widespread around the world (Fig. 4.87). They are not only a medical, but also an

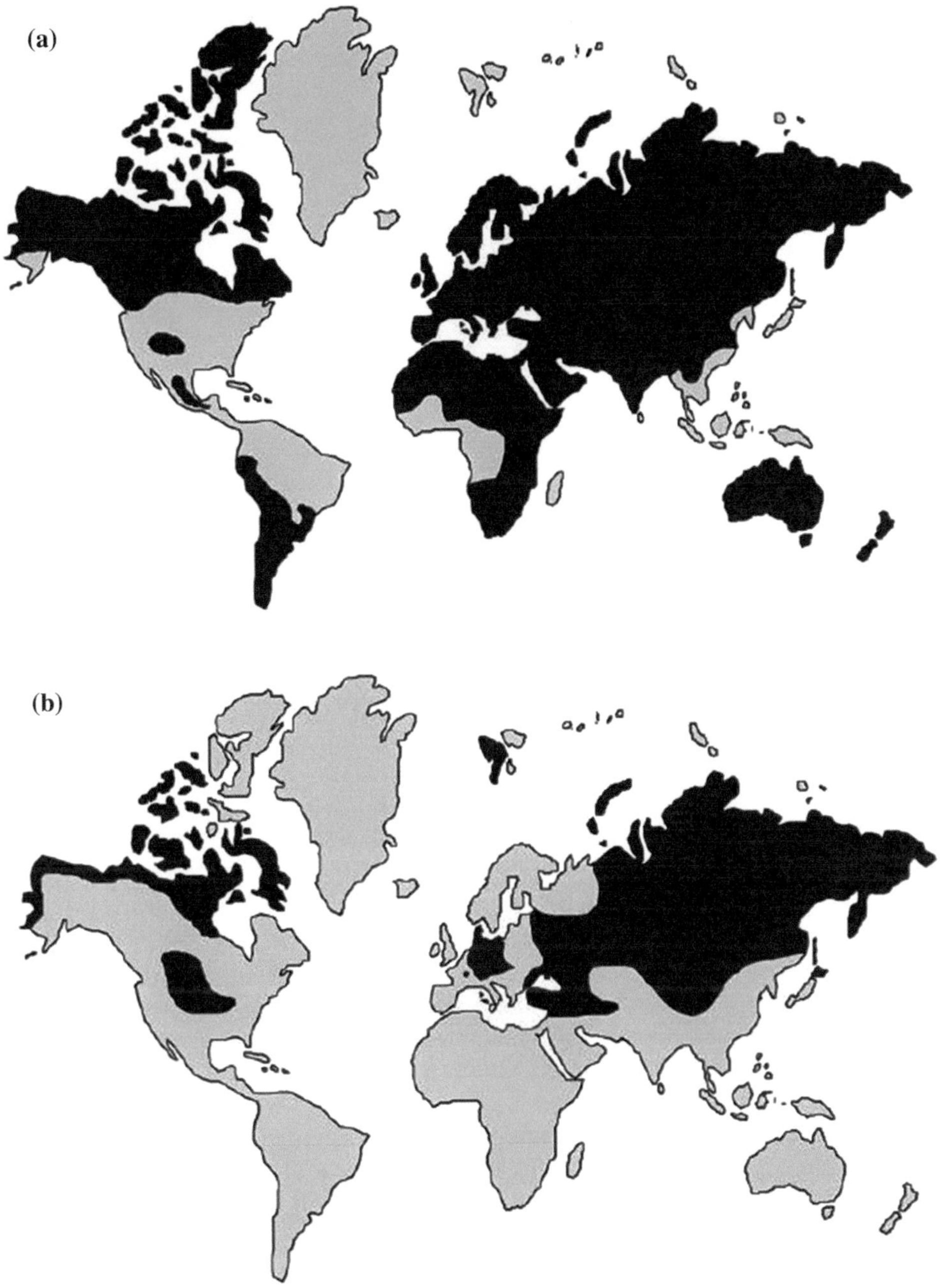

Fig. 4.87 The worldwide distribution of echinococcosis: **a** cystic (hydatid), **b** alveolar. *Credit* Romig et al. (2006)

economic issue, as they affect a large number of livestock. In the USSR, in the 1960s, 12% of sheep, 5% of cattle and 5% of swine were infected. In the mid-1970s, sheep infection rates approached 100% in some regions of Dagestan (Beljaev and Konstantinova 1980). Cystic echinococcosis is widespread in Europe, especially in countries where sheep farming is common (Fig. 4.88). It remains an issue in New Zealand. Moreover, there has been an increase in fox infection rates in rural areas in France, Holland, Germany and Austria. In addition, foxes have been migrating en masse to city parks, creating urban foci (Romig et al. 2006).

Echinococcosis is widespread in poor African and Asian countries. Countries in North and East Africa, the Middle East and Central Asia, as well as Argentina, Chile and Peru in South America are most affected (www.who.int). According to the WHO's estimates, approximately 1 million people in the world is infected with echinococcosis.

Fig. 4.88 Sheep are the main intermediate hosts of *E. Granulosus*. *Credit* A. A. Kadetova (permission obtained)

Alveolar echinococcosis is far less common. Its pathogen is endemic in the Holarctic. The most intensive foci are affiliated with cold regions in Eurasia and North America.

Cystic echinococcosis is ubiquitous in Russia, although the population's morbidity rate is low overall (Map 4.35). However, these statistics are incomplete due to a large number of subclinical and asymptomatic cases. The highest morbidity rates are reported in areas with extensive animal husbandry: Saratov and Orenburg Oblasts and Bashkortostan. Cystic echinococcosis is also found in Dagestan and Karachaevo-Cherkessia in the Northern Caucasus. Alveolar echinococcosis is mainly found in the northern regions of Siberia and the Russian Far East: Yamalo-Nenets Autonomous Okrug, Sakha (Yakutia) and on the Chukotka Peninsula.

About 500 cases of "echinococcosis", meaning cystic and alveolar echinococcosis, are annually reported in Russia. The morbidity rate was 0.36 cases per 100,000 individuals in 2013 (On the Status… 2014). The highest morbidity rates in Russia are found in administrative subjects in which the population hunts extensively or conducts free-range animal husbandry. Overall, 1.3% of all biohelminthiasis morbidity in Russia can be attributed to echinococcosis (On the Status… 2014).

4.21.5 Prevention

Shepherds, individuals living in rural areas with numerous sheep, and people who travel to wild areas and come into contact with infected dogs (whose fur is often contaminated with eggs) or soil contaminated with dog excrement are at a greater risk for contracting echinococcosis. Preventative programs include dog de-worming, improvement of hygiene conditions at slaughterhouses and public education campaigns. Veterinary events that are aimed at slaughterhouse waste disposal and prevent dogs from eating the waste play a key role. Today sheep vaccination is under consideration as another preventative measure.

Alveolar echinococcosis prevention is more difficult. Regular de-worming of domesticated carnivores that have access to wild rodents is recommended. When absolutely necessary, foxes and stray dogs are rounded up, but this measure is not effective. Today a number of countries practice the de-worming of wild, feral and stray host animals with de-worming bait, but these programs are quite costly and difficult to execute.

Maintenance of personal hygiene, limited contact with strange dogs and hand washing after such contact are the foundation of echinococcosis prevention.

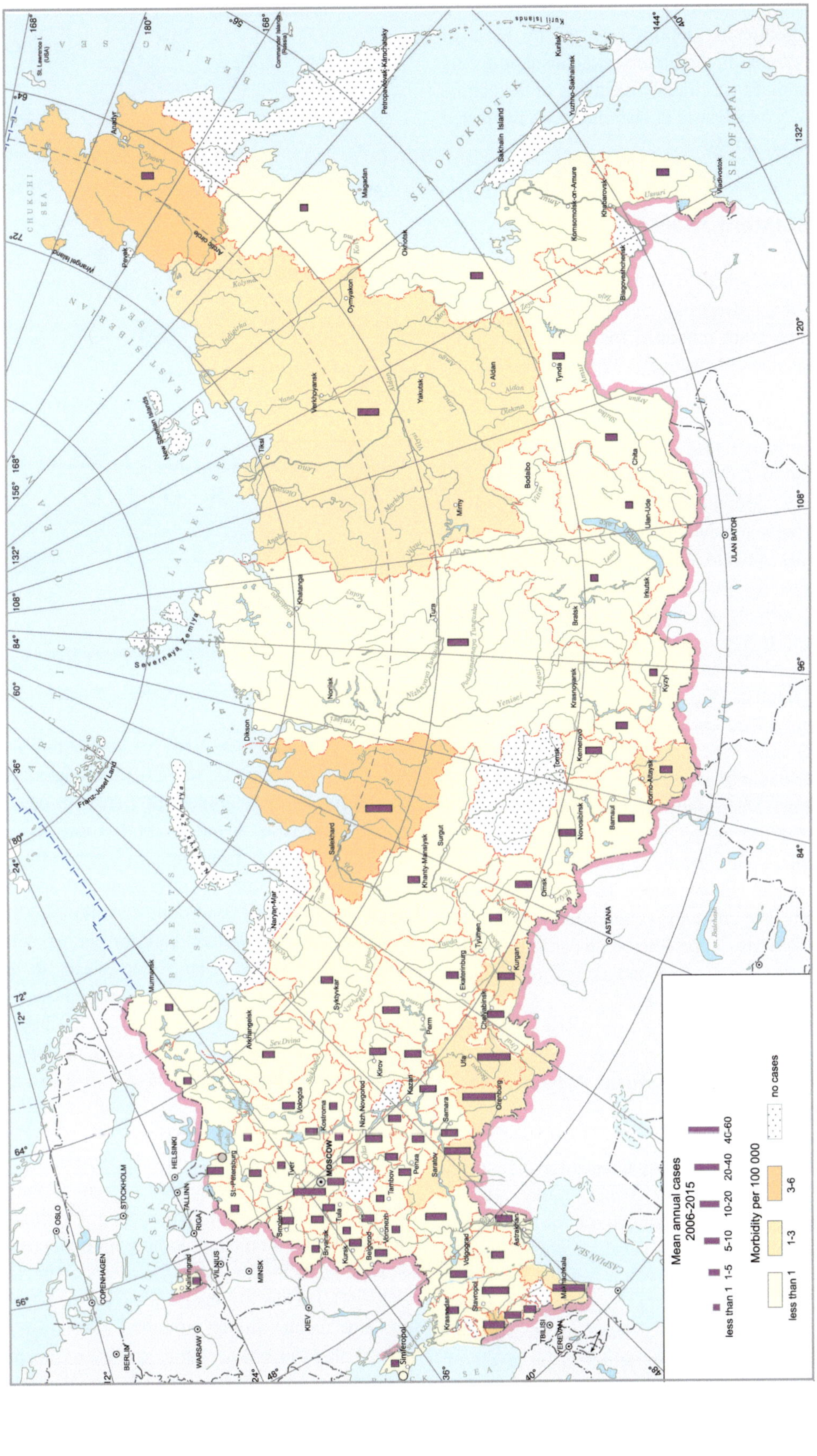

Map 4.35 Echinococcosis morbidity, 2006–2015

4.22 Toxocariasis

Toxocariasis is a helminthic zoonosis caused by larva of nematodes (roundworms) in the *Toxocara* genus that occurs in carnivorous mammals, specifically members of the canine and feline families. Toxocariasis is classified as a neglected disease (Moreira et al. 2014).

4.22.1 General Information

In the wild toxocariasis affects wild canines (foxes, wolves, etc.) and felines (lynxes, leopards, etc.), as well as cats and dogs, whose intestines are inhabited by mature *Toxocara canis* and *T. cati* parasites, respectively. Traditionally the vast majority of toxocariasis cases in humans have been connected to infection with *T. canis*, which has been more extensively studied and is better known to clinicians. As of now, nothing is known of the role of *T. cati* in human pathology, and this issue must be studied (Moreira et al. 2014).

Toxocariasis is a geohelminthasis, or a parasitic disease in which the eggs must undergo a stage of development in the soil in order for the life cycle to be complete. The female helminth excretes millions of unembryonated eggs into the outside environment. In 4–6 weeks, each egg develops into a larva which is capable of infecting the main hosts: wild canines, dogs and humans (Fig. 4.89).

Eggs containing infectious larva end up in the host's intestines, and after some time they penetrate the vasculature and migrate to various organs. Some of the larvae end up back in the intestine, where they develop into the mature stage, re-starting the life cycle. In the organisms of other animals (rodents, birds, etc.), as the human body, migrating larvae never reach the mature stage. Their presence and constant migration give rise to pathologic reactions in humans, characterized by a prolonged, recurring disease course, polymorphous clinical signs and, commonly, attacks on various organs, including the eyes.

4.22.2 Epidemiology and Natural Focus Characteristics

Because the eggs of the toxocariasis pathogen have to spend part of their life cycle in the soil, its foci require specific temperature and humidity conditions to allow for the eggs to develop. Eggs take 5–26 days to develop, based on the temperature of the soil, and they remain infectious for a long period of time (Tumolskaya 1997). Temperature, atmospheric humidity and soil humidity all have an effect on the eggs' viability, which ultimately has an effect on levels of soil contamination at different times of year (Zapariy 2003).

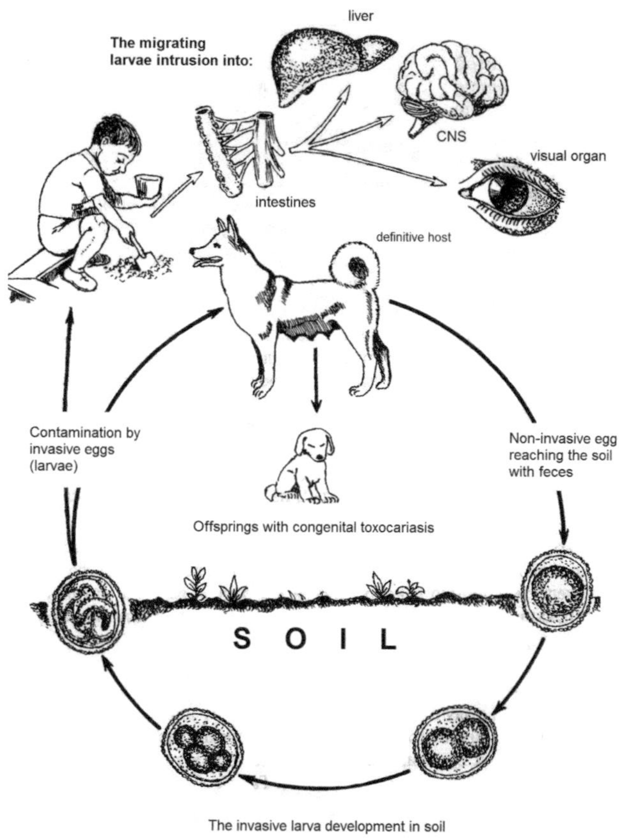

Fig. 4.89 The *Toxocara canis* life cycle. *Credit* Lysenko et al. (2002)

People are at the greatest risk of contracting toxocariasis in the summer and autumn, when meteorological conditions are optimal for the development of *Toxocara* eggs.

Toxocariasis is found in both rural and urban areas. Urban foci, which form in large industrial centres, pose special relevance due to the presence of specific epidemiological and epizootic conditions that allow for the pathogen's wide distribution. Incorrect care for pet and service dogs, as well as the presence of stray dogs, leads to high contamination of the soil in large cities, which causes toxocariasis in all urban dog populations and raises the human morbidity rate (Zapariy 2003). The infection rate of dogs is quite high. Studies in the Altai Republic have shown that approximately 39% of tested animals gave a positive result for *T. canis* (Pautova et al. 2011).

Humans become infected with toxocariasis when they ingest parasitic eggs with food or water, when they inhale dust, or when they come into contact with infected animals. The likelihood of an adult coming into contact with soil that has been contaminated with *T. canis* is about 15% (Lysenko et al. 2002). Due to uncontrolled pet dog walking, which sometimes occurs in playgrounds, as well as a large number of stray dogs in cities, young children who play in public sandboxes are the most vulnerable population (Fig. 4.90).

Fig. 4.90 Stray dogs in a sandbox (www.vk.com). *Credit* http://cs419529.vk.me/v419529283/439a/Rgx7ZW8EL-M.jpg

Many studies have shown that toxocariasis is socially determined, as it is more commonly found in individuals with low economic status (Lysenko et al. 2002).

4.22.3 Historical Background

Toxocariasis was first described in dogs by Werner in 1782: he termed the parasite *Ascaris canis. Toxocara* was identified as a distinct genus in 1905. However, the notion that the nematode can cause pathology in humans was first suspected only in the 1940s. The first clinical case of toxocariasis in a 2-year-old boy was described in the USA in 1950. In the early 1950s, it was established that *Toxocara* larva is the cause for serious pathologies of the eyes and other organs (Despommier 2003; Moreira et al. 2014).

4.22.4 Distribution in Russia and the World

Toxocariasis is a widespread disease that is reported in many countries. The average infection rate in dogs tested on different continents is over 15%, but reaches 93% in some regions. According to seroepidemiological studies, 2–14% of tested practically health individuals in various foci have a positive immune reaction to toxocariasis. Seropositivity in humans ranges from 2.6% in Belgium to 86% on the Caribbean island of St. Lucia (Lysenko et al. 2002). The disease's prevalence in various regions of the world is not known for sure, because toxocariasis is not subject to mandatory reporting. It is clear that its range is quite broad and that the number of infected individuals is far higher than the reported numbers (Tumolskaya 1997). Recently, epidemiologists in a number of countries have begun seriously discussing the issue of toxocariasis. They emphasize its broad prevalence and low public awareness of the disease (Woodhall et al. 2014; Borecka and Klapec 2015).

Toxocariasis is ubiquitous is Russia (Map 4.36), and the highest morbidity rates are found in large cities and industrialized regions. Territories in the Volga Region, the Southern Urals and Siberia (Udmurtia, Perm Krai, Sverdlovsk and Kurgan Oblasts, Altai Republic, Altai Krai, Yamalo-Nenets Autonomous Okrug and Krasnoyarsk Krai) are most affected. All of these subjects have large dog populations, which facilitates the spread of disease.

Today, toxocariasis morbidity is on the rise in Russia, especially in large cities. In part, this increase is related to the broad introduction of diagnostic methods into clinical practices: therefore, the previously underestimated actual morbidity rate is gradually coming into light. 3000–3500 human toxocariasis cases are reported annually in Russia, approximately half of which occur in children (On the Status… 2014).

4.22.5 Prevention

Children between the ages of 3 and 5 who come into close contact with sand and soil, veterinary specialists, dog kennel employees, grocers, garden and orchard owners and pet (dog, cat) owners are at an especially high risk for toxocariasis.

Toxocariasis prevention is both a healthcare and a veterinary issue. It is mostly made up of the following initiatives: (1) dog testing and timely de-worming, (2) limiting the population of stray dogs and creating specified dog parks, (3) hand washing after contact with soil or animals, (4) thorough cleaning of greens, vegetables and other foods that may contain soil particles, (5) preventing animals from entering playgrounds and parks, (6) public sanitary education campaigns, especially those geared at dog owners.

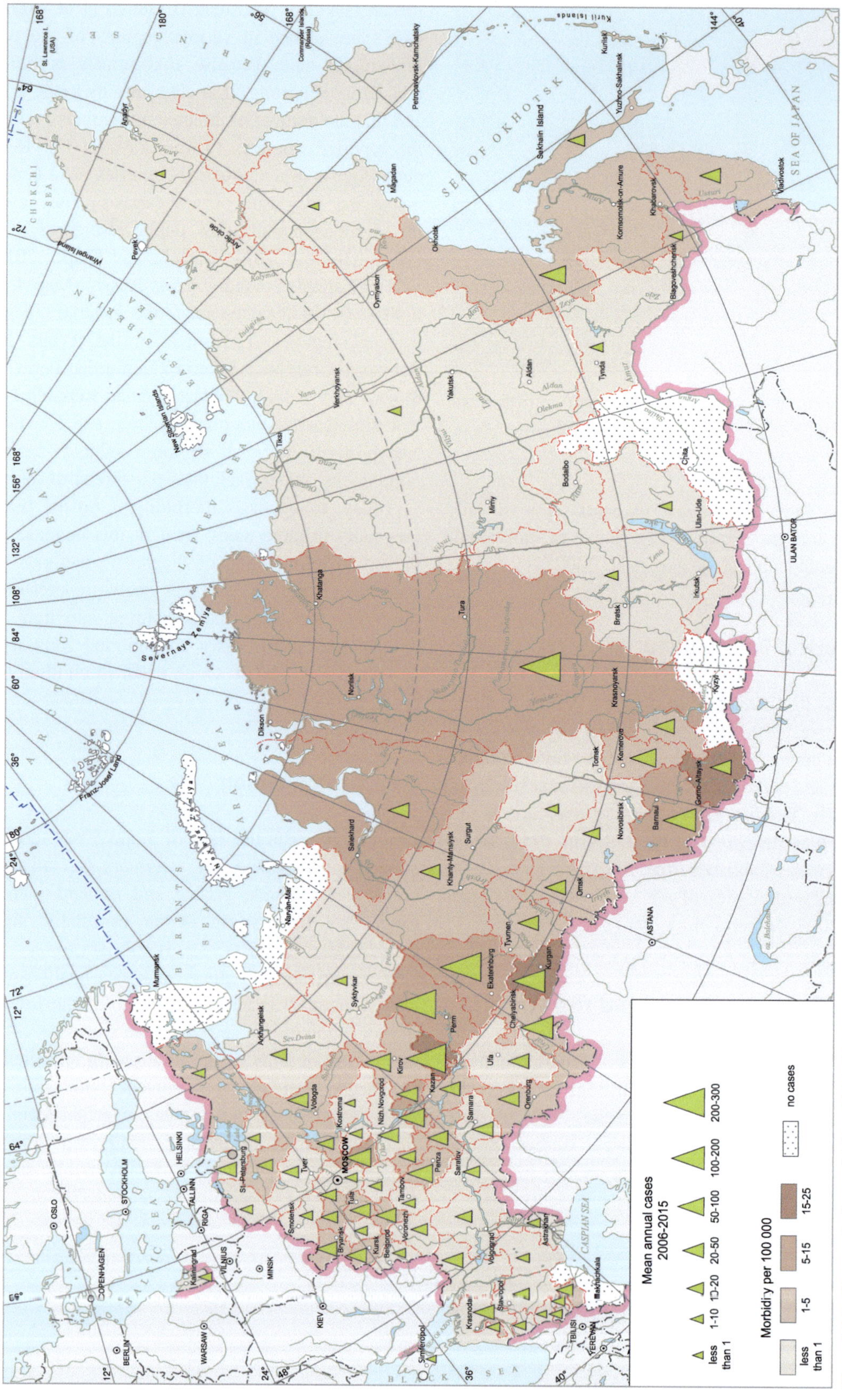

Map 4.36 Toxocariasis morbidity, 2006–2015

4.23 Arboviruses

In 1963, the International Subcommittee on the Taxonomy of Viruses introduced the term "*arbovirus*", which means an ecological group of viruses that are transmitted to susceptible vertebrates by blood-sucking arthropods. Nowadays it is thought that arboviruses are not just transmitted by arthropods, but that they can also be passed on via air-borne and food-borne transmission or other methods (Mahy and Regenmortel 2010).

There are 500–600 arboviruses specified in the world. Many of them are not pathogenic in humans; some were isolated from arthropods, birds and animals, and their public health significance is still unknown. However, about 100 arboviruses can cause disease in humans. Mosquitoes, sandflies, biting midges (Ceratopogonidae) and ticks are the main carriers of arboviruses (Gratz 2004; Kuno and Chang 2005).

4.23.1 General Information

The life cycles and clinical signs of arboviruses vary greatly, as this is a large group of viruses (Monath 2000; Carter and Saunders 2007; International… 2009). The main clinically significant arboviruses belong to the following four families:

Bunyaviridae: viruses in this family cause haemorrhagic fever with renal syndrome (HFRS), Crimean–Congo haemorrhagic fever, Neapolitan fever and pappataci fever.

Flaviviridae: viruses in this family cause tick-borne encephalitis, Japanese encephalitis, Omsk haemorrhagic fever, yellow fever, Kyasanur forest disease, dengue fever and West Nile fever.

Reoviridae: Kemerovo virus, Okhotskiy virus, Aniva virus, epizootic haemorrhagic fever virus and African horse sickness virus are in this family.

Togaviridae: Karelian fever (Pogosta disease) virus, Sindbis virus, Semliki Forest virus, Getah virus, Chikungunya virus, Mayaro virus and equine encephalitis viruses are in this family.

4.23.2 Epidemiology and Natural Focus Characteristics

Arbovirus is a numerous group of viruses, whose members are spread essentially everywhere and have been found in every climate zone except for the polar regions of Earth. At least, 100 known arboviruses have been proven to cause disease in humans, and the list of viruses that are pathogenic to humans and animals is steadily growing (Oganesyan 2002; Medical Virology 2008; Zverev and Boychenko 2010). Blood-sucking arthropods are the specific vectors for the arboviruses that adapt to them. They pass arboviruses to susceptible vertebrates via bites. In addition, arthropods can themselves serve as arbovirus reservoirs in natural foci. Optimal conditions for diseases transmitted primarily by mosquitoes, sandflies and biting midges exist in the equatorial climate zone, due to the constant high temperatures and high humidity. The majority of arboviruses has been isolated from mosquitoes. The closer one moves to the poles, the greater the number of viruses that has adapted to Ixodidae ticks. Argasidae, Gamasina and other ticks can also serve as vectors. Once the virus has entered the arthropod's organism, it begins to reproduce and accumulate in the vector's salivary glands, entering the prey's wound as its blood is sucked. Infection is contingent upon a sufficient concentration of virus in saliva. The virus remains in infected arthropods for their entire lives. Involvement in the vertebrate parasitic system allows the virus to enhance its population's gene pool.

The epizootic and epidemic significance of various vertebrate species is defined by the nature of their contact with blood-sucking arthropods, humans and pets. Birds play a leading role for many arboviruses. They are one of the most ancient reservoirs of pathogens and can move infections across enormous distances, even from one continent to another during seasonal migration (Lvov and Ilyichev 1979). Bats are also epizootically and epidemiologically significant, because viruses can survive in their bodies throughout hibernation. The ubiquitous presence of rodents in every landscape zone and every vertical zone, their high population density, close ties to blood-sucking arthropods and high sensitivity to the pathogens define their role in maintaining natural foci.

Tropical countries are seen as arboviruses' ancestral homeland. Today arboviruses are most prevalent in those countries. The following arbovirus infection outbreaks were reported and decoded in Russia: Karelian fever (Pogosta disease), West Nile fever and Crimean–Congo haemorrhagic fever (Lvov et al. 2000; Moskvitina et al. 2002; On the Status… 2013). The distribution of Japanese encephalitis, dengue fever and West Nile fever has broadened (Botha et al. 2008; World Health Organization 2009).

Bunyaviruses are widespread on every continent. They cause zoonotic natural focal diseases. Like other arboviruses, bunyaviruses can reproduce at two temperature intervals: 36–40 °C and 22–25 °C, allowing them to reproduce in both the vector's and host's bodies. Diseases caused by these viruses tend to be seasonal. In Russia, ticks are the pathogens' main reservoirs and vectors. Rodents, lagomorphs, ruminants, primates and birds are the vertebrate hosts of bunyaviruses. People can become infected not only through vectors after being bitten by blood-sucking arthropods, but also from people via direct contact (Beran 1994; Medical Virology 2008; Zverev and Boychenko 2010). The majority of bunyaviruses causes fevers, haemorrhagic fevers and encephalitis. Crimean–Congo haemorrhagic fever virus (the *Nairovirus* genus) and haemorrhagic fever with renal syndrome virus (the *Hantavirus* genus) are most pathogenic to humans and most medically significant among bunyaviruses commonly found in Russia (Maleev et al. 2003; Smirnova 2007; Tkachenko et al. 2012). It should be noted that these diseases result in stable adaptive immunity (Carter and Saunders 2007; Botha et al. 2008; Klempa et al. 2008; Maes et al. 2009).

Flaviviruses (their name is derived from the Latin *flavus*—yellow, based on the name for yellow fever, caused by a virus in this family) are widespread in Russia and around the world. Flaviviruses cause severe diseases that are accompanied by hepatic and haemorrhagic symptoms (yellow fever, dengue fever, Omsk haemorrhagic fever) or encephalitis (tick-borne encephalitis, Japanese encephalitis) (Medical Virology, 2008). Tick-borne encephalitis, West Nile fever and Omsk haemorrhagic fever are the most relevant of these diseases in Russia (Putintseva et al. 2010; Ruzek et al. 2010; Korenberg et al. 2013; Busygin 2014).

Reoviruses (from the English *respiratory*) consist of 9 genera, including the *Orbivirus* genus, which includes viruses that cause natural focal diseases in humans (Kemerovo virus, Okhotskiy virus, Aniva virus) (Lvov et al. 2014).

Togaviruses (from the Latin *toga*—cape or shawl, reflecting the virion's complex structure) consist of 4 genera while includes viruses in the *Alphavirus* genus that cause natural focal zoonotic infections. Nearly all alphaviruses are ecologically linked to mosquitoes, who act as their vectors and natural reservoirs. Vertebrates—birds, rodents, primates and other mosquito nutrient sources—also serve as their natural reservoirs. At high population densities, combined with large mosquito populations, people become sources of alphaviruses and can transmit them to one another as vectors. Epidemics end when sufficient herd immunity is reached as the result of the development of adaptive immunity during infection and vaccinations (Medical Virology 2008; Zverev and Boychenko 2010). Alphaviruses cause fever, rashes, encephalitis and arthritis in humans. These diseases result in stable adaptive immunity. The main alphaviruses that pose a danger to humans are Sindbis virus, Semliki Forest virus, Chikungunya virus and Karelian Fever (Pogosta disease) virus.

Fig. 4.91 Dmitri Ivanovksy (1864–1920), the founder of virology. *Credit* http://virology.gamaleya.org/index.php?option=com_content&view=article&id=92&Itemid=519

Fig. 4.92 Walter Reed (1851–1902), a military surgeon. *Credit* https://upload.wikimedia.org/wikipedia/commons/4/4e/WalterReed.jpeg **Public domain**

4.23.3 Historical Background

Virology was founded in 1892, when Dmitri Ivanovsky (Fig. 4.91) discovered a unique world of non-bacterial and non-protozoan pathogens, later termed viruses (Mahy and Regenmortel 2010; Ivanovsky Scientific… 2014).

The first human viral disease—yellow fever—was discovered in 1901 by Walter Reed, an American military surgeon (Fig. 4.92) and his colleagues (Carter and Saunders 2007; Mahy and Regenmortel 2010).

In the following years, a variety of viruses causing severe diseases in humans was described. One of these was the West Nile virus, first isolated in 1937 from the blood of an infected woman in the West Nile province of Uganda (Strauss and Strauss 2008). Local doctors wrote the first description of Omsk haemorrhagic fever in Omsk between 1940 and 1945. The Crimean–Congo haemorrhagic fever virus was first found in Crimea in 1944 and isolated from the blood of patients and from its vectors (Ixodidae ticks) in 1945 by Mikhail Chumakov. An outbreak of Karelian fever was reported in Ockelbo, Sweden in 1982. Similar cases were seen in Karelia and Finland (Jones 2012).

4.23.4 Distribution in Russia and the World

Arboviral infections are a relevant issue in the whole world, because they are found nearly everywhere and everyone has been isolated on every continent except Antarctica. As mentioned above, over 500 arboviruses are known today, but thanks to intensive research being done in laboratories in Russia, the USA, Canada, the UK, France, China, South Korea, Japan and in other countries, the list of viruses that are pathogenic to humans and animals and transmitted by arthropods keeps growing.

Russia is characterized by spatial differences in the localization of specific viruses (Lvov et al. 1989; Zverev and Boychenko 2010). The wide prevalence of arboviral infections in Russia (Map 4.37) is related to the diversity of its landscapes, active settlement of new territories, which leads to mass population migration and the high level of anthropogenic influence on ecosystems.

The wide diversity of pathogens, high likelihood of epidemic events and current lack of specific therapies and preventative measures mean that questions pertaining to the geography, epidemiology, clinical signs and diagnosis of arboviral infections are extremely relevant.

4.23.5 Prevention

Vector-borne diseases are combated by fighting vectors by extermination and personal protection. Vectors can be exterminated in the imago stage or in preimaginal stages of development. It is no less crucial to find hatching locations and destroy them or reduce their numbers. Protective measures include preventing vector bites and thus disease transmission. This goal is accomplished with repellents and mechanical protection (nets, canopies, etc.).

4.24 The Morbidity Structure of Natural Focal Infections

In order to analyse the morbidity structure of natural focal infections in the population, three integrated maps were created based on a series of analytical maps: a map of the nosological profiles of Russia's administrative subjects (Map 4.38), a map of the population morbidity due the major natural focal viral and bacterial diseases (Map 4.39), and a map of the population morbidity due to the major natural focal parasitic diseases (Map 4.40).

The integrated map of nosological profiles (Map 4.38) reflects the sets of characteristic natural focal diseases across Russia's administrative subjects. It shows the presence or absence of diseases in the population for every year in the period in question. These sets are shown in the form of matrices with diseases along the vertical axis and years along the horizontal axis. Matrices for small areas are in the legend. The map offers an aggregated map of the distribution of the major natural focal diseases reported in Russia over the course of 10 years (2006–2015) without quantitative morbidity metrics and allows for a number of conclusions of the spread of disease at the federal level. The map demonstrates the presence of nosologic entities and the frequency with which a specific disease appears. It allows for an understanding of annual change in the natural focal disease spectrum in Russia's administrative subjects. The map can help identify the major and most prevalent pathologies. As demonstrated by cartographic analysis, Lyme borreliosis, HFRS and leptospirosis, which are found in the majority of Russia's regions, as well as tick-borne encephalitis and opisthorchiasis, are the leading natural focal diseases in Russia.

The map of population morbidity for a set of natural focal bacterial and viral infections (Map 4.39) allows for the analysis of their morbidity structure in Russia. Using the cartogram method, this map demonstrates the diversity (low,

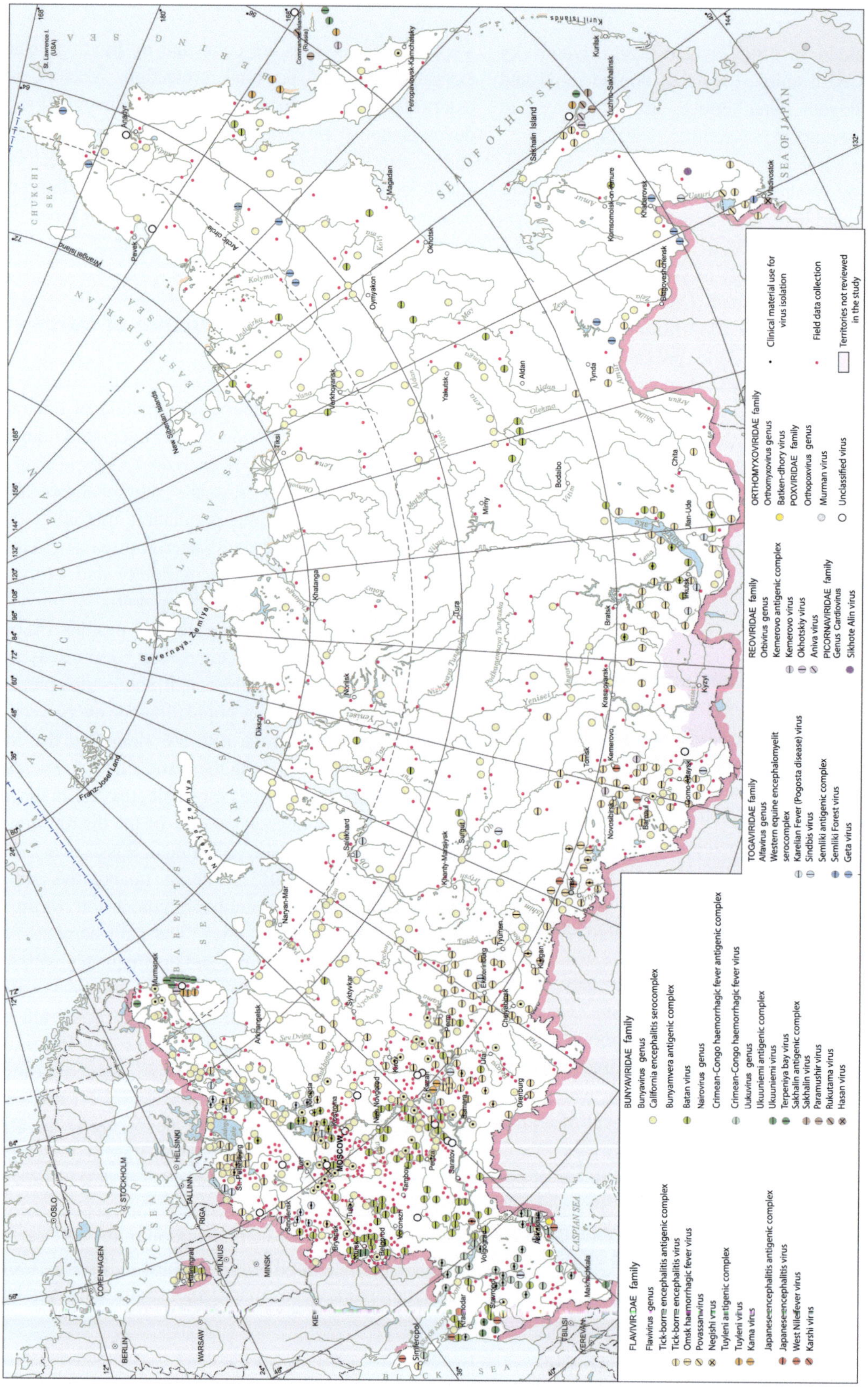

Map 4.37 Arbovirus isolation locations

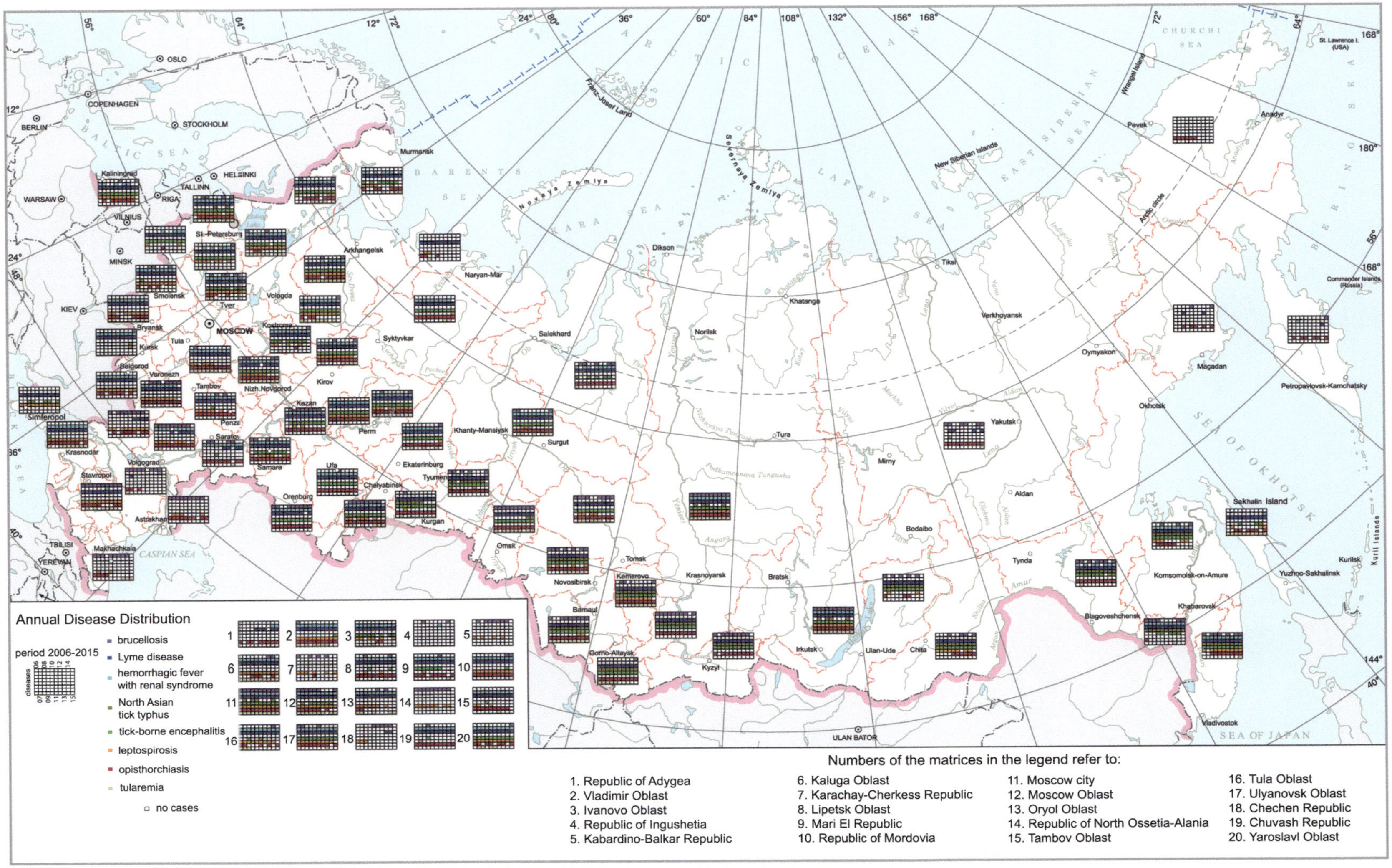

Map 4.38 Nosological profiles

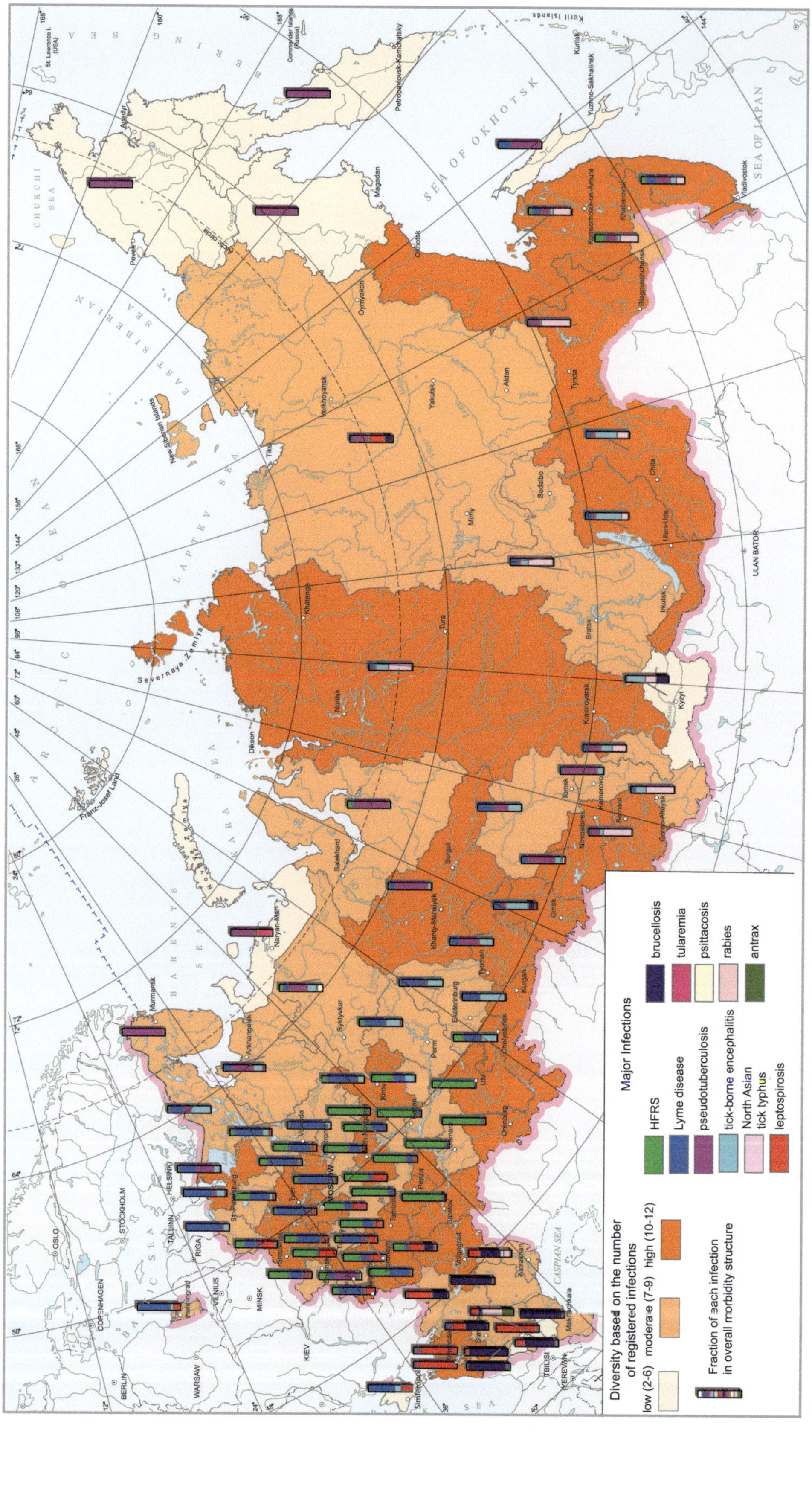

Map 4.39 Morbidity for major natural focal bacterial and viral diseases

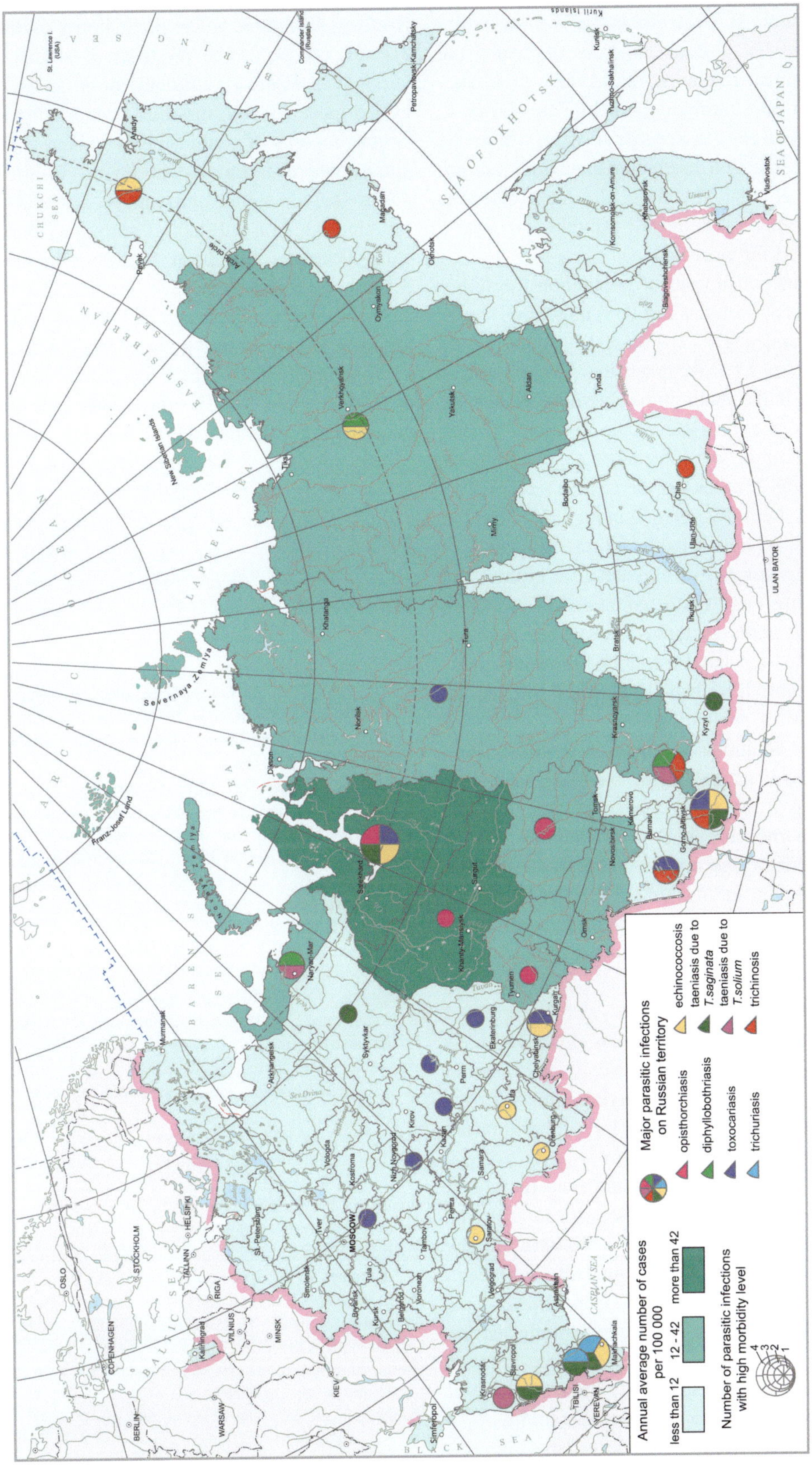

Map 4.40 Morbidity for major natural focal parasitic diseases

moderate, high) of reported infections, while cartodiagrams represent the proportion of each disease in the morbidity structure. The legend list diseases in order of decreasing long-term average annual morbidity rates, while their fractions as part of the overall morbidity structure are shown as bar charts, allowing to place a specific disease within the overall natural focal infection morbidity in various administrative subjects in Russia. Therefore, the main diseases, their diversity and proportion in overall natural focal disease morbidity are reflected for each region.

Analysis of the integrated map of population morbidity for a set of natural focal bacterial and viral infections shows that the greatest diversity (10–12 diseases) is found in the central regions of Russia's European territory, in the south of Western Siberia, and in the south of the Russian Far East. At the same time, the northern regions of the Russian Far East have the lowest number of infections (2–6 diseases). Regionally dominating diseases can be clearly seen. HFRS and Lyme borreliosis are most common in regions in Central Russia; leptospirosis is most significant in the south of Russia's European territory; North Asian tick typhus and tick-borne encephalitis are quite common in southern Siberia; pseudotuberculosis is quite relevant in northern Russia and in the Russian Far East. Tick-borne infections and haemorrhagic fevers have been established to pose the greatest risk for Russia's population.

The map provides additional information on the geography of Russia's main natural focal bacterial and viral infections. This map leads us to the conclusion that the proportion of tick-borne encephalitis is relatively low in Russia's European territory, somewhat higher in the Russian Far East, and quite significant in southern Siberia. HFRS, on the other hand, is mostly reported in Russia's European territory, and its role in the morbidity structure diminishes as we move north.

The map showing parasitic infections with the greatest epidemiological role in Russia is analogous (Map 4.40). As its analysis shows, helminthiases are unevenly distributed in Russia. This unevenness manifests in varying infection rates and numbers of diseases. Regions with high morbidity are primarily found in Western Siberia (over 42.0 average annual cases per 100,000 individuals) and Eastern Siberia (12.2–42.0 cases per 100,000 individuals) respectively. The greatest number of diseases is also found in Western Siberia, some of which are also quite epidemiologically significant. Aside from the commonly reported opisthorchiasis and diphyllobothriasis, the population is severely affected by echinococcosis and a few other parasitic diseases. This is related to the specifics of the traditional culinary culture of ethnic groups living in this area, in which consumption of raw fish and meat is common. The second group of regions with high morbidity and disease diversity (taeniasis, echinococcosis, trichocephalosis) is found in republics of the Northern Caucasus, where the majority of diseases occurs as a result of use of the products of animal husbandry and hunting, as well as traditional consumption of undercooked grilled meat. Finally, the third group is found in the mountainous regions of south Siberia (primarily the Altai Mountains), where morbidity is linked to the popularity of pasture-based animal husbandry (taeniasis) and hunting (trichinosis). In a substantial portion of the Russian Federation (Russia's European territory, the south of Eastern Siberia, the Russian Far East) the annual average number of cases is less than 12.2 per 100,000 individuals. Epidemiologically risky parasitic diseases are not recorded there.

Together the maps allow for an evaluation of the stability of the behaviour of a specific disease and the degree of risk posed by a specific disease in a particular territory and the use of this information for sanitary monitoring and the organization of targeted preventative events and campaigns, especially in areas with heavy recreation and recent development.

References

Act of the chief sanitary inspector of the Russian Federation dated 10.12.2014 No. 81 On the increasing of actions against anthrax in the Russian Federation 2014; http://rospotrebnadzor.ru/upload/iblock/fdb/postanovlenie-_-81-sibirskaya-yazva.pdf. Accessed 10 Jun 2015 (in Russian)

Act of the Russian Federation Government dated 1 December 2004 No. 715 On the approval of the socially significant diseases list and the list of diseases endangering others. http://base.garant.ru/12137881/. Accessed 10 Jun 2015 (in Russian)

Adishcheva OS, Malkhazova SM, Orlov DS (2016) The Distribution of west nile fever in Russia. Vestnik MGU. Series 5. Geography 4:48–55 (in Russian)

Aksoy M, Srigley J, Haider S, Johnstone J (2011) A lethal case of generalized tetanus. CMAJ 183:1045–1048

Ananyina YV (2010) Parasitic and free-living leptospires (leptospiraceae): ecologo-genetic features. Zool Journ 89(1):48–52 (in Russian)

Ananyina YV, Petrov EM (2006) Leptospirosis in Russia: the aetiologic structure and modern epidemiology. RAT-INFO 1:8–10 (in Russian)

Ananyina YuV (2003) Leptospirosis. In: Natural focality of diseases: the Gamaleya RAMN Institute Research. Rusaki, Moscow, pp 122–136 (in Russian)

Ananyina YuV (2009) Patogenic spirochets: ecological and genetic diversity. In: Relevant problems of the natural focality of diseases: from the Russian conference with international participation, dedicated to the 70th anniversary of the theory of Academician EN Pavlovsky on natural focality of diseases (24–25 November 2009, Omsk). Omsk: IC "Omskiy nauchniy vestnik" pp 15–17 (in Russian)

Anina-Radchenko ND (1979) Rabies and the fight against it. Kishinev: Shtiintsa 98p (in Russian)

Anthrax in the Moscow Governorate (1891–1910) (1910) Moscow 4–31 (in Russian)

Arizono N, Shedko M, Yamada M, Uchikawa R, Tegoshi T, Takeda K (2009) Mitochondrial DNA divergence in populations of the tapeworm diphyllobothrium nihonkaiense and its phylogenetic relationship with diphyllobothrium klebanovskii. Parasitol Int 58:22–28. https://doi.org/10.1016/j.parint.2008.09.001

Artemenkov YM, Ostrovskiy GD (2000) Epidemiological difficulties by the example of plague outbreaks in the period from 40-s to 70-s. RAT-info 4:50–56 (in Russian)

Bacot AW, Martin CY (1914) Observations on the mechanism of the transmission of plague by fleas. J Hyg 13.Plague Suppl 3:423–439

Bakulov IA, Gavrilov VA, Seliverstov VV (2001) Anthrax: new pages from the study of an old disease. Vladimir: Izd-vo Posad 283p (in Russian)

Bandara M, Ananda M, Wickramage K, Berger E, Agampodi S (2014) Globalization of leptospirosis through travel and migration. Global Health, 10:61 https://doi.org/10.1186/s12992-014-0061-0

Baryshnikov PI, Fedorova GA, Gatilov IM (2007) Anthrax of animals in altai krai. Altai State Agricultural University, Barnaul, 193p (in Russian)

Baturin AA, Antonov VA, Smelyanskiy VP, Zhukov KV, Chernobay VF, Kolyakina NN (2012) The role of birds being potential reservoirs of the west nile virus in the Russian Federation. Problems of the Specially Dangerous Infections 114:18–21 (in Russian)

Be'er SA (2005) The opisthorchiasis pathogen biology. Moscow: t-vo nauch. izd. KMK, 336p (in Russian)

Begeman L, GeurtsvanKessel C, Finke S, Freuling CM, Koopmans M, Müller T, Ruigrok TJH, Kuiken T (2017) Comparative pathogenesis of rabies in bats and carnivores, and implications for spillover to humans. Lancet Infect Dis. pii: S1473-3099(17) 30574-1. https://doi.org/10.1016/S1473-3099(17)30574-1 [Epub ahead of print]

Beljaev AE, Alekseeva MI (1980) Taeniasis. TSOLIUV, Moscow, pp 3–22 (in Russian)

Beljaev AE, Konstantinova TN (1980) Echinococcosis. TSOLIUV, Moscow, 26p (in Russian)

Beran GW (1994) Handbook of zoonoses. Viral Zoonoses, 2nd edn. CRC Press, 608p

Bernstein AD, Apekina NS, Mikhailova TV, Myasnikov YuA, Khlyap LA, Korotkov YuS, Gavrilovskaya IN (2001) Hantavirus infection of bank voles in the natural focus. Report 1. The features of the infectious process in a vole's body Med Parasitol and the Paras. Diseases 3:22–26 (in Russian)

Bernstein AD, Gavrilovskaya IN, Apekina NS, Dzagurova TK, Tkachenko EA (2010) The features of the natural focality of the hantavirus zoonoses. Epidemiol Vaccination 2(51):5–13 (in Russian)

Bessonov AS (1979) Trichinosis. Zooparasitology (The problems of helminthology) 6:130–208 (in Russian)

Bibikova VA, Klassovsky LN (1974) The transmission of plague by fleas. Medicine, Moscow 188p (in Russian)

Blumer G (1939) Some remarks on the early history of trichinosis (1822–1866). Yale J Biol and Med 11(6):581–588

Bogdanov VD, Golovatin MG (2017) Anthrax in yamal: an ecological view on traditional reindeer husbandry. Russ J Ecol 48:95. https://doi.org/10.1134/S1067413617020059

Borecka A, Klapec T (2015) Epidemiology of human toxocariasis in poland—a review of cases 1978–2009. Ann Agric Environ Med 22 (1):28–31

Botha EM, Mardotter W, Wolfaardt M (2008) Genetic determinants of virulence in pathogenic lineage 2 west nile virus strains. Emerg Inf Dis 14(2):222–230

Botvinkin AD, Poleschuk EM, Kuzmin IV, Borisova TI, Gazaryan SV, Yager P, Rupprecht CE (2003) Novel lyssaviruses isolated from bats in Russia. Emerg Infect Dis 9(12):1623–1625 https://doi.org/10.3201/eid0912.030374

Brantz D (Spring 2005) How parasites make history: on pork and people in nineteenth century Germany and the United States. GHI Bulletin 36:69–79 http:// www.ghi-dc.org/publications/ghipubs/bu/036/36.69.pdf

Briko NI (2010) Chlamydioses. Attending physician 11. http://www.lvrach.ru/2011/10/15435274/. Accessed 20 May 2015 (in Russian)

Bronstein AM, Lucshev VI (2014) Trichinosis. http://www.rmj.ru/artides_2791.htm. Accessed 23 May 15 (in Russian)

Bruschi F, Murrell KD (2002) New aspects of human trichinellosis: the impact of new trichinella species. Postgrad Med J 78(915):15–22

Burgasov PN, Cherkassky BL, Marchuk LM, Shcherbak YuF (1970) Anthrax. Medicine, Moscow, 128p (in Russian)

Burgdorfer W, Barbour AG, Hayes SF, Benach JL, Grunwaldt E, Davis JP (1982) Lyme disease—a tick-borne spirochetosis? Science 216:1317–1319

Busygin FF (2014) Omsk hemorrhagic fever. Izd-vo IP EA Maksheevoy, Omsk, 132p (in Russian)

Butenko AM, Trusova IN (2013) The incidence of crimea-congo hemorrhagic fever across the European, African and Asian Countries (1943–2012). Epidemiol Infectious Diseases 5:46–49 (in Russian)

Carter J, Saunders V (2007) Virology: principles and applications. Wiley, Hoboken, p 358p

Cerqueira GM, Picardeau M (2009) A Century of leptospira strain typing. Infect Genet Evol 9(5):760–768 https://doi.org/10.1016/j.meegid.2009.06.009

Cherkassky BL (1985) The epidemiology and prevention of rabies. Medicine, Moscow, 287p (in Russian)

Cherkassky BL (1994) Infectious and parasitic human diseases. Handbook of an epidemiologist. Medical newspaper, 617p (in Russian)

Cherkassky BL (1996) Especially dangerous infections. Medicine, Moscow, 160p (in Russian)

Cherkassky BL 2002 Epidemiology and prevention of anthrax. Intersan, Moscow, 384p (in Russian)

Cherkassky BL, Zhanuzakov NZh (1980) Anthrax.Kaynar, Almaty, 192p (in Russian)

Cola SG (ed) (1976) Anthrax. Kolos, Moscow, 228p (in Russian)

Cox FE (2002) History of human parasitology. Clin Microbiol Rev 15 (4):595–612

Daiter AB (1985) Natural focal infections in the Urban landscapes. Parasitology 19(3):169–176 (in Russian)

Demina TV, Dzhioev YP, Verkhozina MM, Kozlova IV, Tkachev SE, Plyusnin A, Doroshchenko EK, Lisak OV, Zlobin VI (2010) Genotyping and characterization of the geographical distribution of tick-borne encephalitis virus variants with a set of molecular probes. J Med Virol 82:965–976. https://doi.org/10.1002/jmv.21765

Deschiens V (1901) Atlas de parasitologie. Paris

Despommier D (2003) Toxocariasis: clinical aspects, epidemiology, medical ecology, and molecular aspects. Clin Microbiol Rev 16 (2):265–272

Domaradsky IV (1998) The plague. Medicine, Moscow 175p (in Russian)

ECDC—European centre for disease prevention and control (2016) Annual epidemiological report 2016 Brucellosis. Stockholm: ECDC, http://ecdc.europa.eu/en/healthtopics/brucellosis/Pages/Annualepidemiologicalreport2016.aspx. Accessed 10 Nov 2017

Ecker M, Allison SL, Meixner T, Heinz FX (1999) Sequence analysis and genetic classification of tick-borne encephalitis viruses from Europe and Asia. J Gen Virol 80(1):179–185. https://doi.org/10.1099/0022-1317-80-1-179

Eckert J (1997) Epidemiology of Echinococcus multilocularis and E. *granulosus* in Central Europe. Parasitologia 39(4):337–344

Emond R, Rowland H, Welsby F (2003) Infectious diseases: a colored atlas, 4th edn. Mosby, St.Louis

Encyclopedia of Infectious Diseases (1998) New York, 370p

Epidemiological Monitoring of Tularemia (2005) Moscow: Minzdrav, 59p (in Russian)
Epizootic Situation and Forecast for Rabies in Vladimir Oblast (a Newsletter) (2012) Vladimir, 53p (in Russian)
Fedorov VN, Rogozin II, Fenyuk BK (1955) The plague prevention. Medgiz, Moscow, 230p (in Russian)
Filippova NA (1977) The USSR Fauna. Arachnids. Ixodid ticks (Ixodidae). Leningrad.: Nauka. T.TS. Vol 4, 396p (in Russian)
Fusetti S, Ghiorotto C, Ferronato G (2013) A Case of cephalic tetanus in a developed country. Int J Immunopathology Pharmacool 26 (1):273–277. https://doi.org/10.1177/039463201302600131
Galińska EM, Zagórski J (2013) Brucellosis in humans—etiology, diagnostics. Clin Forms. Ann Agric Environ Med 20(2):233–238
Gamble KC, Clancy MM (Eds) (2013) Infectious diseases of concern to captive and free ranging animals in North America, 2nd ed. Infectious disease committee, American Association of Zoo Veterinarians: Yulee, Florida. 1098p
Gargili A, Thangamani S, Bente D (2013) Influence of laboratory animal hosts on the life cycle of hyalomma marginatum and implications for an in vivo transmission model for crimean-congo hemorrhagic fever virus. Front Cell Infect Microbiol 3:39. https://doi.org/10.3389/fcimb.2013.00039
Ginsburg NN (ed) (1975) Anthrax. Medicine, Moscow, 160p (in Russian)
Godfroid J (2002) Brucellosis in wildlife. Rev Sci Tech Off Int Epiz 21 (2):277–286
Godfroid J, Garin-Bastuji B, Saegerman C, Blasco JM (2013) Brucellosis in terrestrial wildlife. Rev Sci Tech 32(1):27–42
Granwehr BP, Lillibridge KM, Higgs S, Mason PW, Aronson JF, Campbell GA, Barrett AD (2004) West nile virus: where are we now? Lancet Infect Dis 4(9):547–556. https://doi.org/10.1016/S1473-3099(04)01128-4
Gratz N (2004) The Vector-borne human infections of Europe. Their distribution and burden on public health. WHO. 144p
Grizhebovsky G (2006) Rodent and vector surveillance in the northern caucasus. Interregional meeting on prevention and control of plague, WHO, Antananarivo, pp 38–39
Gubler DJ (2007) The continuing spread of west nile virus in the western hemisphere. Clin Infect Dis 45(8):1039–1046. https://doi.org/10.1086/521911
Guide on the Infectious Diseases / Ed. by YuV Lobzin. Saint Petersburg: Foliant; 2000. 936 p. (in Russian)
Guzeeva TM (2011a) Optimization of epidemiological surveillance of biohelminthoses: Auth abstr—Doc. of the med Sciences Moscow, 48p (in Russian)
Guzeeva TM (2011) The Preconditions of tapeworm infections elimination in the Russian Federation. In: Relevant aspects of parasitic diseases nowadays. Minzdravsotsrazvitiya, Rostov-na-Donu p 30–36 (in Russian)
Hampson K, Coudeville L, Lembo T, Sambo M, Kieffer A, Attlan M, Barrat J, Blanton JD, Briggs DJ, Cleaveland S, et al (2015) Estimating the global burden of endemic canine rabies. PLoS Negl Trop Dis 9:e0003709 https://doi.org/10.1371/journal.pntd.0003709
Harding H, Cawich S, Williams E, Crandon I, Williams-Johnson J (2014) Epidemiology of tetanus in Jamaica, 1993–2010. Trop Doct 44:166–168. https://doi.org/10.1177/0049475514526714
Hartskeerl RA, Collares-Pereira M, Ellis WA (2011) Emergence, control and reemerging leptospirosis: dynamics of infection in the changing World Clin Microbiol Infect 17(4):494–501 [https://doi.org/10.1111/j.1469-0691.2011.03474.x]
Hayes EB, Komar N, Nasci RS, Montgomery SP, O'Leary DR, Campbell GL (2005) Epidemiology and transmission dynamics of west nile virus disease. Emerg Inf Dis 11(8):7–15:1167–1173 https://doi.org/10.3201/eid1108.050289a
Hoogstraal H. Ticks of the Sudan. With special reference to equatoria province and with preliminary reviews of the genera *Boophilus, Margaropus* and *Hyalomma.* In: African Ixodoidea. Research Report NM 005 050.29.07. Department of the Navy, Bureau of Medicine and Surgery. Washington DC 1956 pp 436–451
Informational Bulletin of WHO (2014), 99. http://www.who.int/mediacentre/factsheets/fs099/ru/. Accessed 20 Apr 2015 (in Russian)
International Committee on the Virus Taxonomy. http://ictvonlme.org/virusTaxonomy.asp?version=2009. Accessed 12 Jul 2014 (in Russian)
Ivanovsky Scientific Research Institute of Virology. http://www.virology.ru/istoriya-instituta. Accessed 12 Jul 2014 (in Russian)
Jones P (2012) Science foundations: viruses. infobase learning 138p
Kalantarov GS (1912) 12 years of Anthrax in humans in connection with anthrax in animals in the districts of nizhny novgorod governorate (1900–1911) (according to the card registration). Nizhny-Novgorod: Volgar'; 63p (in Russian)
Karan LS, Ciccozzi M, Yakimenko VV, Lo Presti A, Cella E, Zehender G, Rezza G, Platonov AE (2014) The deduced evolution history of omsk hemorrhagic fever virus. J Med Virol 86(7):1181–1187 https://doi.org/10.1002/jmv.23856
Karaseva EV (2000) 50 years (1950–1999) of studying the natural focus of leptospirosis at the nero lake of the yaroslavl region. RAT-info. 4:4–8 (in Russian)
Karaseva EV, Rylnikov VA, Belen'kiy VA, Belen'kaya EI (1986) Some ecological features of brown rats of the central russia in the hollow of the lake nero (Yaroslavl Oblast) and their role in the forming of natural focus of leptospirosis. In: Brown rat. vol. 2. Moscow; pp 3–18 (in Russian)
Kashuba EA, Drozdova TG, Khanipova LV (2014) Ixodid Tick-borne borreliosis (training module). Infectious diseases. 4. http://infection-nmo.geotar.ru/ru/%20jartides_infection/202.html?SSr=140133776808ffffffff27c07df0114080e3767f. Accessed 11 May 2015 (in Russian)
Khatenever LM (ed) (1943) Tularemia infection: a collection of articles. Moscow, Medgiz, 215p (in Russian)
Kiselev LL, Levina ES (2004) Lev Aleksandrovich Zilber (1894–1966): the life in science. Nauka, Moscow, 699p (in Russian)
Klempa B, Tkachenko EA, Dzagurova TK, Yunicheva YV, Morozov VG, Okulova NM, Slyusareva GP, Smirnov AV, Kruger DH (2008) Hemorrhagic fever with renal syndrome caused by 2 lineages of dobrava hantavirus, Russia. Emerg Inf. Dis 14(4). http://www.nc.cdc.gov/eid/article/14/4/07-1310_article https://doi.org/10.3201/eid1404.071310
Knapp W, Masshoff W (1954) Zur Atiologie der abszedierenden retikulozytären Lymphadenitis, einer praktisch wichtigen, vielfach unter dem Bilde einer akuten Appendizitis verlaufenden Erkrankung. Dtsch Med Wochenschr 79(35):1266–1271
Konopatkin AA, Artemov BT, Bakulov IA (1993) Epizootology and the infectious diseases. Kolos, Moscow 687p (in Russian)
Konstantinova TN, Beljaev AE (1997) Trichinoses. RMAPO, Moscow, 24p (in Russian)
Korenberg EI (2001) The study and prevention of the mixed infections transmitted by Ixodid ticks. Vestnik RAMN. 11:41–45 (in Russian)
Korenberg EI, Kovalevsky YuV (1981) Zoning of the Tick-borne encephalitis area. In: The results of science and technology. Ser Med Geography. VINITI, Moscow, vol 11, 148p (in Russian)
Korenberg EI, Kovalevskiy YuV (2000) The Main Features of the eco epidemiology of tick borne encephalitis. In: Problems of tick-borne and parasitic diseases. From the round table of the VI Russian-Italian scientific conf. "Infectious diseases: diagnosis, treatment, prevention" December 14–16th, 2000, St.-Petersburg; pp 13–20 (in Russian)

Korenberg EI, Kalinin MI, Skripnikova IA, Soldatkina LM (1990) The experience of the serologic assay on the Lyme borreliosis endemic territory. Med Parasitology Parasitic Diseases 3:15–16 (in Russian)

Korenberg EI, Vorobyova NN, Sumlivaya ON, Frizen VI, Afanasyeva MV (2007) Tick transmitted diseases in Perm Krai. Methodical recommendations for doctors. GOU VPO PGMA, Perm, 67p (in Russian)

Korenberg EI, Pomelova VG, Osin NS (2013) Infections with natural focality transmitted by Ixodid ticks. Kommentariy, Moscow, 463p (in Russian)

Kozakevich VP, Varshavskiy SN, Nekipelov NV (1981) The natural foci geography of plague in the East Asia (South-west Manchouria). In: Epidemiology and prevention of natural focal infections. Russian Anti-plague Scientific Research Institute "Microbe", Saratov, pp 12–23 (in Russian)

Kucheruk VV (1959) The experience of classifying the natural foci of the extratropical eurasian plague. Report 1. The Principles of typology and classification of the natural foci of infection. Med Parasitology and Parasitic Diseases ; 28(6):658–667 (in Russian)

Kucheruk VV (1972) Structure, typology and regionalization of natural foci of human diseases. In: The results of the development of the natural focality of human diseases theory and further tasks. Coll Med, Moscow, pp 180–212 (in Russian)

Kuchta R, Brabec J, Kubackova P, Scholz T (2013) Tapeworm *Diphyllobothrium dendriticum* (Cestoda)—neglected or emerging human parasite? PLoS Negl Trop Dis 7(12):e2535 https://doi.org/10.1371/journal.pntd.0002535

Kuno G, Chang GJ (2005) Biological transmission of arboviruses: Reexamination of and new insights into components, mechanisms, and unique traits as well as their evolutionary trends. Clin Microbiol Rev 18(4):608–637. https://doi.org/10.1128/CMR.18.4.608-637.2005

Kuranshina NG, Kutilin NG, Nafikova AX, Mannapova RT, Nurtdinova EE (2014) The condition and assessment of natural and anthropogenic foci of rabies in the bashkortostan republic. Ufa, 147p (in Russian)

Kutyrev VV, Popova AYu, Ezhlova EB, Demina YuV, Pakskina ND, Bezsmertniy VE, Toporkov VP, Popov NV, Kabin VV, Yashkulov KB, Bammatov DM, Kovtunov AI, Sandzhiev DN, Zenkevich ES, Grazhdanov AK, Matrosov AN, Kuznecov AA, Sharova IN, Lopatin AA, Grigor'ev MP, Kulichenko AN (2015) Maintaining the epidemiological safety during the worsening of the epizootic situation at the caspian sea natural focus in 2014. Problems of the Especially Dangerous Infections 4:22–29 (in Russian)

Kutyrev VV, Popova AYu, Ezhlova EB, Demina YuV, Pakskina ND, Shchuchinov LV, Mikhajlov EP, Mishchenko AI, Rozhdestvenskiy EN, Bazarova GH, Denisov AV, Sharova IN, Popov NV, Kuznecov AA (2014) Cases of human plague contamination in the gorno-altaysk highland focus in 2014.Report 1. Epidemiological and epizotoological features of the gorno-altaysk natural focus of Plague. Problems of the especially dangerous infections 4:9–16 (in Russian)

Kuzmin IV, Hughes GJ, Botvinkin AD, Orciari LA, Rupprecht CE (2005) Phylogenetic relationships of Irkut and West Caucasian bat viruses within the Lyssavirus genus and suggested quantitative criteria based on the N gene sequence for Lyssavirus genotype definition Virus Res 111(1):28–43. [Epub 2005 Apr 8] https://doi.org/10.1016/j.virusres.2005.03.008

Lee HW (2003) Emergence and control of hantavirus diseases. In: Hantaviruses and hantaviral infections (to the 70th Anniversary of HFRS Study in the Far East Region). Vladivostok, pp 20–42 (in Russian)

Leonova GN, Chentsova IV. Petukhova SA. Somova LM, Belikov SI, Kondratov I.G. et al (2010) First Diagnosed lethal case of lyssavirus infection in Primorsky krai. Pacific Med J 3:90–95

Lindeborg M, Barboutis C, Ehrenborg C, Fransson T, Jaenson TG, Lindgren PE, Lundkvist A, Nystrom F, Salaneck E, Waldenstrom J, Olsen B (2012) Migratory birds, ticks, and crimean-congo hemorrhagic fever virus. Emerg Infect Dis. 18(12):2095–2097 https://dx.doi.org/10.3201/eid1812.120718

Lobzin YuV (2000) A manual on infectious diseases. Foliant, Saint Peterburg, 936p (in Russian)

Lvov DK (ed) (2008) Medical virology. Medical Informational Agency, Moscow, 656p (in Russian)

Lvov DK, Ilyichev VD (1979) Bird migration and pathogen transfer (Ecologico-geographical Relations of Birds with Infectious patogens). Nauka, Moscow, 270p (in Russian)

Lvov DK, Klimenko SM, Gaydamovich SYa (1989) Arboviruses and arboviral infections. Medicine, Moscow, 336p (in Russian)

Lvov DK, Butenko AM, Vyshemirskii OI, Gaydamovich SYa (2000) The West Nile Virus Isolation from the Infected People during the Epidemic Outbreaks in the Volgograd and Astrakhan Oblasts. Problems Virol 3:9–13 (in Russian)

Lvov DK, Kovtunov AI, Yashkulov KB, Gromashevskiy VL, Dzharkenov AF, Shchelkanov MYu, Kulikova LN, Savidzh G, Chimidov NM, Mikhalyaeva LV, Vasilyev AV, Galkina IV, Prilipov AG, Kinney R, Samohvalov EI, Bushkieva BTs, Gablev D, Al'khovskiy SK, Aristova VA, Deryabin PG, Butenko AM, Moskvina TM, Lvov DN, Zlobina LV, Lyapina OV, Sadykova GK, Shatalov AG, Usachev E, Voronina AG, Luneva LI (2004) The features of the circulation of west nile virus (Flaviviridae, Flavivirus) and some other arboviruses across the ecosystems of volga delta, volga-akhtuba floodplain and related aridic landscapes (2000–2002). Questions of Virology, 49(3):45-51 (in Russian)

Lvov DK, Deryabin PG, Shchelkanov MYu (2007) The Dynamics of the populational interactions of the West Nile Virus (WNV) between birds, Mosquitoes, ticks and people in the middle and lower zones of the Volga Delta. In: Proceedings of the 9th National scientific and practice conference of epidemiology and microbiology parasitology. Moscow, pp 199–200 (in Russian)

Lvov DK, Alkhovsky SV, Shchelkanov MYu, Shchetinin AM, Deryabin PG, Gitelman AK, Samokhvalov EI, Botinkov AG (2014) The DNA Analysis of Okhotskiy Virus (OKHV) and Aniva Virus (ANIV) (Reoviridae, Orbivirus), Isolated in Higher Latitudes of Northern Eurasia from the Ixodes Ticks (Ceratixodes uriae White, 1852), the Obligate Ectoparasites of Auks (Alcidae Leach, 1820). Problems of Virology 2:22–27 (in Russian)

Lysenko AYa, Beljaev AE (1987) Epidemiology of helminths: a Textbook. TSOLIUV, Moscow, 36p (in Russian)

Lysenko AYa, Vladimova MG, Kondrashin AV, Majori G (2002) Clinical parasitology. A Guide Geneva: WHO. 752p (in Russian)

Maes P, Clement J, Ranst V (2009) Recent approaches in hantavirus vaccine development. Expert Rev Vaccines 8(1):67–76. https://doi.org/10.1586/14760584.8.1.67

Mahy B, Regenmortel M (eds) (2010) Desk encyclopedia of animal and bacterial virology. Academic Press, Elsevier, 637p

Maleev VV, Galimzyanov HM, Butenko AM, Cherenov IV (2003) Crimea-congo hemorrhagic fever. Izd-vo AGMA, Moscow-Astrakhan, 120p (in Russian)

Marinin LI, Onishchenko GG, Kravchenko TB, Djatlov IA, Tuyrin EA, Stepanov AV, Nikiforov VV (2008) Human anthrax: epidemiology, prevention, diagnostics, treatment. Hygiene, Moscow 416p (in Russian)

Marr JS, Calisher CH (2003) Alexander the great and west nile virus encephalitis. Emerg Inf Dis 9(12):7–15:1599–1603 https://doi.org/10.3201/eid0912.030288

Matrosov AN (2007) Improving the Ecologico-epizootological Monitoring and non-specific prevention inside the natural foci of plague on the territory of the Russian Federation. Auth abstr—Doc of the biol sciences Saratov, 47p (in Russian)

Mead PS, Petersen JM (2009) Tularemia. Bacterial infections of humans. Chapter 42

Meshcheryakova IS (2003) Tularemia. In: Natural focality of diseases: the gamaleya RAMN institute research. Rusaki, Moscow, pp 137–160 (in Russian)

Meshcheryakova IS (2010) Tularemia: current epidemiology and vaccination (for the 80th Anniversary of the Creation of the First Russian Tularemia Laboratory). Epidemiol Vaccination 2:17–22 (in Russian)

Meshcheryakova IS, Dobrovolskiy AA, Demidova TN, Kormilitsyna MI, Mikhailova TV (2014) Transmissive Epidemic Outbreak of Tularemia in Khanty-Mansiysk City in 2013. Epidemiol Vaccination 5(78):14–20 (in Russian)

MG3.5.1.2958-11 (2011) Disinfectant measures against pseudotuberculosis and intestinal yersiniosis. Methodical instructions. Moscow, 32p (in Russian)

Mikhailova TV, Bernstein AD, Nevzorova NV, Apekina NS (2001) The current state of the hantavirus hemorrhagic fevers problem. RAT-info 3:10–13 (in Russian)

Millet NB, Hart GD, Reyman TA (1998) ROMI: Mummification for the common people. In: Cockburn A, Cockburn E, Reyman TA (eds) Mummies, disease and ancient cultures. Cambridge University Press, Ambridge: New York, pp 91–105

Monath TP (ed) (2000) The arboviruses: epidemiology and ecology, vol III, 52p

Moreira GM, Telmo PL, Mendonca M, Moreira AN, McBride AJ, Scaini CJ, Conceicao FR (2014) Human toxocariasis: current advances in diagnostics, treatment, and interventions. Trends Parasitol 30(9):456–464 https://doi.org/10.1016/j.pt.2014.07.003

Moskvitina EA, Vodyanitskaya SY, Lomov YuM, Pichurina NL (2002) Crimean-congo hemorrhagic fever in the rostov oblast: epidemiological zoning and the natural focus activity. Microbiol, Epidemiol and Immunol 6:26–30 (in Russian)

Moskvitina EA, Zabashta MV, Pichurina NL, Orekhov IV, Lomov YuM, Adamenko VI, Feronov DA, Zabashta AV (2011) West nile fever in the rostov oblast: ecologo-epidemiological Features of the 2010 Outbreak. Problems of the Specially Dangerous Infections 4:31–35 (in Russian)

Muratov IV (1993) Carnivorous terrestrial mammals as the final host of diphyllobothrium klebanovskii. Med Parasitol Parasitic Diseases 2:3–5 (in Russian)

Muratov IV, Posokhov PS (1988) The human diphyllobothriasis pathogens. Clin Parasitology 22(2):165–170 (in Russian)

Murrell KD, Pozio E (2011) Worldwide occurrence and impact of human trichinellosis, 1986–2009. Emerg Inft Dis 17:2194–2202. https://doi.org/10.3201/eid1712.110896

Nadelman RB, Wormser GP (1998) Lyme borreliosis. LANCET. 352:557–565. https://doi.org/10.1016/s0140-6736(98)01146-5

Naktin J, Beavis KG (1999) Yersinia enterocolitica and Yersinia pseudotuberculosis. Clinics Lab Medicine 19(3):523–536. https://doi.org/10.4061/2011/735308

Naumov NP, Lobachev VS, Dmitriev PP, Smirin VM (1972) Natural focality of plague in the aral karakum desert. Moscow State University, Moscow 404p (in Russian)

Neronov VM, Malkhazova SM, Tikunov VS (1991) Regional geography of the plague. In: The results of science and technology. Series medical geography. VINITI, Moscow, vol. 17, 229p (in Russian)

Neville J (2004) Rabies in the ancient world. In: King AA, Fooks AR, Aubert M, Wandeler AI (eds) Historical perspective of rabies in Europe and mediterranean basin. OIE (World organisation for animal health), pp 1–12

NIAID Emerging Infectious Diseases/Pathogens, https://www.niaid.nih.gov/research/emerging-infectious-diseases-pathogens). Accessed 21 Dec 2017

Nieto D (1982) Historical notes on cysticercosis. In: Flisser A (ed) Cysticercosis: present state of knowledge and perspectives. Academic Press, Inc., New York, pp 1–7

Oganesyan AS (2002) The tick-borne encephalitis virus circulation in Armenia. Microbiol Epidemiol Immunol 1:107–108 (in Russian)

Olsufiev NG, Dunaeva TN (1970) Natural focality, epidemiology and prevention of Tularemia. Medicine, Moscow, 270p (in Russian)

Olsufiev NG, Rudnev GP (1960) Tularemia. Medgiz, Moscow, 460p (in Russian)

On the morbidity rates of trichinosis, tapeworm infection and taeniasis in the Russian Federation in 2013. A Rospotrebnadzor Letter Dated 30.06.2014. http://rospotrebnadzor.ru/upload/iblock/d3b/2014pismo-trikhinellez.PDF. Accessed 25 Mar 2015 (in Russian)

On the Status and epidemiological wellbeing situation of the population in the Russian Federation in 2012: State report. Moscow: Federal service for surveillance on consumer rights protection and human wellbeing; 2013. 176p (in Russian)

On the Status and epidemiological wellbeing situation of the population in the Russian Federation in 2013: State report. Moscow: Federal service for surveillance on consumer rights protection and human wellbeing; 2014. 191p (in Russian)

On the Status of Sanitary and Epidemiological Wellbeing of the Population in the Russian Federation in 2011: State report. Moscow: The Federal Service for Supervision of Consumer Rights Protection and Human Wellbeing; 2012. 316p (in Russian)

Onischenko GG, Vasiliev NT, Litusov NV, Harechko AT, Vasiliev PG, Sadovoy NV, Kozhukhov VV (1999) Anthrax: the relevant aspects of microbiology, epidemiology, clinical appearance, diagnosis, treatment and prevention. VUNMTS MZ RF, Moscow, 448p (in Russian)

Onishchenko GG, Kutyrev VV (eds) (2004) Natural foci of the plague on the territories of the caucasus, Central Asia and Siberia. Medicine. Moscow, 192p (in Russian)

Otero-Abad B, Torgerson PR (2013) A systematic review of the epidemiology of echinococcosis in domestic and wild animals. PLOS Neglected Tropical Diseases 7(6) http://journals.plos.org/plosntds/article?id=10.1371/journal.pntd.0002249. Accessed 26 May 2015 https://doi.org/10.1371/journal.pntd.0002249

Outbreak Source Investigation *Yersinia pseudotuberculosis.* Ministry of Primary Industries, New Zealand Government; 2014. http://archive.mpi.govt.nz/Portals/0/Documents/food/yersinia/out-break-source-investigation.pdf. Accessed 08 Jun 2015

Pappas G, Papadimitriou P, Akritidis N, Christou L, Tsianos EV (2006) The new global map of human brucellosis. Lancet Infect Dis Feb 6 (2):91–99 https://doi.org/10.1016/s1473-3099(06)70382-6

Pauli G, Bauerfeind U, Blumel J, Burger R, Drosten C, Groner A, Gurtler L, Heiden M, Hildebrandt M, Jansen B, Montag-Lessing T, Offergeld R, Seitz R, Schlenkrich U, Schottstedt V, Strobel J, Willkommen H (2013) West nile virus. Transfus Med Hemother 40 (4):265–284 https://doi.org/10.1159/000353698

Pautova EA, Shchuchinova LD, Shchuchinov LV, Mikhailova SA, Dovgalyov AS, Akbaev MSh (2011) The incidence of toxocariasis in the altai republic. In: Current problems of the epidemiology of infectious diseases. Hygiene, Moscow, pp 763–765 (in Russian)

Pavlenko AL (2013) The features of the leptospirosis epidemiology at the present time. Zaporozhskiy Med J 6(81):63–69 (in Russian)

Pawlowski Z, Schulz MG (1972) Taeniasis and cysticercosis *(Taenia saginata).* Adv Parasitol 10:269–343

Petersen LR, Roehrig JT (2001) West nile virus: a reemerging global pathogen. Emerg Inf Diseases July-August 7(4):611–614. https://doi.org/10.3201/eid0704.010401

Petrisheva PA, Olsufiev NG (1969) Geography of Natural Focal Diseases and the Objectives of Their Prevention. Book of papers. M: Medicina. 340p (in Russian)

Platonov AE, Tolpin VA, Gridneva KA, Titkov AV, Platonova OV, Kolyasnikova NM, Busani L, Rezza G (2014) The incidence of west nile disease in Russia in relation to climatic and environmental factors. Int J Environ Res Public Health 11(2):1211–1232. https://doi.org/10.3390/ijerph110201211

Pokrovskiy VI (1983) Guide to zoonoses. Medicine, Leningrad, 320p (in Russian)

Pokrovsky VI, Onishchenko GG, Cherkasskiy BL (2003) The evolution of infectious diseases in Russia in the 20th Century. Medicine, Moscow 664p (in Russian)

Pokrovsky VI, Pak SG, Briko NI, Danilkin BK (2007) Infectious diseases and epidemiology. A textbook. GEOTAR-Media, Moscow, 816p (in Russian)

Ponomarev NM, Kostuykov MA, Ponomarev AN (2011) The Distribution of animal trichinosis on the altai krai territory. Vesti ASAU 1 (75):69–73 (in Russian)

Popov NV (2002) Discretisation—the main spatio-temporal feature of plague manifestation in ground squirrel-type foci. Saratov State University, Saratov, 192p (in Russian)

Popov VP, Orlov DS, Bezsmertniy VE (2012) Epizootological and epidemiological situation of natural foci of tularemia on the territory of central federal district of the Russian Federation in 1992–2011. In: Problems of the specially dangerous infections. Scientific Research Institute "Microbe", Saratov, 4, pp 1014 (in Russian)

Popov NV, Bezsmertniy VE, Matrosov AN, Knjazeva TV, Kuznecov AA, Fedorov YuM, Popov VP, Verzhuckiy DB, Korzun VM, Chipanin EV, Dubjanskiy VM, Maleckaja OV, Grigor'ev MP, Balahonov SV, Kulichenko AN, Kutyrev VV (2014) Epizootic activity of the natural foci of the plague on the Russian Federation territory in 2014 and a forecast for 2015. Problems of the Especially Dangerous Infections, 1:10–17 (in Russian)

Popov NV, Bezsmertniy VE, Matrosov AN, Knjazeva TV, Kuznecov AA, Fedorov YuM, Popov VP, Verzhuckiy DB, Korzun VM, Kosilko SA, Chipanin EV, Dubjanskij VM, Maleckaja OV, Grigor'ev MP, Zenkevich ES, Toporkov VP, Balahonov SV, Kulichenko AN, Kutyrev VV (2016) Epizootic activity of the natural foci of the plague on the Russian Federation territory in 2015 and a forecast for 2016. Problems Especially Dangerous Infections 1:13–19 (in Russian).

Postovit VA (1997) Infectious diseases. A guide. SOTIS, Saint Petersburg, 502p (in Russian)

Prevention of Ornithosis (2010) The Sanitary-epidemiological rules 3.1.2815–10 (in Russian)

Putinceva EV, Antonov VA, Smeljanskij VP, Pakskina ND, Skudareva ON, Viktorov DV, Tkachenko GA, Pak VA, Zhukov KV, Monastyrskij MV, Borodaj NV, Manankov VV, Pogasij NI, Shpak IM, Savchenko SS, Lemasova LV, Bondareva OS, Zamarina TV, Barkova IA (2014) Features of the epidemic situation of the west nile fever in 2013 worldwide and in the Russian Federation and a forecast for its development in 2014. Problems of the Specially Dangerous Infections 2:33–39 (in Russian)

Putintseva EV, Lipnitsky AV, Alekseev VV (2010) The West nile fever distribution across the world and in the Russian Federation in 2010. Problems of Specially Dangerous Infections, vol 107, pp 38–41 (in Russian)

Rabies in the Russian Federation (a newsletter) (2013) Omsk, 53p (in Russian)

Rall' YuM (1958) The lectures about the Plague's Epizotoology. Stavropol publishing house, Stavropol, 243p (in Russian)

Reisen WK, Fang Y, Lothrop HD, Martinez VM, Wilson J, O'Connor P, Carney R, Cahoon-Young B, Shafii M, Brault A (2006) Overwintering of West Nile virus in southern California. J Med Entomol 43:344–355

Reusken C, Neyman P (2013) Factor driving hantavirus emergence in Europe. Curr Opin Virol 3:92–99

Romig T (2003) Epidemiology of Echinococcosis. Langenbecks Arch Surg 388:209–217. https://doi.org/10.1007/s00423-003-0413-3

Romig T, Dinkel A, Mackenstedt U (2006) The present situation of echinococcosis in Europe. Parasitol Int 55:187–191. https://doi.org/10.1016/j.parint.2005.11.028

Rothschild EV (1978) The spatial structure of the natural focus of the plague and the methods of its study. Moscow, 194p (in Russian)

Rothschild EV (1992) The forecast of the zoonosis foci intensity based on the environmental factors. Nauka Moscow, 184p (in Russian)

Rudakov NV (2009) Modern ideas about Rickettsia and Rickettsiosis. In: Relevant problems of the natural focality of diseases: from the Russian conference with international participation, dedicated to the 70th anniversary of the theory of academician EN Pavlovsky on natural focality of diseases (24–25 November 2009, Omsk). Omsk: IC "Omskiy nauchniy vestnik", pp 21–24 (in Russian)

Rudakov NV, Shpynov SN, Samoylenko IE, Tankibaev MA (2003) Ecology and epidemiology of spotted fever group rickettsiae and new data from their study in Russia and Kazakhstan. Ann N Y Acad Sci 990:12–24

Rudakov NV, Shpynov SN, Samoylenko IE, Kumpan LV, Reshetnikova TA (2006) Modern approaches to the study of rickettsiales. Bullof Siberian medicine, Annex 1:11–116 (in Russian)

Rudakov NV, Shpynov SN, Samoylenko IE, Obert AC (2011) Tick-borne Rickettsiosis and Rickettsias of the tick-borne spotted fever group in Russia. IC "Omskiy nauchniy vestnik", Omsk, 232p (in Russian)

Ruzek D, Yakimenko VV, Karan LS, Tkachev SE (2010) Omsk hemorrhagic fever. Lancet 376(9758):2104–2113. https://doi.org/10.1016/S0140-6736(10)61120-8

Ruzek D, Holbrook M, Yakimenko VV, Karan LS, Tkachev SE (2014) Omsk hemorrhagic fever virus. In: Manual of security sensitive microbes and toxins. Taylor and Francis Group pp 193–200

Rymarenko NV, Chitakova AE, Mazinova ER, Dzhemileva HSh (2014) The first case of west nile fever in a child in the autonomous republic of crimea. Health Child 5:113–115 (in Russian)

Sanitary and Epidemiological Regulations HSR&S 3.1.2.3113–13 dated October 22, 2013 No. 59 Tetanus prevention; 2013. http://rospotrebnadzor.ru/documents/details.php?ELEMENT_ro=3564. Accessed 10 Jun 2015 (in Russian)

Schantz PM, Moore AC, Munoz JL, Hartman BJ, Schaefer JA, Aron AM, Persaud D, Sarti E, Wilson M, Flisser A (1992) Neurocysticercosis in an Orthodox Jewish community in New York City. N Engl J Med 327(10):692–695 https://doi.org/10.1056/nejm199209033271004

Scherret JH, Poidinger M, Mackenzie JS, Broom AK, Deubel V, Lipkin WI, Briese T, Gould EA, Hall RA (2001) The relationships between west nile and kunjin viruses. Emerg Infect Dis 7(4):697–705 https://doi.org/10.3201/eid0704.010418

Scholz T, Garcia HH, Kuchta R, Wicht B (2009) Update on the human broad tapeworm (Genus Diphyllobothrium), Including clinical relevance. Clin Microbiol Rev, pp 146–160 https://doi.org/10.1128/cmr.00033-08

Schuch R, Fischetti VA (2009) The secret life of the anthrax agent *Bacillus anthracis*: bacteriophage-mediated ecological adaptations. PLoS One 4(8):e6532. Published online 2009 Aug 12/ https://doi.org/10.1371/journal.pone.0006532

Sejvar JJ (2003) West nile virus: an historical overview. Ochsner J. Summer- Autumn 5(3):6–10

Selimov MA (1963) The ways of hydrophobia elimination. Medical Literature, Moscow, 294p (in Russian)

Semenova TA, Dovgalyov AS, Avdyukhina TI (2011) The Intermediate hosts of diphyllobothriases pathogens and their infection rate. In: Relevant questions of infectious diseases epidemiology (Coll. of Scientific Works). Hygiene, Moscow, vol. 10, p. 750–754 (in Russian)

Sergiyev VP, Akimova RF, Romanenko NA, Frolova AA (1993) The incidence of diphyllobothriasis and opisthorchiasis in Russia in 1992. Publ Health Env 4:15–17 (in Russian)

WHO Technical Report Series (2013) no. 982. http://apps.who.int/iris/bitstream/10665/85346/1/9789240690943_eng.pdf?ua = 1. Accessed 05 Apr 2015

Sidorov EG (1983) Natural focality of opisthorchiasis Nauka, Almaty 240p (in Russian)

Sidorov GN, Botvinkin AD, Mal'kova MG, Krasil'nikov VR (1992) The distribution, population density, the probability of biocenotic contacts and the degree of synanthropization of the wild canids in the natural foci of rabies across the USSR. Zool Journ 71(4):115–130 (in Russian)

Situation of Tularemia in Khanty-Mansiysk City and Khanty-Mansiysk Region (as on 8.10.2013). http://86.rospotrebnadzor.ru/c/joumal/view_artide_content?groupld=10156&articleld = 118031&version = 1.1. Accessed 10 Jun 2015 (in Russian)

Sjostedt A (2007) Tularemia: history, epidemiology, pathogen physiology, and clinical manifestations. Umea, Sweden, 29p. https://doi.org/10.1196/annals.1409.009

Slonova RA (2003) The History of HFRS study and current situation in the Primorsk Krai. In: Hantaviruses and hantaviral infections (to the 70th Anniversary of HFRS Study in the Far East Region). Vladivostok, p. 4–19 (in Russian)

Sludsky AA (2014) A list of world's vertebrates, potentially hosting a plague pathogen. Problems Especially Dangerous Infections 3:42–51 (in Russian)

Smirnova SE (2007) Crimean-congo hemorrhagic fever (Etiology, Epidemiology, Laboratory Diagnosis). ATISO, Moscow, 304p (in Russian)

Smith KA, Bradley KK, Stobierski MG, Tengelsen L (2005) Compendium of measures to control *Chlamydophila psittaci* (Formerly *Chlamydia psittaci*) infection among humans (Psittacosis) and Pet Birds, 2005. J Am Vet Med Assoc, 532–539

Soldatkin IS, Rudenchik YuV (1988) The epizootical process inside of the natural foci of the plague (a Concept Revision). Ecology of the Sapronoses Patogens, 117–131 (in Russian)

Somov GP, Besednova NN (1981) Hemorrhagic fevers. Medicine, Leningrad, 200p (in Russian)

Somov GP, Pokrovskiy VI, Besednova NN, Antonenko FF (2001) Pseudotuberculosis. Medicine, Moscow, 256p (in Russian)

Souza RA, Imori PFM, Passaglia J, Pitondo-Silva A, Falcao JP (2013) Molecular typing of *Yersinia pseudotuberculosis* strains isolated from livestock in Brazil. Genet Mol Res. 12(4):4869–4878 https://doi.org/10.4238/2013 October.22.6

Steere AC, Malawista SE, Hardin JA, Ruddy S, Askenase W, Andiman WA (1977) Erythema chronicum migrans and Lyme arthritis. The enlarging clinical spectrum. Ann Intern Med 86(6):685–698

Stepanova TF (2001) Epidemiological manifestations of actual risk of public contamination in the combined foci of zoonotic parasitoses. In: Ecological bases of Western Siberian combined natural foci of endemic parasitoses. Tyumen State University, Tyumen pp 129–138 (in Russian)

Stepanova TF (2002) Opisthorchiasis. Tyumen State University, Tyumen, 196p (in Russian)

Strauss JH, Strauss EG (2008) Viruses and Human Disease. 2nd edn. Elsevier 476p. https://doi.org/10.1586/14760584.8.1.67

Suntsov VV, Suntsova NI (2006) The plague. The origin and evolution of the epizootic system. Moscow: T-vo nauch. izd. KMK. 247p (in Russian)

Supotnitsky MV, Supotnitskaya NS (2006) The history of plague in an essay. In 2 volumes. University book, Moscow vol 1, 468p, vol 2, 696p (in Russian)

Suss J (2011) Tick-borne encephalitis 2010: Epidemiology, risk areas, and virus strains in Europe and Asia-an overview. Ticks tick borne dis. 2(1):2–15 https://doi.org/10.1016/j.ttbdis.2010.10.007

Svensson K, Larsson P, Johansson D, Byström M, Forsman M, Johansson A (2005) Evolution of subspecies of *Francisella tularensis*. J Bacteriol 187(11):3903–3908. https://doi.org/10.1128/JB.187.11.3903-3908.2005

Sysuev EB, Pozdnyakov AM, Strygin AV, Shcherbak NP, Iraklionova NS (2013) Natural foci of the dangerous and specially dangerous infectious pathogens. Ornithosis Intern J Appl Fund Res 7:96–97 (in Russian)

Tarasevich IV (2005) Modern ideas about rickettsiosis. Clin Microbial Antimicrob Chemotherapy 7(2):119–129 (in Russian)

Tarasevich IV, Plotnikova LF, Fetisova NF, Makarova VA, Lobanov AV, Dodonov MM, Dayter AB (1990) The epidemiology of rickettsiosis and rickettsia ecology. In: Rickettsias and Rickettsiosis: a collection of articles. Izd-vo AMN USSR, Moscow, pp 21–31 (in Russian)

Tarasov VV (2002) The epidemiology of the vector-borne diseases. Moscow State University, Moscow 336p (in Russian)

Tarshis MG, Cherkassky BL (1997) Animal diseases putting a human in danger. Kolos, Moscow, 298p (in Russian)

Tetanus (2015) Centers for disease control and prevention epidemiology and prevention of vaccine-preventable diseases, pp 341–352

The list of territories endemic for tick-borne encephalitis in 2014. http://www.rospotrebnadzor.ru/deyatelnost/epidemiological-surveillance/?ELEMENT_ID=3039&sphrase_id = 865528. Accessed 20 Mar 2015 (in Russian)

Thompson RCA (2008) The taxonomy, phylogeny and transmission of echinococcus. Exp Parasitol 119:439–446. https://doi.org/10.1016/j.exppara.2008.04.016

Timchenko NF, Popov AF. Past and present of pseudotuberculosis. Infectious diseases 201; 4. http://infection-nmo.geotar.ru/ru/jartides_mfection/201.html?SSr=510133776a01ffffffff27c07df011601332c-3ce). Accessed 05 May 2015 (in Russian)

Timoshenko TM, Majboroda AA, Kupriyanova NYu (1995) The questions of the epidemiology of diphyllobothriasis in the system of ecological monitoring of Baikal. In: Problems of ecology. The 5th International Conference "M. M. Kozhov Memorial Lecture" Materials. Nauka, Novosibirsk, vol. 1, pp 149–152 (in Russian)

Tkachenko EA, Dzagurova TK, Bernstein AD, Okulova NM, Korotina NA, Trankvilevskiy DV, Morozov VG, Yunicheva YuV, Zavora DL, Balovneva MV, Sotskova SE, Mutnykh ES, Smirnova MS, Leonovich OA, Shevelyov AB, Malkin GA (2012) Hemorrhagic fever with renal syndrome in Russia—a XXI century problem. Vestnik RAEN; 1:48–54 (in Russian)

Tkachenko EA, Bernstein AD, Dzagurova TK, Morozov VG, Slonova RA, Ivanov LI, Trankvilevskiy DV, Kruger D (2013) Relevant problems of the hemorrhagic fever with renal syndrome. Microbiol J 1:51–58 (in Russian)

Torgerson PR, Macpherson CNL (2011) The socioeconomic burden of parasitic zoonoses: global trends. Vet Parasitol 182:79–95. https://doi.org/10.1016/j.vetpar.2011.07.017

Tseneva GYa (ed) (2016) Yersinias and yersiniosis. Saint Petersburg, 168p (in Russian)

Tumolskaya NI (1997) Human toxocariasis: a modern approach. The Doctor 9:11–13 (in Russian)

Turnbull PC (2002) Introduction: anthrax history, disease and ecology. Curr Top Microbiol Immunol 271:1–19

Turyanov MH, Tsaregorogsev AD, Lobzin YuV (1998) Infectious diseases. GEOTAR Medicine, Moscow, 319p (in Russian)

Vasilyev KG, Segal AE (1960) The history of the epidemics of Russia. Medgiz, Moscow, 398p (in Russian)

Vodyanitskaya SYu, Moskvitina EA, Pichurina NL, Zabashta AV, Orekhov IV, Mishan'kin BN, Vodopjanov SO, Suchkov IYu (2008) Crimea-congo hemorrhagic fever: ecologo-epizootological value of crows. Problems of the Especially Dangerous Infections 6:12–15 (in Russian)

Volynkina AS, Kotenev ES, Lisitskaja YaV, Maleckaya OV, Shaposhnikova LI, Kulichenko AN (2015) Crimea-congo hemorrhagic fever in the Russian Federation in 2014, the forecast of the epidemiological situation for 2015. Problems of the Especially Dangerous Infections 1:42–45 (in Russian)

Voronov AG (1989) Medical geography. Zoonoses and Sapronoses. Izd-vo MGU, Moscow 107p (in Russian)

Vuitton DA, Zhou H, Bresson-Hadni S, Wang O, Piarroux M, Raoul F, Girau-Doux P (2003) Epidemiology of alveolar echinococcosis with particular reference to China and Europe. Parasitology 127:87–107

Weber K (1974) Erythema-chronicum-migrans-Meningitis—eine bakterielle Infektiontionskrankheit? MMW Munch Med Wochenschr 116(45):1993–1998

WHO (2004) Human plague in 2002 and 2003. Weekly epidemiological record 33(79):301–306

WHO (2007) Guidelines on Tularaemia. Geneva, 115p

WHO (2013) WHO expert consultation on rabies, Second report. In: WHO technical report series 982. World Health Organization, Geneva. Available at http://apps.who.int/iris/bitstream/10665/85346/1/9789240690943_eng.pdf

WHO/FAO/OIE Guidelines for the Surveillance (2005) Prevention and control of taeniosis/cysticercosis, 156p

Wilks D, Farrington M, Rubenstein D (2003) The infectious diseases manual. Blackwell Science, 432p

Woodhall DM, Eberhard ML, Parise ME (2014) Neglected parasitic infections in the United States: toxocariasis. Am J Trop Med Hyg 90(5):810–813. [https://doi.org/10.4269/ajtmh.13-0725]

World Health Organization (2009) Dengue guidelines for diagnosis, treatment, prevention and control, 148p

World Health Organization (2013) Guide for quality control of diphtheria, Tetanus and Pertussis Vaccines, 300p

World Health Statistics 2009. http://www.who.int/gho/publications/world_health_statistics/2009/en/index.html\. Accessed 13 Jun 2012

World Health Statistics 2010. http://www.who.int/gho/publications/world_health_statistics/2010/en/index.html\. Accessed 13 Jun 2012

World Health Statistics 2011. http://www.who.int/gho/publications/world_health_statistics/2011/en/index.html/. Accessed 13 Jun 2012

WHO. Tetanus. http://www.who.int/immunization_monitoring/diseases/tetanus/en/index.html/. Accessed 11 Mar 15

Yastrebov VK, Yakimenko VV (2014) Omsk Hemorrhagic fever: the results of the research (1946–2013). Quest Virol 59(6):5–11 (in Russian)

Yuschenko GV (2000) Pseudotuberculosis and Yersiniosis: epidemiological and ecological aspects. In: Infections caused by Yersinias: An international conference report. Saint Petersburg, p 2 (in Russian)

Yushchuk ND (2014) Bacterial diseases: a textbook. GEOTAR-Media, Moscow, 976p (in Russian)

Zapariy SP (2003) Urban foci of toxocariasis in a large industrial centre of the West Siberia Region Diss—cand of biology Sciences. Tyumen. 122p (in Russian)

Zavoloka AA, Zavoloka AnA (2013) About the Rabies. VetPharma 4 (15):2431 (in Russian)

Zee HW, Zee PW, Johnson KM (1978) Isolation of the etiologic agent of korean hemorrhagic fever. J Infect Dis 137(3):298–308

Zheludkov MM, Tsirelson LE (2010) Reservoirs of *Brucella* infection in nature. Biol Bull Russ Acad Sci 37:709 https://doi.org/10.1134/S106235901007006X

Zilber LA (1966) Operation "The ORE". Sci Life 12:55–56 (in Russian)

Zlobin VI (2006) Epidemiological situation and the problems of tick-borne encephalitis control in the Russian Federation. Bull Siberian Med Annex 1:16–24 (in Russian)

Zlobin VI, Demina TV, Mamaev LV, Butina TV, Belikov SI, Gorin OZ, Dzhioev YuP, Verkhozina MM, Kozlova IV, Voronko IV, Adel'shinRV, Grachev MA (2001) Analysis of genetic variability of TBEV Strains on the primary structure of a cover protein's gene fragment. Vopr Virusol 1:12–16 (in Russian)

Znamenskiy VA, Vishnyakov AK (1967) Etiology of far east scarlet-like fever. Microbiol J 2:125–130 (in Russian)

Zoonotic Diseases: a review of the issue and the information about the project: "The USSR/ UNEP Project "A Training Course on Zoonoses". CMP, Moscow 1981. 22p (in Russian)

Zurek L, Denning SS, Schal C, Watson DW (2001) Vector competence of *Musca domestica* (Diptera: Muscidae) for *Yersinia pseudotuberculosis*. J Med Entomol 38(2):333–335

Zverev VV, Boychenko MN (eds) (2010) In: Medical microbiology, virology and immunology. GEOTAR-Media, Moscow, vol 2, 248p (in Russian)

Spatial Organization of the Sanitary and Epidemiological Service

5

Abstract

The chapter presents a brief characterization of the system of sanitary and epidemiological monitoring in the Russian Federation. Its work is aimed at ensuring epidemiological safety in the environment. A textual description of its structure is provided and illustrated with three maps, one of which reflects border control points that exert sanitary control, while the others show the locations of major sanitary, epidemiological and hygienic institutions in Russia.

5.1 Sanitary and Epidemiological Control

The Federal Service for Surveillance on Consumer Rights Protection and Human Wellbeing (Rospotrebnadzor) provides solutions and answers for public health surveillance issues, the establishment of causality for the rise and spread of infectious disease, public health federal border control, sanitary and epidemiological inspections, and the collection of population public health statistics, as well as a number of other challenges to consumer rights and welfare (Federal Service… 2004).

Rospotrebnadzor oversees hygiene and epidemiology centres and administrative organizations in the administrative subjects of the Russian Federation, anti-plague agencies, organizations and ministries of the sanitary and epidemiological service, as well as public health research institutions (Fig. 5.1).

One of the most important roles fulfilled by Rospotrebnadzor is public health control in air, land and sea border checkpoints in the Russian Federation (Map 5.1). Control is both constant (for routine transport) and temporary, when information about arriving or departing vehicles is received from customs and border control agencies.

One of the main goals of Rospotrebnadzor's anti-plague agencies is sanitary protection of the country's territories against the introduction and distribution of plague and other dangerous infectious diseases. According to data from 2015, anti-plague agencies monitor the 11 plague natural foci that are found in Russia (Map 5.2) (On the Status… 2016).

Rospotrebnadzor's research institutions conduct inquiry into relevant issues in chemistry, biology and radiation safety and contribute to other disciplines that aim to protect population health. The largest experimental laboratories are in Moscow, St. Petersburg, Novosibirsk, Nizhny Novgorod, Khabarovsk and other cities (Map 5.3).

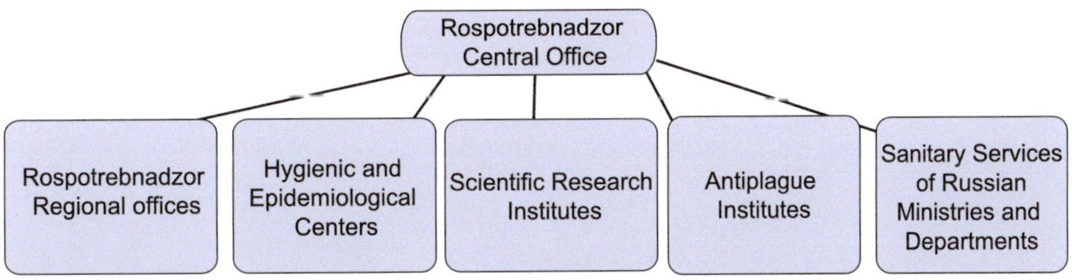

Fig. 5.1 The structure of federal service for surveillance on consumer rights protection and human well-being. *Credit* www.rospotrebnadzor.ru

S. Malkhazova et al., *Mapping Russia's Natural Focal Diseases*, Global Perspectives on Health Geography, https://doi.org/10.1007/978-3-319-89605-2_5

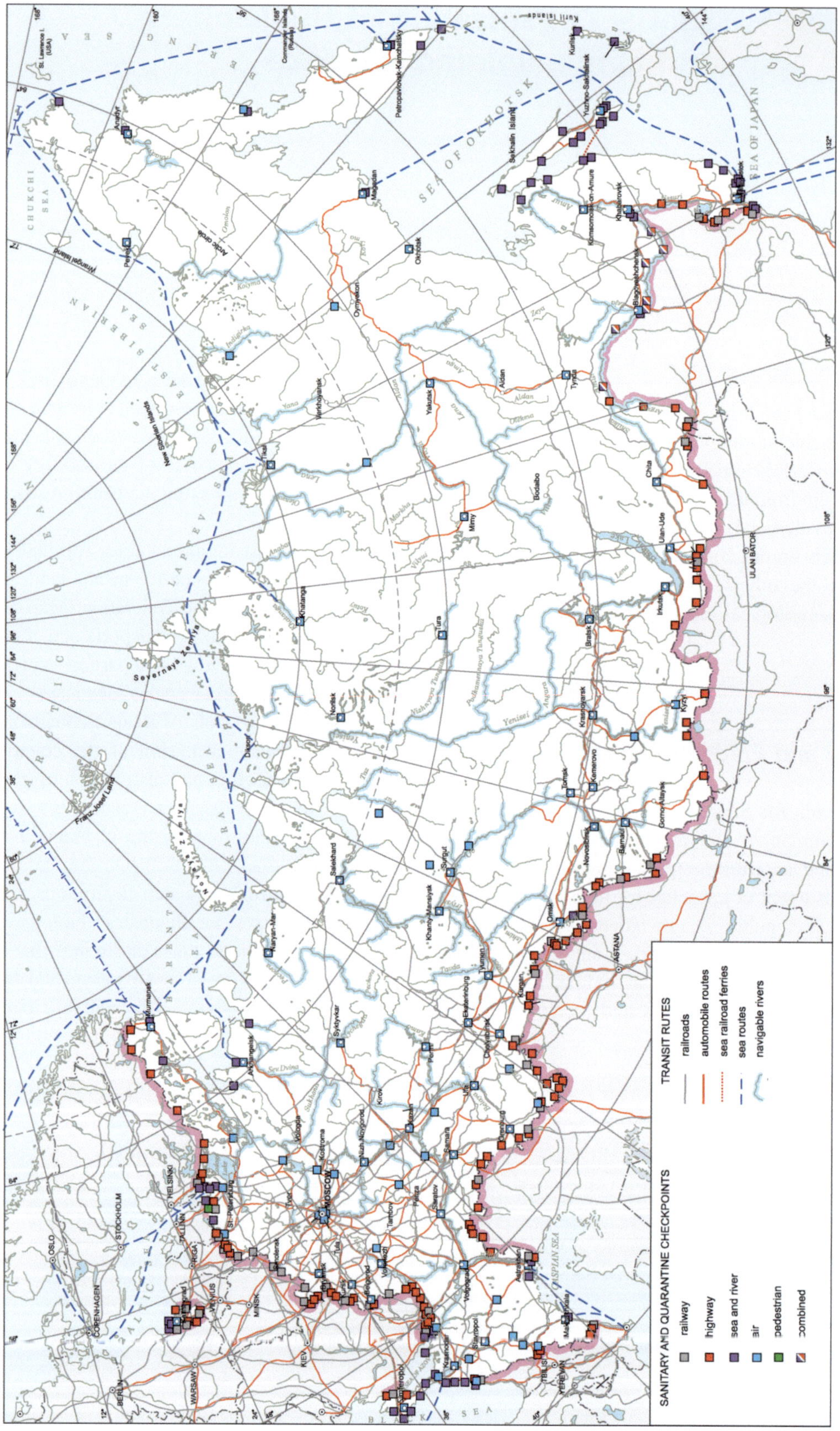

Map 5.1 Federal border checkpoints of the Russian Federation

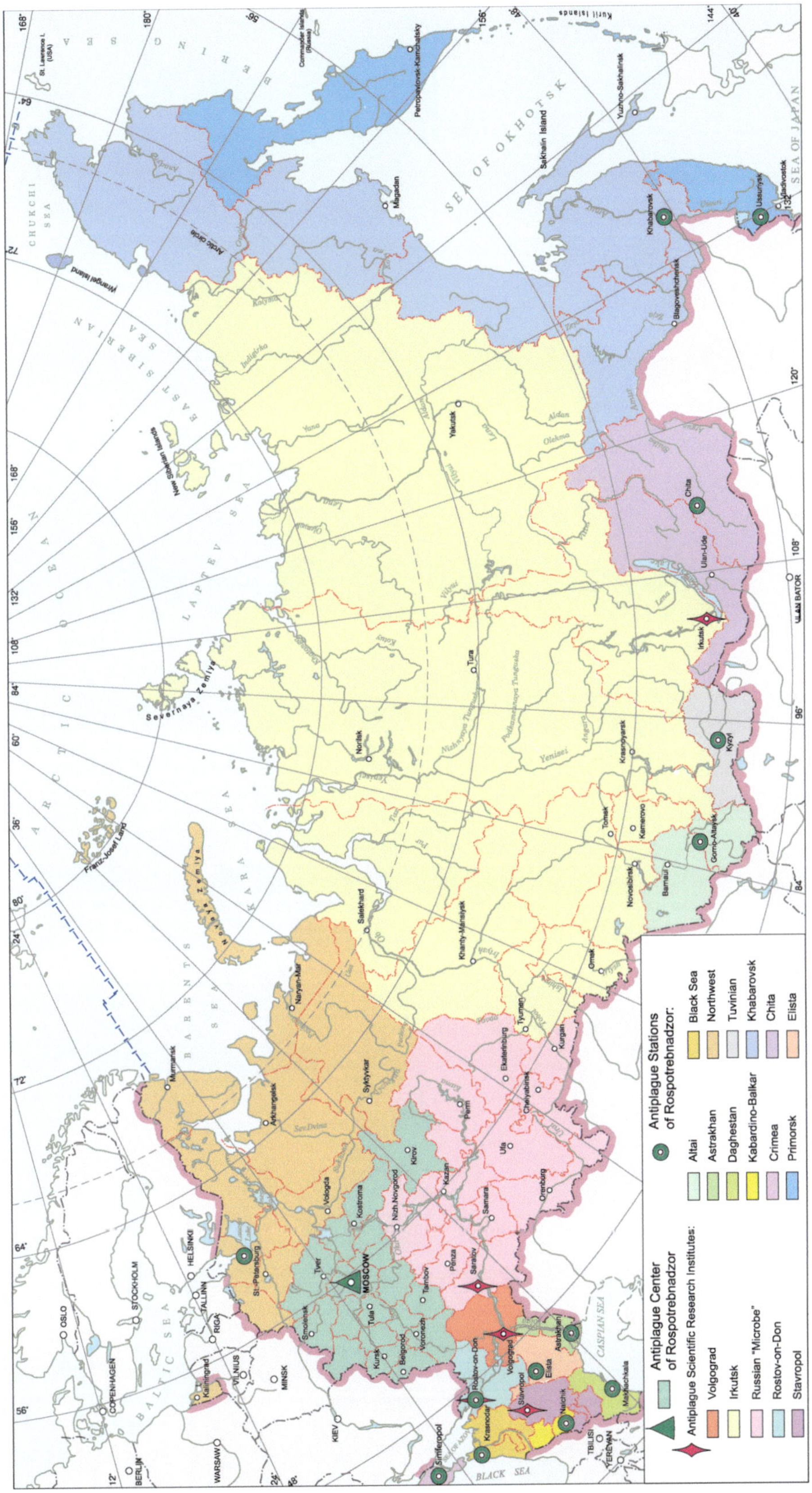

Map 5.2 Anti-plague agencies

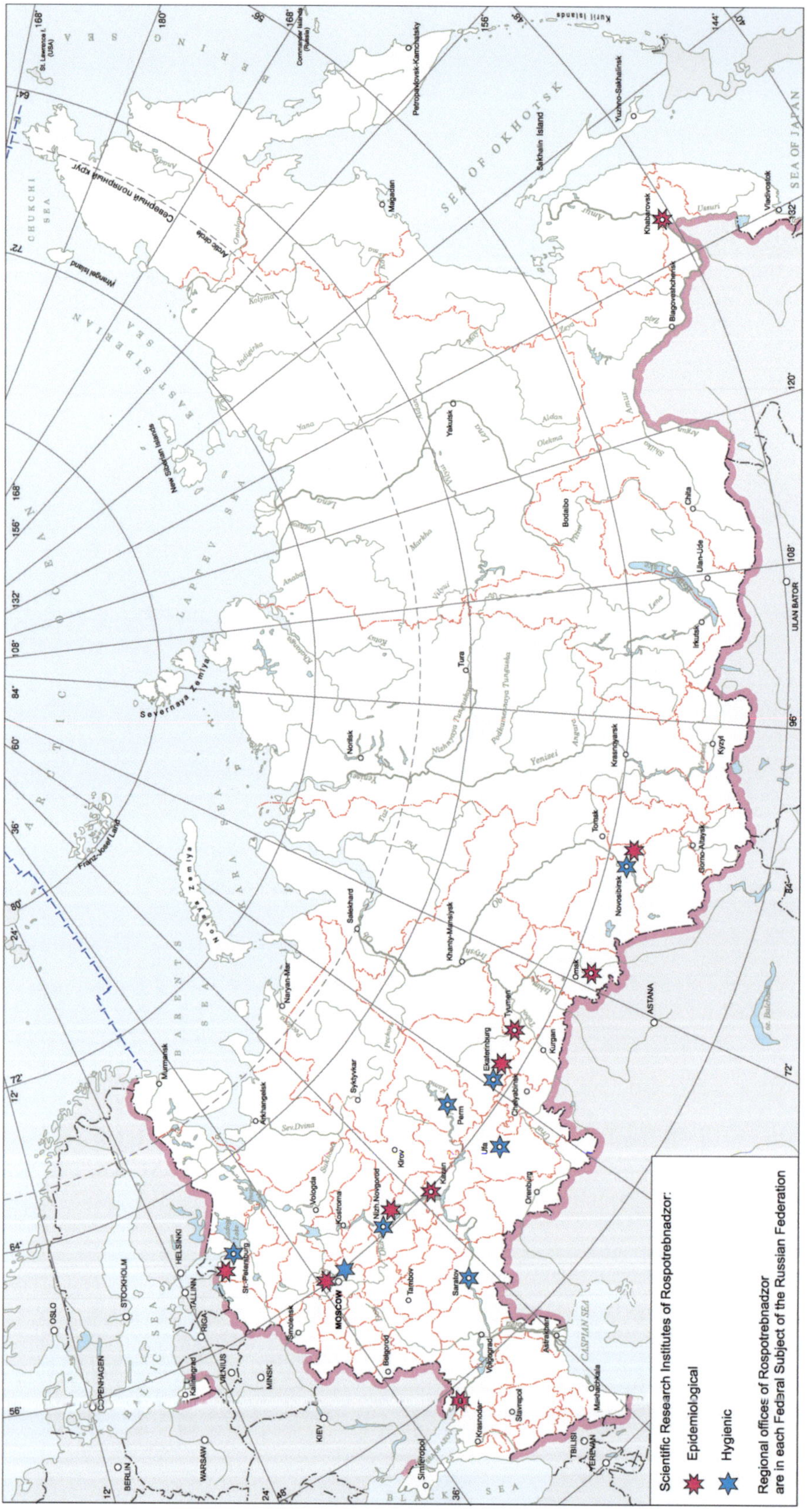

Map 5.3 Federal state public health agencies

Therefore, Rospotrebnadzor's work is aimed at addressing many issues in the sanitary and epidemiological well-being of the population (www.rospotrebnadzor.ru).

References

Federal service for surveillance on consumer rights protection and human wellbeing Regulation No. 322 of 30.06.2004 [electronic resource]. Retrieved from http://base.garant.ru/12136005/. Accessed on 10 June 2015 (in Russian)

On the Status and epidemiological wellbeing situation of the population in the Russian Federation in 2015: State report. Federal Service for Surveillance on Consumer Rights Protection and Human Wellbeing, Moscow, 2016 200 p (in Russian)

Conclusion

The territory of the Russian Federation varies in terms of the medico-geographical conditions regarding natural focal diseases. This variation must be taken into account when new regions are developed, localities are evaluated for their recreational potential and public information campaigns are organized.

This book is a scientific reference publication aimed at a wide audience. The information presented on natural focal diseases may facilitate the work of public health services and sanitary and epidemiological control, thus preventing mass disease in places where the population comes into contact with a natural focus, while the compiled maps can serve as a basis for the development of prognoses on natural focal disease and for the planning of preventative events based on the geographic characteristics of each territory.

Further monitoring and research in medical geography, the elucidation of the tendencies of how modern diseases spread, as well as the creation of more detailed regional atlases (in print and GIS form) will allow for the evaluation of the epidemiological risk posed by natural ecosystems and the creation of scientifically grounded recommendation to public health organizations in order to prevent mass episodes of disease and to treat focal areas. The spread of specific scientific information among is also key, as adherence to sanitary and hygienic norms is crucial for the prevention of the majority of natural focal diseases.

S. Malkhazova et al., *Mapping Russia's Natural Focal Diseases*, Global Perspectives on Health Geography, https://doi.org/10.1007/978-3-319-89605-2

MIX
Papier aus verantwortungsvollen Quellen
Paper from responsible sources
FSC® C105338

If you have any concerns about our products,
you can contact us on
ProductSafety@springernature.com

In case Publisher is established outside the EU,
the EU authorized representative is:
Springer Nature Customer Service Center GmbH
Europaplatz 3, 69115 Heidelberg, Germany

Printed by Libri Plureos GmbH
in Hamburg, Germany